Recent advances in statistics have led to new concepts and solutions in different areas of pharmaceutical research and development. "Springer Series in Pharmaceutical Statistics" focuses on developments in pharmaceutical statistics and their practical applications in the industry. The main target groups are researchers in the pharmaceutical industry, regulatory agencies and members of the academic community working in the field of pharmaceutical statistics. In order to encourage exchanges between experts in the pharma industry working on the same problems from different perspectives, an additional goal of the series is to provide reference material for non-statisticians. The volumes will include the results of recent research conducted by the authors and adhere to high scholarly standards. Volumes focusing on software implementation (e.g. in SAS or R) are especially welcome. The book series covers different aspects of pharmaceutical research, such as drug discovery, development and production.

More information about this series at http://www.springer.com/series/15122

Peter F. Thall

Statistical Remedies
for Medical Researchers

 Springer

Peter F. Thall
Houston, TX, USA

ISSN 2366-8695 ISSN 2366-8709 (electronic)
Springer Series in Pharmaceutical Statistics
ISBN 978-3-030-43716-9 ISBN 978-3-030-43714-5 (eBook)
https://doi.org/10.1007/978-3-030-43714-5

This Springer imprint is published by the registered company Springer Nature Switzerland AG
The registered company address is: Gewerbestrasse 11, 6330 Cham, Switzerland

I am grateful to my medical collaborators, who have taught me so much and provided so many challenging research projects over the years. I also thank Xuemei Wang, Yanxun Xu, and Denai Milton for their generosity in helping with some of the graphics, and Lu Wang and Kelley Kidwell for their constructive comments on my descriptions of bias correction methods.

Preface

Many new medical treatments are developed using powerful technologies, such as microarrays, gene sequencing, or cytometric machines that can rapidly identify dozens of surface markers on millions of cells. The list of innovations is growing all the time, with numerous potential medical applications.

While all this modern machinery may be exciting, the only way to find out how a new treatment for a particular disease works is to give it to people who have the disease and see what happens. An experiment to do this is called a clinical trial. But human beings are much more complex, much less predictable, and much harder to study than gene sequences, cell cultures, or xenografted mice. The main difficulty in the design and conduct of clinical trials is that they must combine scientific method and medical practice. Unfortunately, many clinical trials are designed so poorly that their results are unreliable, misleading, or completely useless. Similar problems arise in analysis of observational data, which may severely misrepresent what one wishes to learn about if flawed statistical methods are used.

The medical research community can do better. The first step is to identify and understand common methodological mistakes. Essentially, this book is a collection of examples showing how things can go wrong with statistical design or data analysis in medical research. For each example, I will provide one or more reasonable alternative statistical methods. The particular topics and examples came from my experiences working as a medical statistician in a large cancer center.

My general aim is to illustrate ideas and statistical methods that will be of interest to a reasonably broad readership, including both statisticians and medical researchers without formal statistical training. In most of the book, I have kept the technical level fairly low. Unavoidably, statisticians may find some of the explanations that I have given at an elementary level to be overly verbose and lacking important technical details. However, a fair number of the examples illustrate serious flaws in methods commonly used by statisticians. Non-statisticians may find some of the formulas to be a bit too technical, especially in the later chapters, although in such cases the reader may simply ignore the formulas. To address both types of reader, I have endeavored to strike a balance without sacrificing important ideas and methods. My overarching goal is to help you avoid shooting yourself in the foot when conducting your medical research.

Houston, USA   Peter F. Thall

Contents

Chapter 1
Why Bother with Statistics?

In the land of the blind, the one-eyed man is king.
Desiderius Erasmus

Contents

Abstract Many statistical practices commonly used by medical researchers, including both statisticians and non-statisticians, have severe flaws that often are not obvious. This chapter begins with a brief list of some of the examples that will be covered in greater detail in later chapters. The point is made, and illustrated repeatedly, that what may seem to be a straightforward application of an elementary statistical procedure may have one or more problems that are likely to lead to incorrect conclusions. Such problems may arise from numerous sources, including misapplication of a method that is not valid in a particular setting, misinterpretation of numerical results, or use of a conventional statistical procedure that is fundamentally wrong. Examples will include being misled by The Innocent Bystander Effect when determining causality, how conditional probabilities may be misinterpreted, the relationship between gambling and medical decision-making, and the use of Bayes' Law to interpret the results of a test for a disease or to compute the probability that a child will have hemophilia based on what has been observed in family members.

© Springer Nature Switzerland AG 2020
P. F. Thall, *Statistical Remedies for Medical Researchers*, Springer Series
in Pharmaceutical Statistics, https://doi.org/10.1007/978-3-030-43714-5_1

1.1   Some Unexpected Problems

The first chapter or two will begin with explanations of some basic ideas in probability and statistics. Readers with statistical training may find this too elementary and a waste of time. But, with each new chapter, problems with the ways that many people actually do medical statistics will begin to accumulate, and their harmful consequences for practicing physicians and patients will become apparent. I once taught a half-day short course on some of these problems to a room full of biostatisticians. They spent an entire morning listening to me explain, by example, why a lot of the things that they did routinely were just plain wrong. At the end, some of them looked like they were in shock, and possibly in need of medical attention.

Many statistical practices commonly used by medical researchers have severe flaws that may not be obvious. Depending on the application, a particular statistical model or method that seems to be the right thing to use may be completely wrong. Common examples include mistaking random variation in data for an actual treatment effect, or assuming that a new targeted agent which kills cancer cells in xenografted mice is certain to provide a therapeutic advance in humans. A disastrous false negative occurs if a treatment advance is missed because an ineffectively low dose of a new agent was chosen in a poorly designed early phase clinical trial. Ignoring patient heterogeneity when comparing treatments may produce a "one size fits all" conclusion that is incorrect for one important patient subgroup, or is incorrect for all subgroups. A numerical example of the latter mistake will be given in Sect. 11.2 of Chap. 11. Comparing data from a single-arm study of a new treatment to historical data on standard therapy may mistake between-study differences for actual between-treatment differences. There are numerous other examples of flawed statistical practices that have become conventions in the medical research community. I will discuss many of them in the chapters that follow.

Bad science leads to bad medicine. Flawed statistical practices and dysfunctional clinical trials can, and often do, lead to all sorts of incorrect conclusions. The practical impact of incorrect statistical inferences is that they can mislead physicians to make poor therapeutic decisions, which may cost patients their lives. Evidence-Based Medicine is of little use if the evidence is wrong, misleading, or misinterpreted.

All statistical methods are based on probability, which often is counterintuitive. This may lead to confusion, incorrect inferences, and undesirable actions. By the same token, correctly following the *Laws of Probability* may lead to conclusions that seem strange, or actions that seem wrong. Here are some examples:

- An experimental treatment that gives a statistically significant test comparing its response rate to the standard therapy rate may only have probability 0.06 of achieving the rate targeted by the test (Chap. 5).
- It may be better for physicians to choose their patients' treatments by flipping a coin than by applying their medical knowledge (Chap. 6).
- A new anticancer treatment that doubles the tumor response rate compared to standard therapy may turn out to only improve expected survival time by a few days or weeks (Chap. 7).

- Comparing the response rates of treatments A and B may show that A is better than B in men, A is better than B in women, but if you ignore sex then B is better than A in people (Chap. 9).
- When administering treatments in multiple stages, it may be best to start each patient's therapy with a treatment that is suboptimal in the first stage (Chap. 12).

While these examples may seem strange, each can be explained by a fairly simple probability computation. I will provide these in the chapters that follow.

1.2 Expert Opinion

Recent advances in computer hardware and computational algorithms have facilitated development and implementation of a myriad of powerful new statistical methods for designing experiments and extracting information from data. Modern statistics can provide reliable solutions to a broad range of problems that, just a few decades ago, could only be dealt with using primitive methods, or could not be solved at all. New statistical methods include survival analysis accounting for multiple time-to-event outcomes and competing risks, feature allocation for identifying patterns in high-dimensional data, graphical models, bias correction methods that estimate causal effects from observational data, clinical trial designs that sequentially evaluate and refine precision ("personalized") medicine, and methods for tracking, predicting, and adaptively modifying multiple biological activities in the human body over time.

But all this powerful new statistical technology has come with a price. Many statistical procedures are so complex that it is difficult to understand what they actually are doing, or interpret their numerical or graphical results. New statistical methods and computer software to implement them have developed so rapidly that modern statistics has become like the federal income tax code. No one can possibly know all of it. For a given experimental design problem or data structure, it is common for different statisticians to disagree about which methodology is best to use. This complexity has led to some severe problems in the scientific community, many of which are not obvious.

Medical researchers must specify and apply statistical methods for the data analyses and study designs in their research papers and grant proposals. They may have attitudes about statistics and statisticians similar to those of college students taking a required statistics course. They do not fully understand the technical details, but they know that, regardless of how painful it may be, statistics and statisticians cannot be avoided. Physicians who do medical research tend to choose statisticians the same way that people choose doctors. In both cases, an individual looks for somebody to solve their problem because they don't know how to solve it themselves. They find someone they think is an expert, and then trust in their expertise. But, like practicing physicians, even the best statisticians make mistakes. Applying statistical methods can be very tricky, and it takes a great deal of experience and a very long time for a statistician to become competent, much less expert, at what they do.

Some medical researchers never bother talking to a statistician at all. They just find a statistical software package, figure out how to run a few programs, and use the computer-generated output to do their own statistical analyses. Scientifically, this is a recipe for disaster. If someone running a statistical software package has little or no understanding of the methods being implemented, this can lead to all sorts of mistakes. What may seem like a simple application of an elementary statistical procedure may be completely wrong, and lead to incorrect conclusions. The reason this matters is that most practicing physicians base their professional decisions on what they read in the published medical literature, in addition to their own experiences. Published papers are based on statistical analyses of data from clinical trials or observational data. If these analyses are flawed or misinterpreted, the results can be disastrously misleading, with very undesirable consequences for patients.

Intelligent people make bad decisions all the time. Based on what they think is information, they decide what is likely to be true or false, and then take actions based on what they have decided. Once someone has jumped to the wrong conclusion, they do all sorts of things that have bad consequences. Coaches of sports teams may lose games that they might have won, portfolio managers may lose money in the stock market when they might have made money, and physicians may lose the lives of patients they might have saved. These are experienced professionals who get things wrong that they might have gotten right. Much of the time, people make bad decisions because they do not understand probability, statistics, or the difference between data and information. Despite the current love affair with "Big Data," many large datasets are complete garbage. Figuring out precisely why this may be true in a given setting requires statistical expertise, and high-quality communication between a competent statistician and whoever provided the data. If a dataset has fundamental flaws, bigger is worse, not better.

If you are a sports fan, you may have noticed that the ways basketball, baseball, and other sports are played have changed radically in recent years. Coaches and team managers did that by hiring sports statisticians. Of course, to be competitive a sports team must have talented athletes, but the team with the most points, or runs, at the end of the game wins. History has shown, again and again, that having the best player in the world on your team does not guarantee you will win a world championship. If you invest money in the stock market, you may have wondered where stock and option prices come from, and how all those computer programs that you never see make so much money. Statisticians did that. If you are a medical researcher, you may have wondered why so much of the published literature is so full of statistical models, methods, and analyses that seem to get more complicated all the time. Many physicians now talk about "Evidence-Based Medicine" as if it is a radical new idea. It is just statistical inference and decision-making applied to medical data.

Unfortunately, in this era of clever new statistical models, methods, and computational algorithms, being applied to everything and anything where data can be collected, there is a pervasive problem. It is very easy to get things wrong when you apply statistical methods. Certainly, the complexity of many statistical methods may make them difficult to understand, but the problem actually is a lot deeper.

1.3 The Innocent Bystander Effect

Making a conclusion or deciding what you believe about something based on observed data is called *statistical inference*. A lot of incorrect inferences are caused by "The Innocent Bystander Effect." For example, suppose that you and a stranger are waiting at a bus stop. A third person suddenly walks up, calls the stranger some nasty names, knocks him out with a single punch, and runs away. You stand there in shock for a minute or two, looking at the man lying on the pavement in a growing pool of blood, and wondering what to do. A minute or two later, people who happen to be walking by notice you and the bleeding man, a crowd forms, and a police car soon arrives. The police see you standing next to the unconscious man on the sidewalk, and they arrest you for assault, which later turns out to be murder, since the stranger hit his head on the sidewalk when he fell and died of a brain hemorrhage. You are tried in a court of law, found guilty of murder by a jury of your peers, and sentenced to death. As you sit in your jail cell pondering your fate, you realize your problem was that neither the people walking by nor the police saw the assailant or the assault. They just saw you standing over the man on the sidewalk, and they jumped to the conclusion that you must have killed him. This is an example of "Evidence-Based Justice."

What does this have to do with medical research? Plenty. Replace the dead stranger with a rapidly fatal subtype of a disease, the unknown assailant with the actual but unknown cause of the disease, and you with a biomarker that, due to the play of chance, happened to be positive (+) more often than negative (−) in some blood samples taken from some people who died of the disease. So, researchers are ignorant of the actual cause of the disease, but they see data in which being biomarker + and having the rapidly fatal subtype of the disease are strongly associated. Based on this, they jump to the conclusion that being biomarker + must increase the chance that someone with the disease has the rapidly fatal subtype, and they choose treatments on that basis. To make matters worse, even if being biomarker + and having the rapidly fatal version of the disease actually are positively associated, it still does not necessarily imply that being biomarker + *causes* the rapidly fatal subtype. A common warning in statistical science is that *association does not imply causality*. Just because two things tend to occur together does not imply that one causes the other. What if the biomarker is just an Innocent Bystander? There may be a third, "latent" variable that is not known but that causes both of the events [biomarker +] and [rapidly fatal subtype]. A simple example is leukemia where the cause of the rapidly fatal subtype is a cytogenetic abnormality that occurs early in the blood cell differentiation process, known as hematopoiesis. The biomarker being + is one of many different downstream consequences, rather than a cause. So giving leukemia patients a designed molecule that targets and kills the biomarker + leukemia cells will not stop the rapidly fatal leukemia from continuing to grow and eventually kill patients. The cells become leukemic before the biomarker has a chance to be + or −, and enough of the leukemia cells are biomarker − so that, *even if you could kill all of the biomarker + leukemia cells, it would not cure the disease.* But people often ignore this sort of thing, or may be unaware of it. In this type of setting, they may invest a great deal of time and money developing a treatment that targets the biomarker, with the goal to treat the rapidly fatal subtype. Unfortunately, the treatment is doomed to failure before it even is developed, since targeting only the + biomarker cannot cure the disease.

The above example is a simplified version of much more complex settings that one sees in practice. For example, there may be multiple disease subtypes, some of which are not affected by the targeted molecule, or redundancies in the leukemia cell differentiation process so that knocking out one pathway still leaves others that produce leukemia cells. It really is not a mystery why so many "targeted therapies" for various cancers fail. The truth is that many diseases are smarter than we are. Still, we can do a much better job of developing and evaluating new treatments.

A different sort of problem is that people often incorrectly reverse the direction of a conditional probability. To explain this phenomenon, I first need to establish some basic concepts. In general, if E and F are two possible events, the *conditional probability of E given that you know F has occurred* is defined as

$$\Pr(E \mid F) = \frac{\Pr(E \text{ and } F)}{\Pr(F)}.$$

This quantifies your uncertainty about the event E if you know that F is true. Two events E and F are said to be *independent* if $\Pr(E \mid F) = \Pr(E)$, which says that knowing F is true does not alter your uncertainty about E. It is easy to show that independence is symmetric, that is, it does not have a direction, so $\Pr(E \mid F) = \Pr(E)$ implies that $\Pr(F \mid E) = \Pr(F)$. A third equivalent way to define independence is $\Pr(E \text{ and } F) = \Pr(E) \Pr(F)$.

For example, consider the cross-classified count data in Table 1.1. Based on the table, since the total is $20 + 60 + 40 + 80 = 200$ patients,

$$\Pr(\text{Treatment } A) = \frac{20 + 60}{200} = 0.40$$

and

$$\Pr(\text{Response and Treatment } A) = \frac{20}{200} = 0.10,$$

so the definition says

Table 1.1 Counts of response and nonresponse for treatments A and B

Treatment	Response	Nonresponse	Total
A	20	60	80
B	40	80	120
Total	60	140	200

$$\text{Pr(Response } | \text{ Treatment } A) = \frac{0.10}{0.40} = 0.25.$$

In words, 25% of the patients who are given treatment A respond. Since the unconditional probability is Pr(Response) $= (20 + 40)/200 = 0.30$, conditioning on the additional information that treatment A was given lowers this to 0.25. Similarly, Pr(Response $|$ Treatment B) $= 0.20/0.60 = 0.33$, so knowing that treatment B was given raises the unconditional probability from 0.30 to 0.33. This also shows that, in this dataset, treatment and response are not independent.

Remember, Pr(Response | Treatment A) and the reverse conditional probability Pr(Treatment A | Response) are very different things. For the first, you assume that treatment A was given and compute the probability of response. For the second, you assume that treatment was a response, and compute the probability that treatment A was given. To see how confusing the two different directions of conditional probabilities can lead one astray, consider the following example. Suppose that a parent has a child diagnosed as being autistic, and the parent knows that their child received a vaccination at school to help them develop immunity for various diseases like measles, chicken pox, and mumps. Since many schoolchildren receive vaccinations, Pr(vaccination) is large. While the numbers vary by state, year, and type of vaccination, let's suppose for this example that Pr(vaccination)$=0.94$, and that this rate is the same for any selected subgroup of schoolchildren. Therefore, for example, the conditional probability Pr(vaccinated | autistic)$=$Pr(vaccinated)$=0.94$. In words, knowing that a child is autistic does not change the probability that the child was vaccinated. That is, being autistic and getting vaccinated are independent events.

But suppose that the parent makes the mistake of reversing the order of conditioning and concludes, incorrectly, that this must imply that Pr(autistic | vaccinated)$=0.94$. This is just plain wrong. These two conditional probabilities are related to each other, by a probability formula known as either *Bayes' Law* or *The Law of Reverse Probabilities*, which I will explain later. But these two conditional probabilities are very different from each other.

- Pr(vaccinated | autistic) quantifies the likelihood that a child known to be autistic will be vaccinated.
- Pr(autistic | vaccinated) quantifies the likelihood that a child known to have received a vaccination will be diagnosed with autism.

Suppose that the parent does not understand that they have made the error of thinking that these two conditional probabilities that go in opposite directions are the same thing, and they find out that autism is very rare, with Pr(autistic)$=0.015$, or 15 in 1000. Then the extreme difference between the number 0.015 and Pr(vaccinated | autistic)$=0.94$, which they mistakenly think is Pr(autistic | vaccinated), may seem to imply that receiving a vaccination at school *causes* autism in children.

The parent's incorrect reasoning may have been a lot simpler. Not many people ever bother to compute conditional probabilities like those given above. They may just know that autism is rare, noticed that their child received a vaccination at some point before being diagnosed as autistic, and concluded that the vaccination was the cause. In this case, vaccination was just an Innocent Bystander. Using the same reasoning, they might as well have inferred that going to school causes autism.

This fallacious reasoning actually motivated scientists to look for biological causes or associations that might explain how vaccination causes autism. The evidence leads in the opposite direction, however. For example, a review of multiple studies of this issue, given by Plotkin et al. (2009), showed that, in a wide variety of different settings, there is no relationship between vaccination and subsequent development of autism. Knowing that a child has been vaccinated does not change the probability that the child is, or will be diagnosed as, autistic. Moreover, on more fundamental grounds, no biological mechanism whereby vaccination causes autism has ever been identified.

Unfortunately, given the ready ability to mass mediate one's fears or opinions via the Internet or television, and the fact that many people readily believe what they read or see, it is easy to convince millions of people that all sorts of ideas are true, regardless of whatever empirical evidence may be available. Very few people read papers published in scientific journals. This is how very large numbers of people came to believe that vaccinations cause autism in children, drinking too much soda pop caused polio, and breast implants caused a wide variety of diseases in women. All of these beliefs have been contradicted strongly by scientific evidence. But once a large number of people share a common belief, it is very difficult to convince them that it is untrue.

Takeaway Messages About the Innocent Bystander Effect

1. The Innocent Bystander Effect occurs when some event or variable that happened to be observed before or during the occurrence of an important event of primary interest is mistakenly considered to be the cause of the event.
2. The Innocent Bystander variable may be positively associated with the event of primary interest, but not cause the event. This is the case when both the Innocent Bystander variable and the event of interest both are likely to be caused by a third, unobserved or unreported "lurking" variable. In the soda pop consumption—polio incidence example, higher rates of both were caused by warmer temperatures during the summer, which was the lurking variable.
3. If the lurking variable is unknown or is not observed, it is impossible to know that it affected both the Innocent Bystander variable and the event of interest. Consequently, it may be believed very widely that the Innocent Bystander variable causes the event of interest.
4. Mass mediation of incorrect causal inferences due to The Innocent Bystander Effect has become a major problem in both the scientific community and modern society.
5. A general fact to keep in mind to help guard against being misled by The Innocent Bystander Effect is **association does not imply causation**.

1.4 Gambling and Medicine

A *gamble* is an action taken where the outcome is uncertain, with a risk of loss or failure and a chance of profit or success. Buying a lottery ticket, matching or raising a bet in a poker game, or buying a stock at a particular price all are gambles. For each, the most important decision actually is made earlier, namely, whether or not to make the gamble at all. The decision of whether to sit down at a poker table and say "Deal me in" is more important than any decisions you make later on while you're playing. We all make plenty of gambles in our day-to-day lives, by choosing what clothes to wear, what foods to eat, or what sort of over-the-counter pain medicine to take. Most of the time, our decisions do not seem to have consequences large enough to matter much, but sometimes a lot may be at stake.

If a physician has a patient with a life-threatening disease for which two treatments are available, say *A* and *B*, then the physician has the choices [treat the patient with *A*], [treat the patient with *B*], or [do not treat the patient]. This is a greatly simplified version of most medical decision-making, where there may be lots of possible treatments, the "loss" may be some combination of the patient suffering treatment-related adverse effects or death, and the "profit" may be some combination of reducing their suffering, curing their disease, or extending their life. For example, if the disease is acute leukemia, *A* may be a standard chemotherapy, and *B* may be an allogeneic stem cell transplant. The option [do not treat the patient] may be sensible if the patient's disease is so advanced, or the patient's physical condition is so poor, that any treatment is very likely to either kill the patient or be futile. Given that the physician has taken on the responsibility of making a treatment choice for the patient, or more properly making a treatment recommendation to the patient, the gamble must be made by the physician. For the patient, given that they have the disease, whatever the physician recommends, the patient must make the gamble of either accepting the physician's recommendation or not. This is not the same thing as deciding whether to fold a poker hand or match the most recent bet, since one can simply choose not to sit down at the poker table in the first place. I will not get into detailed discussions of gambling or decision analysis, since this actually is an immense field that can get quite complex and mathematical. The journal *Medical Decision Making* is devoted entirely to this area. Some useful books are those by Parmigiani (2002), Felder and Mayrhofer (2017), and Sox et al. (2013), among many others.

Suppose that, based on a blood test, your doctor tells you that you have a Really Bad Disease, and that your chance of surviving 3 years is about 20%. After getting over the shock of hearing this news, you might ask your doctor some questions. Where did the number 20% come from? What are the available treatment options? For each treatment, what is the chance of surviving 3 years, and what are the possible side effects and their chances of occurring? What effects may your personal characteristics, such as how advanced your disease is, your age, or your medical history, have on your survival? You also might ask whether your doctor is absolutely sure that you have the disease, or if it is possible that the blood test might be wrong, and if so, what the probability is that the diagnosis actually is correct. You might ask how long you can expect to survive if you do not take any treatment at all. If your doctor can't answer your questions convincingly, you might talk to another doctor to get a second opinion. You might take the time to read the medical literature on the Really Bad Disease yourself, or even find whatever data may be available and look for a biostatistician to analyze or interpret it for you. After all, your life is at stake.

For a physician, providing reliable answers to these questions requires knowledge about statistical analyses of data from other people previously diagnosed with the disease, the treatments they received, their side effects, and how long they survived. This is the sort of thing that is published in medical journals. Any competent doctor already has thought about each of these questions before making a treatment recommendation, and may suggest two or more treatments, with an explanation of the potential risks and benefits of each. If you are a practicing physician, then you know all about this process. If you treat life-threatening diseases, then gambling with people's lives is what you do routinely in your day-to-day practice when you make treatment decisions. You probably are familiar with the published medical literature on the diseases that you treat. If you want to be good at what you do, inevitably you must rely on the statistical data analyses described in the papers that you read. But, in any given area of medicine, there often are so many papers published so frequently that it is difficult or impossible to read them all. To make things even harder, sometimes different papers contradict each other. It is not easy being a physician in the modern world.

Reading the medical literature can be perilous. A lot of published medical papers contain imprecise, misleading, or incorrect conclusions because their statistical methods are flawed in some way. There may be problems with study design, assumed statistical models, data analyses, or interpretation of results. But detecting such flaws can be very difficult, even for a statistician. Because journal space is limited, many medical papers provide only a brief sketch of the statistical methods that were applied. They often do not provide enough detail for you to figure out exactly what was done with the data, precisely how the data were obtained, or how each variable was defined. Sometimes, the statistical models and methods are conventions that you have seen many times before but, for reasons that are not obvious, they are used in a way that is inappropriate for the particular dataset being analyzed. So, even if a physician reads the medical literature very carefully, and uses it as a guide to practice so-called *evidence-based medicine*, the physician may be misled unknowingly by statistical errors. Often, the authors themselves are not aware of statistical errors that appear in their papers.

For example, in survival analysis, there are many examples of statistical errors that may lead to incorrect conclusions. One common error occurs when comparing two competing treatments, say E and S, where the Kaplan and Meier (1958) (KM) estimates of the two survival probability curves cross each other, as illustrated in Fig. 1.1. The two KM plots are estimates of the survivor probability functions $\Pr(T > t \mid E)$ and $\Pr(T > t \mid S)$ for all times $t > 0$, where T denotes survival time. The proportional hazards (PH) assumption, which underlies the commonly used Cox (1972) regression model, says that the E versus S treatment effect is constant over

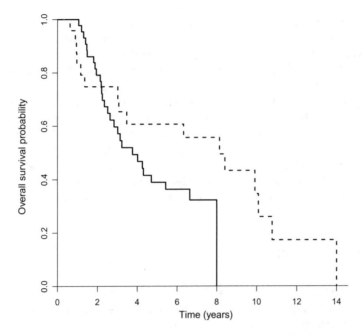

Fig. 1.1 Two Kaplan–Meier survival distribution estimates that cross each other

time. But the fact that the two KM curves cross each other implies that the PH assumption cannot be correct. Survival is better for E compared to S up to the time, slightly after 2 years, when the curves cross, and after that, the effect is reversed so that S is superior to E. So the PH assumption cannot be true. Since two estimated hazard ratio (HRs) computed from the KM curves at two different times give values <1 for a time before 2 years and >1 thereafter, it makes no sense to talk about one parameter called "the hazard ratio." The HR actually is a function that changes over time. In this kind of setting, any single numerical estimate of one nominal "treatment effect" expressed as either an HR or log(HR) in a table from a fitted Cox model makes no sense.

When it first was published, Cox's PH survival time regression model was a big breakthrough. It provided a very useful model that accounts for effects of both treatments and patient prognostic variables on event time outcomes, like death or disease progression, subject to administrative right-censoring at the end of follow up. A statistician named Frank Harrell wrote an easy-to-use computer program for fitting a Cox model, so at that point medical researchers had a practical new tool for doing regression analysis with survival time data. As with any regression model, you could even include treatment–covariate interactions, if you were interested in what later would be called "precision medicine."

Unfortunately, despite its widespread use in medical research, there are many time-to-event datasets that the Cox model does not fit well because its PH assumption is not met. This is a big problem because, more generally, when a statistical regression model provides a poor fit to a dataset, inferences based on the fitted model may be misleading, or just plain wrong. In statistics, "model criticism" or "goodness-of-fit analysis" is a necessary exercise when fitting a model to data, since it may be the case that the assumed model does not adequately describe the data, so a different model is needed. People who do not know this, or who are aware of it but choose to ignore it, are not statisticians. They just know how to run statistical software packages.

The good news is that, since 1972, the toolkit of statistical models and computer software for dealing with a wide variety of complex time-to-event data structures with multiple events of different types has grown tremendously. A useful book on assessing goodness-of-fit for the Cox model, giving extended versions of the model that can be used when the PH assumption is violated, is Therneau and Grambsch (2000). A book covering a wide array of Bayesian models and methods for survival analysis is Ibrahim et al. (2001). I will discuss some examples in Chaps. 5, 9, and 11.

1.5 Testing Positive

Suppose that there is a very reliable blood test for a disease. No test is perfect, but suppose that, if someone has the disease, then there is a 99% chance that the test will come out +. If someone does not have the disease, there is a 95% chance that the test will come out −. To make this precise, denote $D = $ [have the disease] and $\overline{D} = $ [do not have the disease]. The disease is rare, since only 0.1% of people, or 1 in 1000, have the disease, written as $\Pr(D) = 0.001$.

Now, suppose that you are tested, and it comes out +. Given this scary test result, what you are interested in is the probability that you actually have the disease. It turns out that, based on your + test, there is only about a 2% chance that you have the disease or, equivalently, a 98% change that you do not have it. If this seems strange, you might just chalk it up to the fact that probability is nonintuitive. But here is a formal explanation.

The probability computation that explains this strange result can be done by applying *Bayes' Law*. This is an extremely powerful tool that shows how to reverse the direction of a conditional probability, which quantifies how likely an event is, given that you know whether some other event has occurred. The information that I gave above can be formalized as follows. The *sensitivity* of the test is the conditional probability $\Pr(\text{test} + \mid D) = 0.99$. This says that, if someone has the disease, there is a 99% chance that the test will correctly come out +. The *specificity* of the test is $\Pr(\text{test} - \mid \overline{D}) = 0.95$. This says that, if someone does *not* have the disease, there is a 95% chance that the test will correctly come out −. The third quantity that we need in order to do the key computation is the *prevalence* of the disease, which is $\Pr(D) = 0.001$ in this example. But the probability that you actually want to know is $\Pr(D \mid \text{test} +)$, which is the *reverse* of the sensitivity probability, $\Pr(\text{test} + \mid D)$.

Bayes' Law sometimes is called *The Law of Reverse Probability*. But whatever you call it, it's The Law, and those who disobey it may suffer terrible consequences. Suppose that E and F are two possible events. Denote the complement of E, the event that E does not occur, by $\overline{E}$. The *Law of Total Probability* implies that

$$\Pr(F) = \Pr(F \text{ and } E) + \Pr(F \text{ and } \overline{E})$$
$$= \Pr(F \mid E)\Pr(E) + \Pr(F \mid \overline{E})\Pr(\overline{E}).$$

Bayes' Law is given by the following Magic Formula:

$$\Pr(E \mid F) = \frac{\Pr(F \mid E)\Pr(E)}{\Pr(F \mid E)\Pr(E) + \Pr(F \mid \overline{E})\Pr(\overline{E})}.$$

Notice that the denominator on the right-hand side is just the expanded version of $\Pr(F)$. To apply Bayes' Law to the disease testing example, replace E with D, and F with [test +]. This gives

$$\Pr(D \mid \text{test } +) = \frac{\Pr(\text{test } + \mid D)\Pr(D)}{\Pr(\text{test } + \mid D)\Pr(D) + \Pr(\text{test } + \mid \overline{D})\Pr(\overline{D})}.$$

Of course, this formula is only magic if you haven't seen it before. Let's plug in the numbers that we have and find out what the answer is. Since (test +) and (test −) are complementary events, their probabilities must sum to 1, so $\Pr(\text{test} -) = 1 - \Pr(\text{test} +)$. This is *The Law of Complementary Events*. You probably should keep track of these laws, to avoid being sent to Statistics Jail. This law is true for conditional probabilities, provided that they condition on the same event. So,

$$\Pr(\text{test } + \mid \overline{D}) = 1 - \Pr(\text{test } - \mid \overline{D}) = 1 - 0.95 = 0.05.$$

In words, this says that the test has a 5% false positive rate. *The Law of Complementary Events* also says that $\Pr(\overline{D}) = 1 - \Pr(D) = 1 - 0.001 = 0.999$. Plugging all of these numbers into the Magic Formula gives the answer:

$$\Pr(D \mid \text{test } +) = \frac{0.99 \times 0.001}{0.99 \times 0.001 + 0.05 \times (1 - 0.001)} = \frac{0.00099}{0.00099 + 0.04995} = 0.0194,$$

or about 2%. Although this is a pretty small percentage, what your + test actually did was increase your probability of having the disease from the population value of $\Pr(D) = 0.001$, which was your risk before you were tested, to the updated value $\Pr(D \mid \text{test } +) = 0.0194$. That is, the + test increased your risk of actually having the disease by a multiplier of about 19, which is pretty large. The updated value is obtained by using the information that your test was +, and applying Bayes' Law. Another way to think about this process is to say that the prevalence $\Pr(D) = 0.001$ was your *prior probability of having the disease*, and then you applied *Bayes' Law* to incorporate your new data, that you tested +, to learn that your *posterior probability of having the disease* was 0.0194. So, applying *Bayes' Law* can be thought of as a way to learn from new information.

By the way, notice that $\Pr(\text{test } + \mid D) = 0.99$ while $\Pr(D \mid \text{test } +) = 0.0194$. So, these two reverse probabilities, while related, have very different meanings, and are very different numbers. You need to apply Bayes' Law to get from one to the other. Remember the vaccination and autism example?

The prevalence of the disease in the population, which is the probability that any given individual has D, actually plays a very important role in this computation. The conditional probability $\Pr(D \mid \text{test} +)$ is sometimes called the *positive predictive value* (PPV) of the test. This can be written as

$$PPV = \frac{0.99 \Pr(D)}{0.99 \Pr(D) + 0.05\{1 - \Pr(D)\}}.$$

For example, if $\Pr(D) = 0.01$ rather than 0.001, so the disease is 10 times more prevalent, then the previous computation becomes

$$\Pr(D \mid \text{test} +) = \frac{0.99 \times 0.01}{0.99 \times 0.01 + 0.05 \times (1 - 0.01)} = \frac{0.0099}{0.0099 + 0.0594} = 0.143.$$

So for a disease with prevalence 0.01, a + test says that you have about a one in seven chance of having the disease, which is more worrying than the one in fifty chance obtained for a disease with prevalence of 0.001. If the disease is a lot more common, with $\Pr(D) = 0.05$, or 1 in 20, then the PPV $= 0.51$. So, the prevalence of the disease in the population matters a lot since it can have a large effect on the PPV. If you do not know the prevalence, then you can't do the computation, so the main effect of being told that you tested positive is to scare you. Given the above computations, you might hope that the disease is rare but, again, hope is not a useful strategy. Instead, just get onto the Internet, look up the disease prevalence, and apply *Bayes' Law*.

Another practical question is, what if, after testing +, you get tested a *second* time and it also comes out +? Assuming that the two test results are independent, it turns out that you can apply *Bayes' Law* a second time, using your updated probability 0.0194 from the first computation in place of the population prevalence 0.001 that you started with before the first test. This is yet another magic formula that you may derive, if you like doing probability computations. Or you can just take my word for it. The updated formula for your probability of D, given that both the first and second tests were +, is

$$\Pr(D \mid \text{test} + \text{twice}) = \frac{0.99 \times 0.0194}{0.99 \times 0.0194 + 0.05 \times (1 - 0.0194)}$$

$$= \frac{0.019206}{0.019206 + 0.04903} = 0.281.$$

Uh oh. Depending on what the disease is, and how big you think a 28% chance of having the disease is, now you may have a problem that needs medical attention. If the disease is gonorrhea, which actually has a prevalence of about 0.001 in the US, it might be a good idea to see a doctor and get some antibiotics. Of course, a doctor probably ordered the tests in the first place.

This is an example of how *Bayes' Law* can be used repeatedly as a learning process. At each stage after the first, the posterior is used as the prior for the next stage. In the above example, the prior was $\Pr(D) = 0.001$, and this gave the posterior $\Pr(D \mid \text{test } +) = 0.0194$. Using this posterior as your new prior value, to incorporate the new data that the second test was positive at the second stage, you do the computation by just replacing the 0.001 with 0.0194 in the formula to get the new posterior value 0.281. In practice, say in some complex process where you repeatedly acquire new information, this can be repeated over hundreds or thousands of stages.

To complete this disease testing example, it also is useful to know how likely it is that you have the disease if you test $-$. Bayes' Law gives this as

$$\Pr(D \mid \text{test } -) = \frac{\Pr(\text{test } - \mid D)\Pr(D)}{\Pr(\text{test } - \mid D)\Pr(D) + \Pr(\text{test } - \mid \overline{D})\Pr(\overline{D})}$$

$$= \frac{(1 - 0.99) \times 0.001}{(1 - 0.99) \times 0.001 + 0.95 \times 0.999} = \frac{0.00001}{0.00001 + 0.94905} = 0.00001,$$

which is 1 in 100,000. So, if you test negative, it is great news, and you can go out and celebrate. Isn't *Bayes' Law* wonderful?

Takeaway Messages About Testing for a Disease

1. Being able to attach probabilities to events in order to quantify your uncertainty about them can be extremely useful. If you are a physician, it can help you make decisions like whether to give a particular antibiotic to a patient who tested +.
2. A positive test does not necessarily mean that you have the disease, that is, it does not mean that the probability you have the disease equals 1.
3. The three key pieces of information needed to apply Bayes' Law to evaluate the disease test are (i) the test's sensitivity, (ii) the test's specificity, and (iii) the prevalence of the disease in the population.
4. The numerical answers often are not what one might expect based on intuition.
5. You can apply the formula for Bayes' Law a second time by just using $\Pr(D \mid \text{test } +)$ from a first test that was + in place of $\Pr(D)$ in the formula. Actually, you can repeat this sort of update as many times as you are willing to be tested.
6. You should never confuse $\Pr(D \mid \text{test } +)$ with its reverse probability $\Pr(\text{test } + \mid D)$, since they have very different meanings.

By the way, reverse probability aside, here is a bit of practical advice. It is based on the idea of just thinking about what you might do or not do in terms of conditional probabilities. It often is very useful to think about whether you have control over the event that is to the right of the conditioning bar. In some cases, there are things that you can control.

Let's go back to the disease testing example. What if the D in the example is gonorrhea? The numerical results suggest that it is not a good idea to have unprotected sex with a stranger you just met. Think about the two conditional probabilities

Pr(gonorrhea or another venereal disease | you had unprotected sex)

versus

Pr(gonorrhea or another venereal disease | you had sex and used a condom).

The two points are that (1) these conditional probabilities quantify the risks of an extremely undesirable event, and (2) you have control over the event to the right of the conditioning bar. It also applies to things like whether you look both ways before you cross the street, limit your consumption of sugar and alcohol, or find a competent statistician to design your experiments and analyze your data. The simple point is that, *from a practical perspective, sometimes it is very useful to think about conditional probabilities by reading them from right to left.* That is, first think about the event to the right of the conditioning bar and, given that, think about the probability of the event on the left. If you have some control over what goes to the right of the conditioning bar, this little exercise can be extremely useful. The issue is not only that any given action may have several potential consequences, but that, in many cases, you can choose between possible actions by thinking about their possible consequences in terms of conditional probabilities. This may seem pretty simple minded, but it is amazing how many people do not perform this sort of thought experiment. Statistics has lots of thought experiments that can be used to make better decisions in real life.

1.6   Bayes' Law and Hemophilia

It is useful to see how Bayes' Law can be applied to take advantage of knowledge about a genetic abnormality. Hemophilia is a genetic disease carried on the human X chromosome. Denote the defective chromosome that causes hemophilia by X_h. People with this disease have blood with a severely reduced ability to clot, which may cause severe bleeding from a slight injury. Here are some basic facts:

1. Males have one X and one Y chromosome, denoted by XY, while females have two X chromosomes, XX. Everybody gets one chromosome from their mother and one from their father. Since the mother must contribute an X chromosome, while the father contributes either an X or a Y, the father's contribution determines whether the child will be male or female.
2. A female with chromosomes $X_h X$ does not have hemophilia, since the genetic information on the normal X chromosome provides what is needed for normal blood clotting.
3. A male hemophiliac has chromosomes $X_h Y$. Since he must have received the Y chromosome from his father, he must have received the defective X_h chromosome from his mother.

4. A female hemophiliac has chromosomes $X_h X_h$, with one affected X_h each from her mother and her father. This is extremely rare, since the X_h chromosome is rare.

If a woman, Louise, has an affected brother, with chromosomes $X_h Y$, but she, her mother, and her father all are normal, then her mother must have chromosomes $X_h X$, that is, her mother is physically normal but she is a carrier of the disease. Since Louise must have received a normal X from her father and the probability that she received the X_h chromosome from her mother is 1/2, the probability that Louise also is a carrier must be $\Pr(X_h X) = 1/2$. So the probability that Louise is not a carrier is $\Pr(XX) = 1 - 1/2 = 1/2$.

Now, let's think about possibilities and risks for Louise's male children. Let $Y_1 = 1$ if her firstborn son is a hemophiliac, and 0 if not. The probability that her firstborn son is a hemophiliac is

$$
\begin{aligned}
\Pr(Y_1 = 1) = {} & \Pr(Y_1 = 1 \mid \text{Louise is } X_h X)\,\Pr(\text{Louise is } X_h X) \\
& + \Pr(Y_1 = 1 \mid \text{Louise is } XX)\,\Pr(\text{Louise is } XX) \\
= {} & 1/2 \times 1/2 + 0 \times (1 - 1/2) = 1/4.
\end{aligned}
$$

The 0 entry quantifies the fact that, if Louise has two normal X chromosomes, then her firstborn son could not possibly be a hemophiliac. So, *knowing that her brother is a hemophiliac implies that there is a 1 in 4 chance that her firstborn son will be a hemophiliac*.

Next, suppose you are told that Louise's first son is normal. What is the probability that she is a carrier? Before we had this knowledge about her firstborn son, we knew that the probability she was a carrier was 1/2. To see how this new information changes this probability, we can apply Bayes' Law. Denote H = [Louise is a carrier] and $\overline{H}$ = [Louise is not a carrier]. Recall that

$$
\Pr(Y_1 = 1 \mid H) = 1/2 = \Pr(Y_1 = 0 \mid H).
$$

It also must be the case that

$$
\Pr(Y_1 = 1 \mid \overline{H}) = 0 \quad \text{and} \quad \Pr(Y_1 = 0 \mid \overline{H}) = 1.
$$

That is, if Louise is not a carrier, then her son cannot be a hemophiliac. Bayes' Law says that the probability Louise is a carrier, given that her firstborn son is normal, must be

$$
\begin{aligned}
\Pr(H \mid Y_1 = 0) &= \frac{\Pr(Y_1 = 0 \mid H)\,\Pr(H)}{\Pr(Y_1 = 0 \mid H)\,\Pr(H) + \Pr(Y_1 = 0 \mid \overline{H})\,\Pr(\overline{H})} \\
&= \frac{1/2 \times 1/2}{1/2 \times 1/2 + 1 \times (1 - 1/2)} = 1/3.
\end{aligned}
$$

So, *knowing that Louise's firstborn son is normal decreases the probability that she is a carrier from 1/2 to 1/3.*

Louise plans to have a second child. If that child is male, what is the probability that he will be a hemophiliac? Denote $Y_2 = 1$ if her secondborn son is a hemophiliac, 0 if not. Since you know that her firstborn son was normal, the quantity that should be computed is the predictive probability $Pr(Y_2 = 1 \mid Y_1 = 0)$. Since we can partition the event $(Y_2 = 1)$ as

$$(Y_2 = 1) = (Y_2 = 1 \text{ and } H) \cup (Y_2 = 1 \text{ and } \overline{H}),$$

the Law of Total Probability says that

$$Pr(Y_2 = 1) = Pr(Y_2 = 1 \text{ and } H) + Pr(Y_2 = 1 \text{ and } \overline{H}).$$

So, the predictive probability for Louise's second son can be computed as

$$\begin{aligned}
Pr(Y_2 = 1 \mid Y_1 = 0) &= Pr(Y_2 = 1 \mid H) \times Pr(H \mid Y_1 = 0) \\
&\quad + Pr(Y_2 = 1 \mid \overline{H}) \times Pr(\overline{H} \mid Y_1 = 0) \\
&= 1/2 \times 1/3 + 0 \times (1 - 1/3) = 1/6.
\end{aligned}$$

This says that *knowing that Louise's firstborn son is normal reduces the probability that her secondborn son will be a hemophiliac from 1/4=0.25 to 1/6=0.17.*

These examples of Bayesian learning show how one's uncertainty about important events can be quantified by their conditional probabilities given your current knowledge. They also show how new information can be used to change those probabilities, by using the process of Bayesian learning.

Chapter 2
Frequentists and Bayesians

The most misleading assumptions are the ones you don't even know you're making.
Douglas Adams

Contents

Abstract This chapter explains some elementary statistical concepts, including the distinction between a statistical estimator computed from data and the parameter that is being estimated. The process of making inferences from data will be discussed, including the importance of accounting for variability in data and one's uncertainty when making statistical inferences. A detailed account of binomial confidence intervals will be presented, including a brief history of how Gauss, DeMoivre, and Laplace established important ideas that still are relevant today. The relationship between sample size and statistical reliability will be discussed and illustrated. I will introduce Bayesian statistics, which treats parameters as random quantities and thus is fundamentally different from frequentist statistics, which treats parameters as fixed but unknown. Graphical illustrations of posterior distributions of parameters will be given, including posterior credible intervals, illustrations of how a Bayesian analysis combines prior knowledge and data to form posterior distributions and make inferences, and how reliability improves with larger sample size. An example will be given of how being biomarker positive or negative may be related to the probability that someone has a particular disease.

© Springer Nature Switzerland AG 2020   19
P. F. Thall, *Statistical Remedies for Medical Researchers*, Springer Series
in Pharmaceutical Statistics, https://doi.org/10.1007/978-3-030-43714-5_2

2.1 Statistical Inference

All statistical methods rely on underlying probability models, which describe uncertainty about possible events that may or may not happen. Statistical inferences are based on observed data. They may take the form of statements about the probabilities of certain events of interest. For example, when applying Bayesian statistics, an inference might take the form "Given the observed data, the probability that treatment A is superior to treatment B is 0.85." This simple statement actually is very different from the way that treatments are compared based on statistical tests of hypotheses, which are what researchers use most commonly. In Chap. 5, I will discuss some ways to quantify the strength of scientific evidence that are alternatives to conventional tests of hypotheses.

If you are engaged in medical research, then you already know that statistics plays a central role in how you design studies, analyze data, and reach conclusions about treatment effects. If you are not a statistician, then you must rely on statisticians to get things right. But we do not live in a perfect world. Human beings make mistakes and, unfortunately, this occurs more frequently than many people realize when applying statistical methods during the research process. What may be surprising is that highly educated, intelligent, dedicated researchers may make serious statistical errors. It is not easy being a medical statistician in the modern world.

Statistics, done sensibly, provides a rational foundation for scientific method. One might even say that statistics *is* scientific method. The two main objects in statistics are data, which can be observed, and parameters, which are conceptual objects that usually can't be observed. Still, you know what a parameter represents. In medical research, a parameter may be the probability that a patient will respond to a particular treatment, the effect of a patient's age on their expected survival time, or mean daily pain score during the week following a particular surgery. Parameters can be thought of as describing or quantifying what can be expected to happen to a population of future patients who will receive the treatment.

Although you don't know the numerical values of parameters, you can take a statistical approach to learn about them from observed data. This is done by performing a medical experiment in which patients are treated and their outcomes are observed, that is, by conducting a *clinical trial*, and using the resulting data to make statistical inferences about the parameters that you are interested in. Planning how to do this is called *experimental design*, which can take a myriad of different forms, depending on the medical setting, what you want to find out, and the available resources. Or you might just find some data on what happened in the past to patients who received the treatment you want to evaluate, and analyze it statistically to draw your conclusions. This is called *observational data analysis*, or *historical data analysis*. There are so many ways to do experimental design or data analysis incorrectly, someone could write a book about them. If you got this far, you are reading that book.

It is common to hear or read that someone "fed the data into the computer" and came up with some sort of magical answer to an important question. This implies that a computer can think, and that whatever it outputs must be correct. For most people, any technology that they do not understand looks like magic. Actually, a computer is a tool, designed to follow instructions which are codified in computer programs, also known as software. Recent advances in artificial intelligence notwithstanding, computers are only as smart as the programmed instructions that they receive. Of course, modern computers are very fast, and a gang of computers that receive programmed instructions over the Internet can be quite a powerful tool, as we now are learning. When operating systems learn to function autonomously and reprogram themselves and each other, much of what I have written here may seem archaic and even quaint. But we are not there, at least not yet.

When someone says that they "crunched the numbers," what they actually are talking about, most of the time, is that some sort of statistical method was applied to some dataset, and the necessary computations were done using a computer program. The program, along with the data, were input into a computer, and the output was the "crunched numbers." What this really means is that the statistical analysis reduced what may have been a large dataset to a small number of statistics that represent the dataset, in some summary fashion. For example, if each of 200 patient's ages is recorded, then this may be reduced by computing two statistics, the sample mean $\overline{age}$ and the standard deviation s, which characterizes variability. So, 200 numbers were crunched into two numbers. Under reasonable assumptions, if your goal is to estimate the mean age in the population that the sample of 200 ages came from, then $(\overline{age}, s)$ actually contains all the information that you need. The formal name for this is that these are "sufficient statistics." For example, the statistics $(\overline{age}, s) = (40, 10)$ might be used to compute a 95% confidence interval $[L, U]$ for the mean age of the population from which the data were sampled, using the formulas $L = \overline{age} - 1.96s/\sqrt{200} = 38.6$ and $U = \overline{age} + 1.96s/\sqrt{200} = 41.4$. But if $s = 40$, then the 95% confidence interval becomes $[33.1, 46.9]$, which is a lot wider than the interval $[38.6, 41.4]$ if $s = 10$. So the numerical value of s, which quantifies the variability in the data, matters a lot. This sort of number crunching is done all the time, but the real question is how such a confidence interval may be interpreted to make an inference about the mean age of people that the sample represents. This turns out to be a nontrivial issue, since confidence intervals often are misinterpreted.

This brings to mind a consulting job that I once did for a law firm. It turned out that one of the junior lawyers had hired a company that did data analysis to "crunch some numbers" and come up with the sample mean, $\overline{X}$, of an important variable in a law case. The people at the company had tried to provide both $\overline{X}$ and s, but the junior lawyer said that he did not want s. He just took the value of $\overline{X}$, told the company to destroy their copy of the data, and paid them. As the case proceeded, the numerical value of $\overline{X}$ turned out to be the bone of contention, and the opposing side asked for a 95% confidence interval to quantify the variability of $\overline{X}$. That is when they hired me. I had a brief but pithy conversation with a senior partner, just the two of us, in a big room with a huge wooden table. When he asked me if there was some magical statistical way to compute a confidence interval based on $\overline{X}$ alone, I explained that s was needed. The senior partner thanked me, asked me to send him an invoice, shook my hand, and I left. I never heard from or saw any of them again. The moral to this story is that a statistical estimator is of little use unless one also has some index of the variability in the data, to quantify the reliability of the estimator.

Going back to the earlier example of estimating the mean age, suppose that each patient's data are more complex, which usually is the case. Suppose the data consist of (Age, Sex, Disease Severity, Treatment, Outcome), where "Outcome" is the four-level ordinal disease status variable Y taking on the possible values {Complete Response, Partial Response, Stable Disease, Progressive Disease}, abbreviated as {CR, PR, SD, PD}. Suppose that you have a computer program that does the computations needed to fit a statistical regression model, say with response as the outcome and the vector $X =$ (Age, Sex, Disease Severity, Treatment) of possible predictive variables, or "covariates." So now age is not the only variable, and its main role is as a possible predictive variable for response level. One may be interested in, say, estimating $\Pr(Y = PR$ or $CR)$ for a given X. A numerical summary of a fitted regression model describing how Y varies with X that is output may have about 20 statistics that, together, describe the relationships between these variables. Thus, $200 \times 5 = 1000$ numbers or categories in the dataset have been "crunched" to about 20 summary statistics. These statistics typically appear in the output from a statistical software package that was used to fit the model. To understand what they mean, you must understand the regression model that was fit to the data. In this sense, fitting a statistical model typically performs a dimension reduction, in this example from the 1000 raw data values to the 20 statistics that characterize a fitted model. What actually is going on, inferentially, is that the fitted regression model may reveal important patterns in the data, including the effects of age, sex, disease severity, and treatment on response level.

Any given data analysis may have been done extremely well, reasonably well, or so poorly that it produced an incorrect or misleading inference. Since this book is about flawed statistical practices, and remedies in the form of practical alternative statistical methods that address these flaws, it is important to first establish what is meant by "statistical inference." Statistics, as it is today, is a rapidly developing field with a myriad of different models, methods, and computational algorithms, many of which are astonishingly powerful. But the basic ideas are pretty simple. Statistics gives us computational methods to use observed data to make conclusions about the phenomenon that produced the data. These conclusions are called statistical inferences.

A statistical inference may be a parameter estimate like $\overline{\text{age}}$ accompanied by a confidence interval, a statement about how likely it is that one treatment is better than another, or a prediction about the future survival time of an individual with a certain disease and covariates who is given a particular treatment. For example, an inference might be based on a sample of 50 patients with some disease who received a drug, with the inference having to do with how the drug will behave, on average, in any future patients with that disease who are given the drug. It might be the predicted survival time of a single patient given that patient's characteristics, disease severity, and history of previous treatments. Another type of inference, to compare drugs A and B, might refer to the three possibilities, {A is better than B}, {B is better than A}, or {A and B are equivalent} in terms of some key parameter like mean survival time.

A statistical inference might use observed data to assign probabilities to these three events, which necessarily must sum to 1 since these are the only three possibilities. Depending on the particular treatments, an inference might take the form "Based on the available data, there is a 92% chance that A is better than B."

Epistemology is an area of philosophy defined as the theory of knowledge, also known as "The Basis for Assent." It deals with how one may distinguish justified belief from pure opinion. This does not always involve the sort of empirical evidence that is the focus of statistical science. For most people, including much of the scientific community, statistical inference, which is based on observable data, is pretty exotic stuff. There are lots of other, much more popular ways to decide what is true and what is false, or to take the intermediate approach of assigning probabilities to possible events. If you are a religious fundamentalist, then you can deduce anything you need to know from your religion's book, by praying to your God in search of the answer, or just asking one of your religion's local priests. If you are a narcissist, then you will just conclude whatever your intuition tells you must be true, since you know that you always are right. If you are a hedonist, there is no reason to bother thinking about what is or is not true, since you can just have fun doing whatever you enjoy. If you are a politician or a compulsive liar, which might be a difficult distinction to make, you will always say whatever you think sounds convincing, no matter how ridiculous it may appear to be.

This book is for the tiny subgroup of the global population who care about applying statistical science to design experiments and make sensible inferences from observed data. These are people who have chosen to undertake the painful task of thinking, and possibly even abandoning conventional behavior that they have discovered is foolish.

2.2 Frequentist Statistics

Classical "frequentist" statistical methods treat parameters as fixed but unknown quantities. Here is a simple example of how frequentist statistics may be used to estimate a parameter. Suppose that you give a new drug to $n = 50$ patients with a particular disease and observe $R =$ the number who responded out of the 50. Then $\hat{\pi} = R/n = R/50$ is the usual frequentist statistical estimator of the parameter $\pi =$ probability that any given patient with the disease will respond to this treatment. Usually, this is given as a percentage $(100 \times R/50)\%$. But the statistic R is a random quantity, since before the experiment you do not know which of the possible values 0, 1, ..., 50 it will take on. That is, you have uncertainty about R. The binomial probability distribution gives explicit formulas for the probabilities that R will take on each of these 51 possible values. I will give the binomial probability formula below, but you can find it in any introductory statistics book. The binomial distribution depends on the parameter π, which you do not know. So, the statistical problem is to use the data $(R, 50)$ to estimate π. That is a formal description of what people are doing when they compute a percentage. Most researchers agree that $\hat{\pi} = R/50$, the

sample proportion, is a reasonable statistical estimator of π. If you observe $R = 30$
responses, then $\hat{\pi} = 30/50 = 0.60$, equivalently a 60% response rate. But since the
value 0.60 is based on data, and data are random, if you did the experiment a second
time you would be very likely to obtain a different value of R and thus a different $\hat{\pi}$. Moreover, if you observed $R = 6$ responses in $n = 10$ patients, or $R = 180$
responses in $n = 300$ patients, you would get the same answer of 60%. Obviously,
"60%" based on $n = 10$, 50, or 300 are not equally reliable. But in any case, you
cannot conclude with certainty that $p = 0.60$, only that your estimator $\hat{\pi}$ of π *based*
on your sample data was observed to be 0.60.

The number of responders, R, follows a *binomial distribution* with parameters
(n, π), written $R \sim binom(n, \pi)$. The statistic R can take on any integer value
$0, 1, \ldots, n$, and its mean, or "expected value" is πn. For example, if $n = 10$ then
you can possibly observe any of the values $0, 1, \ldots, 10$ for R, but you expect to observe
$\pi 10$ responses, which may be written as $E(R) = \pi 10$. The binomial distribution of
R is given by the formula

$$\Pr(R = r \mid n, \pi) = \frac{n!}{r!(n-r)!} \pi^r (1 - \pi)^{n-r} = \binom{n}{r} \pi^r (1 - \pi)^{n-r}$$

for each possible value $r = 0, 1, \ldots, n$. This formula is valid if the n trials (or
patients) that each resulted in response or nonresponse are independent of each other.
For example, if patient #1 has a response, independence implies that it has no effect on
whether any other patent will respond. The notation $k! = k \times (k - 1) \times \cdots \times 2 \times 1$
is the "factorial" of k for any positive integer k. The factorial terms are included
to account for all possible orderings of the r responses and $n - r$ nonresponses for
each r, which ensures that the binomial probabilities sum to 1. For example, if $n = 5$
and $r = 2$, then which of the 10 possible orderings of the two responses and three
nonresponses occurs does not matter. For $r = 2$,

$$\Pr(R = 2 \mid n = 5, \pi) = \frac{5!}{2!\,(5-2)!} \pi^2 (1 - \pi)^{5-2} = 10\pi^2 (1 - \pi)^3.$$

Whatever the response probability π may be, since the only possible values of R are
$0, 1, 2, 3, 4,$ or 5, it must be the case that

$$\sum_{k=0}^{5} \Pr(R = k \mid n = 5, \pi) = \sum_{k=0}^{5} \frac{5!}{k!\,(5-k)!} \pi^k (1 - \pi)^{5-k} = 1.$$

If n is larger, you do not need to compute these probabilities by hand using the
binomial formulas, since there are lots of computer programs that do this sort of
tedious work for you. For example, in Splus, $dbinom(r, n, \pi)$ does the numerical computation of $\Pr(R = r \mid n, \pi)$ for the binomial(n, π) for any sample size
n and probability π, and $pbinom(r, n, \pi) = \Pr(R \leq r \mid n, \pi)$. So you can obtain
probabilities like $dbinom(2, 5, 0.40) = 0.3456$ in a few seconds. But you also can

compute $dbinom(10, 30, 0.40) = 0.1152$ and $pbinom(10, 30, 0.40) = 0.2915$ just as quickly, despite the fact that these would take a very long time to compute by hand. Of course, in practice, you never know the value of π.

As a thought experiment to see how the binomial distribution behaves, assume that, in fact, $\pi = 0.70$ and you treat $n = 30$ patients. Then $R \sim binom(30, 0.70)$. Although you might possibly observe anywhere between 0 and 30 responses, you can expect that $0.70 \times 30 = 21$ patients will respond and 9 will not respond. In this experiment, the usual binomial estimator of the parameter π is the statistic $\hat{\pi} = R/30$, so if you observe $R = 18$ then $\hat{\pi} = 18/30 = 0.60$. But the computed statistic $\hat{\pi} = 0.60$ is smaller than the true value $\pi = 0.70$ that it is estimating. How can this be? It is because R is a random variable, following a $binom(30, 0.70)$ distribution, so $\hat{\pi} = R/30$ also must be random, and it has a distribution on its possible values $0, 1/30, \ldots, 29/30, 1$. For the $binom(30, 0.70)$ distribution, $\Pr(R \leq 18) = 0.16$, so it should not be surprising that the observed value is $R = 18$, giving $\hat{\pi} = 0.60$, was observed.

What I have done here with this "What if you know the actual value of π" computation is to reverse the inferential process that actually is done in practice, which is to learn about π from data. The scientific process may proceed by deciding to treat $n = 30$ patients, writing down the assumed binomial model [number of responders] $= R \sim binom(30, \pi)$, treating 30 patients, observing R, and using the observed data to make inferences about the unknown parameter π. So, observing $R = 18$ gives the usual estimated parameter value $\hat{\pi} = 0.60$. Now, remember the idea that $n = 300$ should give more reliable inferences, in particular a more reliable estimator of π, than $n = 30$. So, computing $\hat{\pi}$ is not enough, since you cannot be certain that $\hat{\pi} = \pi$. *A statistical inference based on a computed estimate does not say that one is certain the numerical estimate equals the parameter value.* Saying "If you give a patient this treatment, there is a 60% chance they will respond" is incomplete and misleading, because one cannot be certain that the response rate in the patient population of interest equals the statistic 60% computed from the sample. It is only a statistical estimate of the response rate. In general, *any statistical estimate must be accompanied by additional statistical information that quantifies one's uncertainty about the estimator.*

To quantify your uncertainty about your estimate $\hat{\pi} = 0.60$, you might apply a famous theorem attributed to two French mathematicians, de Moivre (1718) and Laplace (1812), translated into English in Laplace (1902). This is an example of a "Central Limit Theorem." There are lots of Central Limit Theorems, but the DeMoivre–Laplace Theorem was the first. The French nobility loved people like DeMoivre and Laplace, mainly because a lot of these noble people had a lot of free time and were really into gambling, especially card games, and probability computations could help them win. The DeMoivre–Laplace Theorem provides a tool to formally quantify one's uncertainty about $\hat{\pi} = R/n$. It says that, for a sufficiently large sample n, the distribution of $\hat{\pi}$ is, approximately, a bell-shaped curve with mean π and variance $\pi(1 - \pi)/n$. It also says that the approximation gets better as n grows larger, hence the name "Central Limit Theorem." The phrase "Bell-Shaped Curve" actually is a slang term for "normal distribution" or "Gaussian distribution."

Gauss was a brilliant German mathematician who was interested in all sorts of things, including statistics, physics, and astronomy. Although there is some argument about whether Laplace or Gauss invented the normal distribution first, Gauss derived it while he was inventing least squares regression.

The DeMoivre–Laplace Theorem implies that, approximately, for a binomial sample of sufficiently large size n, a 95% confidence interval (ci) for π will have lower and upper limits

$$L = \hat{\pi} - 1.96\sqrt{(\hat{\pi}(1 - \hat{\pi})/n} \quad \text{and} \quad U = \hat{\pi} + 1.96\sqrt{(\hat{\pi}(1 - \hat{\pi})/n}. \qquad (2.1)$$

The approximation gets closer to 95% as the sample size n increases, so the formula isn't of much use for small sample sizes, say $n = 10$. The closeness of the approximation to 95% also depends on the values of both n and π, and it does not work well for π near 0 or 1. But of course, you do not know π; otherwise, you would not be estimating it. You also can compute 90% approximate confidence intervals by replacing 1.96 with 1.645. There is a large published literature on how close various approximate confidence intervals are to their nominal level. But these days most statisticians study and validate such things by computer simulation. It is important to keep in mind that, since L and U are computed from data, they are statistics. The famous formula (2.1) has been presented in a vast number of introductory statistics text books. As far as I can tell, in modern times, its two primary uses are to analyze data and to torture college students. For example, suppose that you observe there were $R = 30$ responses in a binomial sample of size $n = 50$. If you substitute the estimator $\hat{\pi} = R/n = 30/50 = 0.60$ obtained from the data into the formula (2.1), this gives the approximate 95% ci limits $L = 0.46$ and $U = 0.74$. Instead of bothering to solve the formulas, you might just use a computer program to obtain the numerical values of L and U and proudly declare "I used The Computer to crunch the numbers." Referring to The Computer, which is a technology that very few people really understand, often helps to make people believe what you are telling them. But remember, scientifically, the whole point of applying this statistical formula, provided to us long ago by French and German mathematicians and immortalized in countless textbooks, is to quantify uncertainty about the estimate $\hat{\pi} = R/n$. If you have an extremely large sample, then for all practical purposes the issue of uncertainty about the estimator goes away. For example, if you observe 6000 responses in a sample of 10,000 patients, then $\hat{\pi} = 0.60$ as before, but the approximate 95% confidence interval becomes 0.59–0.61. The point is, if your sample size is 10 or 50, don't pretend that it is 10,000.

But, once the numbers have been crunched, the question is, exactly what do the lower limit $L = 0.46$ and upper limit $U = 0.74$ actually mean? This same question of interpretation is what we ended up with after computing the 95% ci limits $L = 38.6$ and $U = 41.4$ years for mean age. The interpretation of a confidence interval requires some explanation, since one needs to know how it is defined. If you haven't seen this before, you might make sure that you are comfortably seated, and take a deep breath. Here is the definition of a 95% confidence interval. Suppose that, hypothetically, you were to repeat the experiment described above a very large number of times, always with the same sample size $n = 50$, observe R for each repetition, and compute the 95% ci limits $[L, \ U]$ each time using the above formula. So, you need to imagine that you could repeat your 50-patient experiment, say, 1000 times, or maybe 1,000,000 times. Remember, L and U are statistics, so you will get different $[L, \ U]$ pairs as you perform the many imaginary repetitions of the experiment. But throughout all this, π is some fixed but unknown number. The DeMoivre–Laplace theorem ensures us that you should expect, in about 95% of the many imaginary repetitions of the experiment, e.g., approximately 950 out of 1000, or 950,000 out of 1,000,000, that the pair of statistics L and U computed using the above formulas would have the fixed but unknown value of π between them. In the other 5% of the hypothetical repetitions, either $\pi < L$ or $U < \pi$. So, the definition of a confidence interval requires you to think about conducting an arbitrarily large number of imaginary replications of the experiment. Of course, the experiment was only conducted once, and you do not know whether or not the one random interval $[L, \ U]$ that was computed from the one observed dataset contains the fixed but unknown value of π.

Actually, it is fairly easy to repeat this sort imaginary experiment a very large number of times by writing and running a simple computer program. To do this, you may assume particular numerical values of π and n, and for each computer experiment generate n random 1's and 0's where a 1 has probability π and a 0 has probability $1 - \pi$, record $R =$ the total number of 1's, and compute $\hat{\pi} = R/n$, L, and U. If you repeat this, say, 1000 times and draw a histogram of all the $\hat{\pi}$ values, it should look approximately bell-shaped and be centered around the numerical value of π that you assumed, unless you made a mistake. The larger the number of repetitions, the more bell-shaped the curve will look, and the closer the middle will be to the value of π that you assumed. Since you can do this exercise on a computer fairly easily and very quickly for a very large number of replications, say, 10,000, or even 1,000,000, it is fun to have the computer program draw the histogram of $\hat{\pi}$ values repeatedly for successively larger values of the sample size n, and watch the histogram becoming more and more bell-shaped. Hundreds of years after the DeMoivre–Laplace theorem was proved mathematically, it is easy for someone with a laptop computer and a little programming skill to illustrate it graphically. Figure 2.1 gives histograms of $\hat{\pi}$ values for sample size values $n = 100$, 500, 1000, and 2000, to illustrate what the DeMoivre–Laplace Theorem looks like.

Even if someone accepts the statistical approach to estimation, they might argue "Why do I need to think about parameters at all? If I treat 30 patients and 18 of them respond, then the response rate is $(100 \times 18/30)\% = 60\%$." I didn't need a parameter to compute that. A probability is a percentage divided by 100, so the corresponding probability is always between 0 and 1. The above statement is equivalent to saying that $\pi = \Pr(\text{Response}) = 0.60$. Since you never know for sure whether a given patient will respond to a given treatment, saying $\pi = 0.60$ quantifies your uncertainty. But the question is, do you really know for sure that $\pi = 0.60$, based on the data from your 30-patient experiment? *While the parameter π quantifies your uncertainty about whether a given patient will respond, in practice you also will always have uncertainty about π. The best that you can do is estimate π from data.*

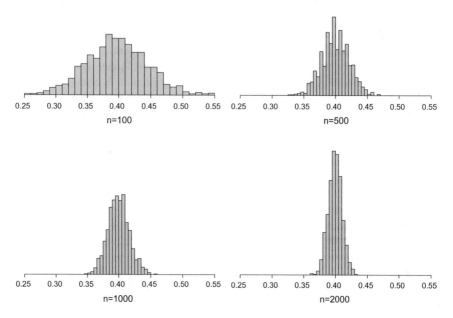

Fig. 2.1 Histograms of sample proportions $\hat{\pi} = X/n$ for sample sizes $n = 100, 500, 1000$, and 2000

Although a confidence interval quantifies uncertainty, we are still left with the problem that the definition of confidence interval is pretty technical, and many people find it hard to follow or remember. So, at this point, the whole statistical process of estimating parameters may seem like a mess. But do not despair. There is a very sensible way to estimate a parameter and also quantify your uncertainty about your estimator that is easy to understand. It is called *Bayesian Statistics*.

2.3   Bayesian Statistics

Bayesian statistics treats parameters as random quantities and endows them with probability distributions to quantify uncertainty about them. This is very different from frequentist statistics, which considers parameters to be fixed but unknown. The difference is more than just philosophical, since the two approaches can lead to very different inferences based on the same dataset. There will be many examples of this in the chapters to follow. Bayesian statistics is named after Thomas Bayes, a minister living in England during the eighteenth century whose notes on a special case of what is now known as *Bayes' Law*, also called *The Law of Reverse Probability* or *The Magic Formula*, were discovered and published after his death. It turns out that Laplace actually derived the general form of this law, and spent many years refining and applying this new way of thinking. This may seem unfair, and perhaps it should be called "Laplacian Statistics." But maybe Bayes considered what he had derived to be so unimportant or arcane that he tucked away his notes in a desk drawer. Or maybe he was just a humble Man of God. If you are interested, Stigler (1986) provides a history of all that, and a technical review of Laplace's 1774 paper. In the twentieth century, Jerome Cornfield (1969) was one of the early pioneers in advocating application of Bayesian statistics in medical research and clinical trials. This is remarkable because, at that time, none of the modern computational tools that now make Bayesian methods so easy to apply and thus so powerful had yet been developed. A brief history is given by Wittes (2012). If you want to have some fun, you might read a very informative, nontechnical book by Mc Grayne (2011), who gives an extensive history of Bayesian statistics and how it has been used over the years to solve a wide array of practical problems.

I have discussed the need to quantify one's uncertainty when using data to estimate a parameter, and I have suggested that one may use a confidence interval to do this. But the definition of "confidence interval" is a bit strange, and requires you to imagine repeating the experiment that generated your data many times, which may be unappealing or even confusing, since the experiment actually was done only once. Here is a different way to go about describing uncertainty.

First of all, most non-statisticians misinterpret the definition of a 95% confidence interval. When questioned, based on the confidence interval computed from the data $(R, n) = (30, 50)$ in the example given earlier, most people say

"There is a 95% probability that π is between 0.46 and 0.74."

I have asked many of my physician researcher colleagues how they interpret a 95% confidence interval and, aside from the particular numerical confidence limits obtained from their data, almost invariably this is the form of their answer. Interestingly, I also have asked many of my younger statistician colleagues the same question, and quite a few of them give exactly the same incorrect answer.

The above statement, or misstatement depending on your philosophy, may be written formally in terms of the Bayesian posterior probability equation

$$\Pr(0.46 \leq \pi \leq 0.74 \mid 30 \text{ responses in 50 patients}) = 0.95.$$

The vertical line "|" is called the "conditioning bar" since what goes to the right of it is assumed to be known, which in this example is the observed data, while what goes to the left of it is the event about which you are uncertain. Reading the above equation from left to right, words for the conditioning bar are "given that." *Under the frequentist approach to statistics, this probability equation cannot be correct, since making a probability statement about π treats it like a random quantity.* This contradicts the assumption in frequentist statistics that any unknown parameter, like π, is a fixed value. So, in frequentist statistics, the above probability equation makes no sense. When confronted with this dilemma, many non-statisticians become confused, essentially because they actually think of π as a random quantity, since they have uncertainty about it. The above probability statement makes perfect sense to them. Such people are known as "Bayesians." In my experience, whether they know it or not, almost all physicians are Bayesians.

The good news is that the probability equation given above actually is an example of a *Bayesian 95% posterior credible interval*, although the numbers are not quite correct since they were computed using the 95% confidence interval formula. But there is no need to despair, since there are plenty of Bayesian statisticians who are happy to compute the correct credible interval limits, and also plenty of easy-to-use computer programs that you can run to do the numerical computations yourself.

For example, taking a Bayesian approach, suppose that you begin with the prior assumption that the unknown *random* parameter π is distributed uniformly between 0 and 1. This reflects the belief that no value of π is either more or less likely than any other value. Then, given that you observe 30 responses in 50 patients, the 95% posterior credible interval for π is [0.46, 0.72]. These numbers are not very different from the frequentist confidence limits [0.46, 0.74] in this simple example. The point is that, with a Bayesian approach, you have a coherent way to describe your uncertainty about π once you have observed your data. Conceptually, a credible interval is a very different way to quantify uncertainty than a confidence interval. Credible intervals make perfect sense. You do not need to think about a huge number of imaginary experiments, only the experiment that actually was done and the data that actually were observed. This simple example notwithstanding, later on I will give examples where a Bayesian analysis gives a coherent answer in settings where a frequentist design or data analysis gives a very different inference that is either useless or dangerously misleading. The Bayesian approach to statistical inference quantifies uncertainty in a way that is very different from conventional frequentist statistics, which leads to problematic things like confidence intervals and p-values. I have not yet talked about the horrors of p-values, which is quite a painful topic. I will discuss that scientific train wreck in Chap. 5.

Here are some formalities about Bayesian statistical inference. It begins with the key assumption that all model parameters, θ, are random, since you have uncertainty about them. Consequently, a Bayesian model must include an assumed prior distribution $p(\theta \mid \phi)$ on the parameters, to describe how much or how little you know about θ before data are observed. I have denoted the *prior hyperparameters* by ϕ, which are actual numerical values that quantify your prior uncertainty about θ. For observable data Y and parameter θ, the likelihood function, $\mathcal{L}(Y, \theta)$, is the probability distribution of Y given θ. This describes uncertainty about the data before it is observed. Once Y is observed, Bayesian inference is done by using the assumed prior and the likelihood function to compute the *posterior distribution* of the parameter,

$$p(\theta \mid Y, \phi) = c \, \mathcal{L}(Y, \theta) \, p(\theta \mid \phi).$$

The multiplicative constant c is included to ensure that the posterior $p(\theta \mid Y, \phi)$ is a probability distribution on θ. Inferences about θ are based on its posterior.

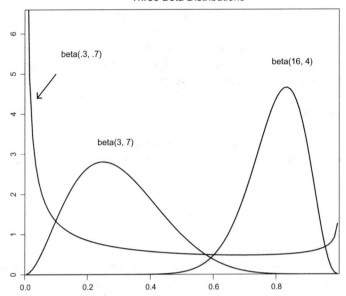

Fig. 2.2 Probability densities of three different beta distributions, corresponding to a non-informative prior with mean 0.3 and effective sample size ESS = 1, a second with mean 0.3 and ESS = 10, and a third with mean 0.8 and ESS = 20

Mathematically,

$$c = \frac{1}{\int_\theta \mathcal{L}(Y, \theta)\, p(\theta \mid \phi) d\theta},$$

although in practice c often does not need to be computed to obtain a posterior. Thus, the observed data Y are used to turn the prior $p(\theta \mid \phi)$ into the posterior $p(\theta \mid Y, \phi)$. This sometimes is called *Bayesian Learning*.

As an example of how this works, let's revisit the simple experiment where n patients are treated, the parameter is $\theta = \pi = \Pr(\text{response})$, and you observe $Y = R =$ number of responses out of the n patients. Recall that R follows the binomial distribution, $R \sim binom(n, \pi)$. To estimate the response probability, $\theta = \pi$, the most common Bayesian approach is to assume a $beta(a, b)$ prior on p, which has hyperparameters $\phi = (a, b)$, where $a > 0$ and $b > 0$, mean $\mu = a/(a + b)$, and variance $\mu(1 - \mu)/(a + b + 1)$. The sum $a + b$ is the prior effective sample size (ESS). For example, the uniform distribution for π is a $beta(1, 1)$, which has $\mu = 0.50$ and ESS = 2. A more informative $beta(5, 5)$ distribution also has mean 0.50, but ESS = $5 + 5 = 10$, reflecting prior knowledge corresponding to 10 observations. As an illustration, Fig. 2.2 gives three different beta distributions for a probability π.

A nice property of the beta prior and binomial likelihood, known as a beta-binomial model, is that, once one has observed the data (R, n), the posterior also is a beta distribution,

$$\pi \mid (R, n), (a, b) \sim beta(a + R, b + n - R).$$

That is, to obtain the two parameters of the beta posterior of π, all you need to do is add the observed number of responses R to a and add the number of observed nonresponses (failures) $n - R$ to b. So, the posterior mean of π is $\mu = (R + a)/(n + a + b)$ and the ESS $= n + a + b$. The technical name for this is that, for a binomial distribution with probability π, the beta prior on π is *conjugate*, since the posterior is also a beta distribution.

An interesting property of an assumed beta prior $\pi \sim beta(a, b)$ is that, once the binomial data (R, n) are observed, the posterior mean of π can be written in the following form:

$$\frac{R + a}{n + a + b} = \frac{R}{n} \times \frac{n}{a + b + n} + \frac{a}{a + b} \times \frac{a + b}{a + b + n}.$$

That is, the posterior mean of π can be expressed as a weighted average of the conventional estimator R/n and the prior mean $a/(a + b)$ of π. As the sample size n grows larger, the weight $n/(a + b + n)$ given to R/n increases toward 1 and the weight $(a + b)/(a + b + n)$ given to the prior mean $a/(a + b)$ decreases toward 0. Put another way, as the sample size becomes larger, the influence of the prior on posterior inferences becomes smaller. So, with larger sample sizes, the data dominate posterior inferences. For example, if you start with a moderately informative beta prior based on earlier data with three responses in 10 patients, formally $\pi \sim beta(3, 7)$, then the posterior beta distribution is $[\pi \mid R, n] \sim beta(3 + R, 7 + n - R)$. So, starting with this prior, observing $R = 30$ responses in $n = 50$ patients would give posterior mean

$$\frac{30 + 3}{50 + 3 + 7} = \frac{30}{50} \times \frac{50}{3 + 7 + 50} + \frac{3}{3 + 7} \times \frac{3 + 7}{3 + 7 + 50}$$

$$= 0.60 \times 0.833 + 0.30 \times 0.167 = 0.55.$$

One way of describing this is that the Bayesian model *shrinks* the usual frequentist estimator $30/50 = 0.60$ toward the prior mean $3/10 = 0.20$. In general, the amount of shrinkage depends on the weights $n/(a + b + n)$ for the sample proportion R/n and $(a + b)/(a + b + n)$ for the prior mean. This illustrates the general property that the Bayesian posterior mean of a model parameter θ can be thought of as an average of the prior mean and the usual frequentist estimator of θ.

If you are very uncertain *a priori*, you might assume a prior equivalent to having information on only one observation, i.e., $\pi \sim beta(0.50, 0.50)$, which has ESS $= 0.5 + 0.5 = 1$. This gives the posterior $[\pi \mid R, n] \sim beta(R + 0.50, n - R + 0.50)$, which has mean

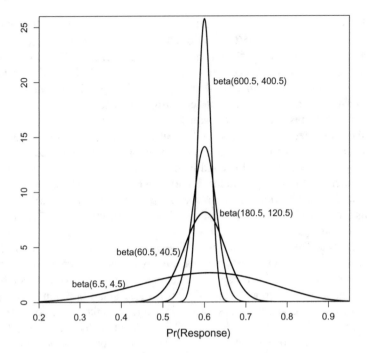

Fig. 2.3 Four beta distributions obtained, starting with a beta(0.50, 0.50) prior, as posteriors from samples of size 10, 100, 300, or 1000, all having empirical proportion 0.60

$$\frac{R+0.50}{R+0.50+n-R+0.50} = \frac{R+0.50}{n+1}.$$

For the data $(R, n) = (18, 30)$, the posterior is beta(18.50, 12.50). This posterior has mean $18.50/(18.50+12.50) = 0.597$ and 95% credible interval (ci) $[0.42, 0.76]$. That is,

$$\Pr(0.42 < \pi < 0.76 \mid 18/30 \text{ responses}] = 0.95.$$

Notice that the posterior mean 0.597 is nearly identical to the conventional estimate $18/30 = 0.60$, since the $beta(0.5, 0.5)$ prior has an effective sample size of only $0.5 + 0.5 = 1$, so it contains very little information relative to the sample size $n = 30$. Expressing prior and posterior uncertainty this way shows that, even after observing the data $(R, n) = (18, 30)$, you still don't know much about π, since the 95% posterior ci $[0.42, 0.76]$ has width $0.76 - 0.42 = 0.34$, and it contains both 0.50 and 70. If the data were $(R, n) = (6, 10), (60, 100), (180, 300),$ or $(600, 1000)$ these all give the same empirical proportion of 0.60, but the corresponding posteriors of π would be a beta(6.5, 4.5), beta(60.5, 40.5), beta(180.5, 120.5), or beta(600.5, 400.5) which give very different posterior 95% credible intervals for π. These four posteriors are illustrated in Fig. 2.3.

With the posterior based on $n = 300$, you have a lot less uncertainty about π, since its ESS $= 301$ and the 95% posterior credible interval for π is a lot smaller. In particular, for $n = 300$ the posterior ci does not contain 0.70, and $\Pr(\pi > 0.70 \mid R = 180, n = 300)) = 0.0001$. You can never be sure that $\pi = 0.60$, but if your data are based on $n = 300$ rather than $n = 10$ or 100, then, as Fig. 2.3 illustrates, the larger sample provides a much more reliable posterior inference. The point is that not all reports of a "60% response rate" are equally credible.

By the way, there is nothing sacred about the probability 0.95 for computing posterior credible intervals, although it is very popular in practice. You could use other coverage probabilities if you like, and it sometimes is useful to compute posterior credible intervals for multiple values. For example, given 18/30 responses, starting with a beta(0.50, 0.50) prior on π, the respective posterior 80, 90, 95 and 99% credible intervals for π are [0.48, 0.70], [0.45, 0.74], [0.42, 0.76], and [0.37, 0.77]. As should be expected, for the given data 18/30 responses, a higher coverage probability gives a wider interval. These four posterior credible intervals are illustrated in Fig. 2.4.

With very small sample sizes, making inferences about an unknown parameter or parameters may be a bit tricky. For example, suppose that you are studying a new, very expensive drug, and you only have a very small preliminary sample of $n = 3$ patients. The conventional estimator $\hat{\pi} = R/3$ can take on the possible values $0/3 = 0, 1/3 = 0.33, 2/3 = 0.67$, and $3/3 = 1$. A bit of thought shows that the estimator $\hat{\pi} = 0/3 = 0$ does not make much sense, since this says "Since I observed 0 responses in 3 patients, I estimate that the probability of response is 0. That is, my estimate says that *response is impossible with this drug.*" Similarly, if you observe $R = 3$ then $\hat{\pi} = 1$, the estimated response probability ia 1, and you get the equally silly inference "Since I observed 3 responses on 3 patients, I conclude that *response is certain with this drug.*" But we can avoid making such silly inferences since we know that reporting only the point estimate $\hat{\pi} = R/3$ does not account for uncertainty about this statistical estimator. It may appear that a sensible way out is to compute and report a 95% confidence interval. But the usual formula for a 95% confidence interval for π, which I gave above, gives [L, U] $=[0, 0]$ if $R = 0$ and [1, 1] if $R = 3$, which both are silly answers. So, you need some other way to quantify your uncertainty about $\hat{\pi}$. To avoid this sort of problem, many different methods for computing exact or approximate confidence intervals for π based on binomial samples have been proposed. A very thorough review is given by Piret and Amado (2008).

But why bother with any of this? Problems with the conventional estimator R/n and confidence interval formula for small sample sizes disappear if one takes a Bayesian approach. The beta-binomial model always gives a coherent, easily interpretable answer, since by definition the 2.5th and 97.5th percentiles of the posterior of π are a 95% posterior credible interval for π. It can be obtained in a few seconds using any statistical software package that computes beta distribution percentiles. For example, the R formulas $L = qbeta(0.025, \alpha, \beta)$ and $U = qbeta(0.975, \alpha, \beta)$ give the 2.5th and 97.5th percentiles of a $beta(\alpha, \beta)$ distribution, which are the 95% credible interval limits. In the beta-binomial computation, plugging in the values $\alpha = a + R$ and $\beta = b + n - R$ gives lower and upper limits of a 95% posterior credible interval. Similar percentile computations can be done very easily using R

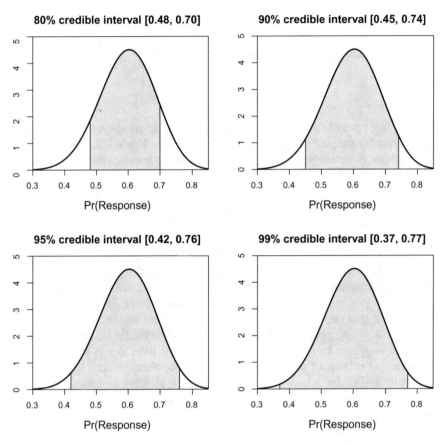

Fig. 2.4 Four posterior credible intervals having different coverage probabilities for $\pi = $ Pr(Response), each obtained from a sample with 18/30 responses, assuming a beta(0.50, 0.50) prior

or other languages for numerous other standard distributions, including the normal, Poisson, gamma, and exponential.

In the above example where $n = 3$, if we start by assuming a non-informative prior $\pi \sim beta(0.50, 0.50)$, which has mean 0.50 and effective sample size $0.50 + 0.50 = 1$, then the posterior is $\pi \sim beta(R + 0.50, 3 - R + 0.50)$. Of course, regardless of what statistical method is used, such a tiny sample size gives very little information about π. This is quantified by the respective posterior 95% credible intervals for π, which are [0, 0.54] for $R = 0$, [0.04, 0.82] for $R = 1$, [0.18, 0.96] for $R = 2$, and [0.46, 1.00] for $R = 3$. These wide 95% credible intervals quantify how little a sample size of $n = 3$ can tell you about π. But for point estimation of π, the posterior mean is $\hat{\pi} = (R + 0.50)/4$, which takes on the values $\hat{\pi} = 0.125, 0.375, 0.675, 0.875$ for $R = 0, 1, 2, 3$, respectively. So using the posterior mean as a point estimator of π never gives a silly value. Essentially, this is because the Bayesian approach computes the posterior mean by shrinking the conventional estimator of the mean, which is R/n in this example, toward the prior mean, which is assumed to be 0.50. Applying the general formula given earlier, the posterior mean is

$$\frac{R + 0.50}{4} = \frac{R}{3} \times \frac{3}{4} + 0.50 \times \frac{1}{4}.$$

In words, the posterior mean is a weighted average of the sample mean and the prior mean, with the weights depending on the sample size. If you like, you can repeat these computations using different priors, to see how the posterior means and credible intervals vary as you change the prior.

So, a sample of size 3 reduces your uncertainty about π, but not by much. This example is a good illustration of the important fact that looking at a point estimate without some quantification of uncertainty can be very misleading. For example, the sample proportions 1/3 and 100/300 both equal 0.33, but the corresponding 95% posterior credible intervals are [0.04, 0.82] and [0.28, 0.39]. Just saying "A 33% response rate was observed" is bad scientific practice, since it conveys no information about the reliability of the reported estimate. A similar frequentist analysis based on confidence intervals would reach the same conclusion. The point is that tiny sample sizes give very unreliable inferences. If this example seems artificial, you might bear in mind that dose escalation or de-escalation decisions in phase I clinical trials, where people's lives are at stake, typically are based on cohorts of size 3, and statistics of the form [# dose limiting toxicities at dose d]/n for $n = 3$ or $n = 6$ are used for making decisions about what dose to give to the next cohort of patients. In this setting, no matter what you observe, the data from a given cohort treated at a given dose tell you very little about the probability of toxicity at that dose. An absolutely terrible convention in phase I dose-finding trials is to report two toxicities in six patients treated at a particular dose of an experimental agent as "a 33% toxicity rate" without noting that it is based on a tiny sample of size 6. I will discuss this sort of lunacy in more detail in Chap. 9, which describes severe problems with conventional methods used in early phase clinical trials. In particular, I will discuss the use of very small samples to make unreliable inferences that affect whether people will live or die.

Here is another example of how you might use posterior credible intervals to quantify uncertainty. Let's go back to the blood test for the disease, but think about it statistically. So I will assume that we have some data and want to learn about the reliability of the test. Suppose you have a sample of 200 patients with D, and you test each for the presence of a biomarker that you think might be associated with the disease. The biomarker was identified by complex genomic or proteomic analyses or some other laboratory studies, and you wish to determine whether it might be used as an aid in diagnosing whether patients have the disease. If you observe 180 patients who are biomarker $+$ and 20 who are biomarker $-$, then the percentages are $100 \times 180/200\% = 90\%$ biomarker $+$ and $100\% - 90\% = 10\%$ biomarker $-$. You might say "I estimate that 90% of people with the disease are biomarker positive, with a 95% posterior credible interval 0.85–0.94 for Pr(biomarker + | D)." This sort of inference may seem simple, but there are some important issues to think about.

To make this example a bit more concrete, suppose that the disease is an infection, and the above estimate is used as a basis for the policy of giving antibiotics to anyone who tests biomarker +. One key thing that is missing here is an estimate of the percentage of people without the disease who test biomarker +. Suppose that a second sample of 200 noninfected people also is 90% biomarker +. Then the biomarker does not seem to be associated with the disease at all, since the rates of being biomarker + in infected and noninfected people are identical. So, in this case, the policy does not appear to make much sense.

But if the rate among non-diseased people is $100 \times 160/200\% = 80\%$ biomarker +, then the two rates are slightly different, so a question then is whether the difference is large enough to matter. This gets at the general problem of estimating the association between the biomarker being + or − and whether someone has the disease or not. Of course, if you want to use the biomarker as a prognostic tool, then what you really are interested in is the reverse conditional probability, Pr(D | biomarker +), that somebody who is biomarker + actually has the disease.

A key statistical issue is estimating and comparing the two conditional probabilities Pr(D | biomarker +) and Pr(D | biomarker −), since these are what matter most to someone who has just been tested. As mentioned above, these probabilities condition the events in the reverse order of what was given above, which could be written as Pr(biomarker + | D)=0.90 and Pr(biomarker + | $\overline{D}$)=0.80. Remember that it is extremely important not to confuse Pr(biomarker + | D) with the reverse conditional probability Pr(D | biomarker +) since they mean very different things. Let's think about how reliable the 0.90 and 0.80 are as estimators of the population rates of being biomarker + in people who do or do not have the disease. Since the values 0.90 and 0.80 are really statistical estimators of the two subpopulation probabilities, this could be written more formally as $\widehat{Pr}$(biomarker + I disease)=0.90 and $\widehat{Pr}$(biomarker + I no disease)=0.80. The hat symbol ⌢ over the probabilities can be read as "is a statistical estimator of the thing underneath the hat."

If we take a Bayesian approach to the biomarker data by assuming that

$$\pi_{+|D} = Pr(biomarker \ + \ | \ D)$$

and

$$\pi_{+|\overline{D}} = Pr(biomarker + \ | \ No \ D)$$

are random quantities that both follow non-informative beta(0.50, 0.50) priors then, given the observed data, we have the posterior probabilities

$$Pr(0.85 < \pi_{+|D} < 0.94 \ | \ 180 + in \ 200 \ with \ D) = 0.95$$

and

$$Pr(0.74 < \pi_{+|\bar{D}} < 0.85 \ | \ 160 + in \ 200 \ without \ D) = 0.95.$$

In other words, based on what has been observed in 200 patients with and 200 patients without the disease who were tested, a posterior 95% credible interval for $\pi_{+|D}$ is [0.85, 0.94] and a posterior 95% credible interval for $\pi_{+|\bar{D}}$ is [0.74, 0.85]. Since these two 95% credible intervals do not overlap, the data seem to provide reasonably convincing evidence that the probabilities of being biomarker + actually are different in people who do or do not have the disease. So now it seems likely that biomarker status and disease status may be related in some way.

These days, many statisticians who were trained as frequentists are quietly starting to use Bayesian methods. Essentially, this is because Bayesian methods work well in practice, are easy to interpret and explain to non-statisticians, and often have better properties than their frequentist competitors. Using Bayesian credible intervals rather than frequentist confidence intervals is just one example. There will be many more in the chapters that follow.

Chapter 3
Knocking Down the Straw Man

Everybody Else Is Doing It, So Why Can't We?
Title of The Cranberries' first album

Contents

Abstract An introduction to clinical trials is given, including a list of things to consider when designing a clinical trial. An extensive discussion is given of the Simon (1989) two-stage phase II clinical trial design, because it is used very commonly, and it very often is misunderstood or applied inappropriately. This design provides a useful illustration of more general problems with the way that the frequentist paradigm of testing hypotheses is misused or misinterpreted, and problems that may arise when making inferences based on single-arm trials. These include (1) treatment-trial confounding, (2) making biased treatment comparisons, (3) the common misunderstanding of what an alternative hypothesis means, (4) how rejection of a null hypothesis often is misinterpreted, (5) consequences of ignoring toxicity in phase II trials, (6) assuming incorrectly that a null response probability estimated from historical data is known with certainty, and (7) logistical problems that may arise when making interim adaptive decisions during a trial. Two alternative Bayesian methods will be described. The first is posterior evaluation of binomial data from a phase II trial based on a binary response variable, leading to a conclusion that is substantively different from that based on a test of hypotheses. The second is a practical Bayesian phase II design that monitors both response rate for futility and toxicity rate for safety.

© Springer Nature Switzerland AG 2020   39
P. F. Thall, *Statistical Remedies for Medical Researchers*, Springer Series
in Pharmaceutical Statistics, https://doi.org/10.1007/978-3-030-43714-5_3

3.1 Designing Clinical Trials

Before talking about specific clinical trial designs in detail, I will provide a general framework for how I think one ought to go about constructing a statistical design for a trial. My perspective was developed during three decades designing hundreds of clinical trials. First of all, one should keep in mind the overriding fact that a clinical trial is a medical experiment with human subjects. It has the following two main purposes.

The Main Purposes of a Clinical Trial

1. Treat the patients in the trial.
2. Obtain data about treatment effects that may benefit future patients.

In some early phase trials, an additional goal is to modify the treatment or treatments being studied, usually by changing dose or schedule of administration adaptively as patients are treated and data are acquired. A good design should do a reasonable job of serving all of these goals, despite the fact that they may be at odds with each other. The first goal often surprises statisticians the first time they see it, since many statisticians think of clinical trials as abstractions, mainly requiring a sample size computation using an out-of-the-box method based on a test of hypotheses. The second goal may seem obvious, but a surprisingly large number of clinical trials do not produce useful data, for a variety of reasons. Some trials produce data so unreliable or misleading that it would have been better to not conduct the trial at all. In some cases, this can be determined by a careful examination of the statistical design before the trial is begun. This may show that the trial will produce data that are not useful or are fundamentally flawed, or that the designs will not adequately protect patient safety. Some examples of this will be given in Chap. 10.

To achieve the two major goals, both medical and statistical thinking must be applied carefully while constructing a design. This process should begin with the statistician(s) determining key elements from the physician(s). A list of these elements is as follows:

Some Things to Think About When Designing a Clinical Trial

1. The disease(s) to be treated, and the trial entry criteria.
2. The treatments, doses, schedules, or treatment combinations to be evaluated.
3. Any within-patient multistage adaptive rules that will be applied to modify dose, change treatment, or deal with adverse events.
4. Any existing standard treatment or treatments that the patients would receive if they were not enrolled in the trial.
5. The main clinical outcome or outcomes that will be used to evaluate treatment effects, including possible severe adverse events, and the possibility of multiple co-primary outcomes.
6. The main goals of the trial, including how the trial results may be used for planning future studies or changing clinical practice.
7. Additional secondary goals of the trial.

 8. Ranges of anticipated accrual rates and feasible sample sizes.
 9. Whether the trial will involve one or several institutions.
10. Monetary costs, funding sources and limitations, and human resources.
11. Regulatory issues and requirements that must be satisfied.

Item 4 deals with the key question "Compared to what?" since, ultimately, all clinical trials are comparative. Even in settings where there is no safe and minimally effective treatment for a disease, such as so-called phase IIA trials, the comparator is to not treat the patient at all. In some instances, I have found that an investigator has started the process of developing a clinical trial design by thinking about the last item on the list, after assuming that their experimental treatment must be safe and effective because the preclinical results in cells and mice were so exciting. They often consider a statistician's main purpose to be doing a sample size and power computation, ideally to obtain the smallest sample size needed to obtain regulatory approval. It is good to have a rich fantasy life, but this seems to be more useful in creative endeavors like making films or writing fiction, rather than in evaluating new medical treatments. Optimism may feel good, but it is a poor scientific strategy.

In my experience, the physician-statistician conversation may lead the physician(s) to rethink and modify some aspect of the therapeutic process, the experimental treatments to be studied, what outcomes to evaluate, or how to allocate resources to the trial. It also may motivate the statistician(s) to develop a new design methodology. Since accounting for everything on the list in some ideal way usually is impossible, actual clinical trial design and conduct always must be a reasonable compromise between a theoretical ideal and practical reality.

Long before clinical trials became commonplace in medical research, Ronald Fisher invented a lot of the major ideas that now make up the infrastructure of scientific experimentation. Expressing his opinion about the need for sound experimental design in science, he stated (1938)

> To consult the statistician after an experiment is finished is often merely to ask him to conduct a *post mortem* examination. He can perhaps say what the experiment died of.

Clinical trials are much harder to plan and carry out per design than agricultural or laboratory experiments. There are many reasons for this, but the most prominent among them is the fact that a clinical trial involves medical treatment of human beings. To follow the protocol's design, treatment must be done per some specified set of rules that, ideally, reflect what physicians actually do, and what they are willing to do for patients enrolled in the trial. Here are some simple realities about clinical trials that should be kept in mind when planning or conducting a trial.

1. Most clinical trials are designed at least twice. Many are redesigned several times, either before the trial is begun or during conduct if something unexpected has occurred. This may include severe adverse treatment effects, a sponsor withdrawing funding, insurance companies unwilling to fund a certain treatment or treatments, or logistical problems with data observation and storage.
2. Many clinical trials are not conducted exactly as designed. This may be due to the problem that the design's requirements do not realistically reflect what physicians, nurses, and patients are willing to do. Patient noncompliance is quite common.

3. Applying outcome-adaptive decision rules, as done, for example, in early phase dose-finding or trials with early stopping rules for treatment futility or superiority, may be difficult logistically.

4. Due to limitations imposed by a trial's entry criteria, its resulting data may represent only a subpopulation of the patients who have a particular disease.

3.2   A Common Phase II Design

Here are some examples of what can go wrong when statistical hypothesis testing is used for making inferences or taking actions in a clinical trial. To do this, I first will focus on the Simon two-stage phase II design (Simon 1989). I chose this design because the technical basis for its construction is hypothesis testing, due to its simplicity it is used very commonly in oncology, and it shares several properties with other hypothesis test-based designs. There is a huge published literature on phase II clinical trial designs, mostly written by statisticians. After the basic ideas on phase II designs where established in the 1980s and 1990s, there followed an explosion of papers providing technical modifications or improvements. I will not attempt to review the phase II design literature here. If you are interested, you can find that sort of thing in other books, such as those by Brown et al. (2014), Jung (2013), or Ting et al. (2017).

When Simon published the two-stage design in 1989, his main motivation was the practical problem that, at the time, there were no easy-to-use single-arm phase II designs that included interim rules to stop early for futility. Simon's solution to this problem now may seem obvious, but when he introduced the design it was quite a clever solution to a problem that had been largely ignored by people who conducted phase II trials. Simon recognized that, because many new agents were ineffective, people who ran phase II trials needed a practical tool that would stop the trial early if the interim data showed that the agent had an unacceptably low response rate, compared to a prespecified fixed standard response probability. The advantage of this is that, if you can determine that a new agent is ineffective at some interim point in a phase II trial, then by stopping the trial early you can avoid wasting time and resources. From an ethical perspective, there is no reason to continue treating patients with an experimental therapy that doesn't provide an improvement over current standard treatment. This was, and still is, very important, because most new treatments do not provide a substantive improvement over existing standard therapies. Since the Simon two-stage design includes an early stopping rule for futility at the end of its first stage, in 1989, it was a big step forward. If the design does not stop the trial early, then its final data are used to decide whether the new treatment is sufficiently promising to warrant further study a large confirmatory phase III trial. This usually requires patients to be randomized between the new treatment and standard therapy, in order to obtain an unbiased comparison. Bias and randomization are major issues that I will discuss in Chap. 6.

After reviewing the main elements of the Simon two-stage design, I will explain why, based on this design, it is very easy to make decisions or take actions that don't really make sense. This applies more generally to any hypothesis test-based clinical trial design. There are several inferential problems, some of which may have very undesirable consequences, that may not be obvious to people who use this design routinely.

As a temporary digression, it is worth listing some general problems with many phase II trial designs. For a single-arm phase II trial, serious problems may arise from any of the following:

1. Assuming that one knows the exact value of a standard or null parameter that actually is not known, and either has been estimated from previous data or specified arbitrarily.
2. Bias and confounding due to lack of randomization.
3. Assuming that a clinical outcome that is used for interim decision-making is observed quickly, when in fact it requires a nontrivial amount of time to observe.
4. Ignoring toxicity.

Later on, I will give examples of each of these, their consequences, and some alternative methods that avoid these problems.

The two main reasons that the Simon two-stage design is used so often seem to be that it is very simple, and there is free online software that requires very little input and computes design parameters very quickly. Over the years, this design has become a convention, which allows people to use it without being questioned or having to suffer through the painful process of thinking about the issues that I listed above. The aim of the design is to evaluate an experimental treatment, E, based on an early binary "response" outcome that is assumed to characterize the anti-disease effect of E. The design is constructed based on a hypothesis testing framework that involves three fixed response probabilities. The first is the unknown response probability, p_E, of E. The other two probabilities are numerical values that one must specify, since they are used to define hypotheses about p_E and construct the design. The specified null response probability, p_0, usually corresponds to the best available standard therapy. If $p_E \leq p_0$, then E is considered not promising. The third probability is a desirable alternative target p_a, which is larger than p_0.

If the trial's decision rules conclude that E is promising, this motivates a confirmatory trial in which E is randomized against some standard therapy, S, and the outcome is longer term, usually survival time or progression-free survival time. Formally, given prespecified p_0 and $p_a > p_0$, the Simon design is based on a one-sided test of the null hypothesis $H_0 : p_E \leq p_0$ versus the alternative hypothesis $H_a : p_E > p_0$. So one must specify p_0 and p_a as design parameters, and also two probabilities, α and β, of making incorrect decisions. The parameter α is the Type I error probability, that the design's test incorrectly rejects H_0 when $p_E = p_0$, that is, when H_0 is true, a "false positive" decision. The parameter β is the Type II error probability, that the design's test incorrectly accepts H_0 when $p_E = p_a$, i.e., when E achieves the targeted alternative, a "false negative" decision. The value $1 - \beta$ is called the "power" of the test if in fact $p_E = p_a$, that is, if the response probability with E is the targeted alternative. In most applications, the values $\alpha = 0.05$ or 0.10, and $\beta = 0.05, 0.10$, 0.15 or 0.20 are used. This is an example of a classical frequentist test of hypotheses.

The algorithm for constructing a two-stage Simon design is as follows. First, specify the four values p_0, p_a, α, and β. The difference $\delta = p_a - p_0$ often is called *the targeted improvement*. The design then is determined by four other parameters that must be computed. These are the sample sizes n_1 for stage 1 and n_2 for stage 2, and two integer-valued decision cutoffs, r_1 for stage 1 and r_2 for stage 2. A freely available online computer program *SimonsTwoStageDesign.aspx* that does this is available at http://cancer.unc.edu/biostatistics/program/ivanova.

Since the design may stop the trial early after stage 1, the actual sample size is either n_1 or the maximum $N = n_1 + n_2$. To use the online computer program, one inputs the four values $(p_0, p_a, \alpha, \beta)$ and it quickly outputs the design parameters (n_1, r_1, n_2, r_2). There actually are two versions of the design, the "optimal" design that minimizes the expected sample size, and the "minimax" design that makes the value of the maximum sample size N as small as possible if H_0 is true. Denote the observed data by $R_1 = $ the number of responses seen in the n_1 patients treated in stage 1, and, if stage 2 is conducted, $R_2 = $ the number of responses seen in the n_2 treated in stage 2. Here are the decision rules:

Stage 1: Treat n_1 patients. If $R_1 \leq r_1$, then stop the trial early for futility and accept H_0. If $R_1 > r_1$, then continue to stage 2.

Stage 2: Treat n_2 additional patients. If $R_1 + R_2 \leq r_2$, then accept H_0. If $R_1 + R_2 > r_2$ then reject H_0.

This all looks pretty nice. It controls the Type I and Type II error probabilities. It has simple decision rules that seem to make perfect sense. It stops early if the response rate is too low. Maybe best of all, using the free online program, once you input the four values $(p_0, p_a, \alpha, \beta)$, you can obtain design parameters almost instantly. So it makes the problem of designing a phase II trial extremely easy to solve. Unfortunately, if you use this design there is a good chance that you will shoot yourself in the foot. I will explain this sad reality below.

Here is a typical application. Suppose that $p_0 = 0.30$, you target response probability $p_a = 0.45$ that increases the null value by 50%, and you control the false positive (Type I) and false negative (Type II) error probabilities so that $\alpha = \beta = 0.10$. The optimal two-stage design has $n_1 = 30, r_1 = 9, n_2 = 52$, and $r_2 = 29$. So, it says to treat 30 patients in stage 1, and stop and accept the null hypothesis $H_0 : p_E \leq 0.30$, i.e., conclude that E is not promising, if there are nine or fewer responses in those 30 patients. Otherwise, if there are 10 or more responses, then treat 52 more patients in stage 2. At the end of stage 2, accept H_0 if there are 29 or fewer responses in the total of $30 + 52 = 82$ patients, or reject H_0, i.e., declare E to be "promising," if there are 30 or more responses in the 82 patients.

3.3 A Common Misinterpretation

The main problem with the Simon design is not with the design itself, which does exactly what it claims to do. The problem often arises due to confusion about the more general issue of what statistical hypothesis testing actually is doing. If this design ends up rejecting H_0 at the end of the trial, many people would conclude that $p_E \geq 0.45$, that is, that the response probability p_E of E is at least as large as the targeted value $p_a = 0.45$. This is a very common misinterpretation. Here is the key point.

Rejecting a null hypothesis $p_E = p_0$ does not imply that the alternative hypothesis $p_E \geq p_a$ has been accepted.

To see what this does and does not mean, first notice that the smallest empirical response rate required to reject H_0 at the end of stage 2 is $30/82 = 0.366$. But this sample proportion is far below the fixed target value $p_a = 0.45$. So, if you wrongly conclude that a "significant" test implies $p_E \geq 0.45$, this final decision rule doesn't seem to make sense. *How can you conclude that E has response probability $p_E \geq 0.45$ if your data have empirical response rate 0.366?* This conundrum arises from a very common misunderstanding of what the hypothesis test's decision rule actually is doing. The hypothesis test either accepts H_0 or rejects H_0 in favor of H_a. But H_a *says that $p_E > p_0 = 0.30$, not that $p_E \geq p_a = 0.45$. That is, in this conventional, frequentist one-sided hypothesis test, "Reject the null hypothesis" only concludes that p_E is larger than the null value $p_0 = 0.30$, not that it is at least as large as the targeted alternative value $p_a = 0.45$.*

So, the question then becomes, what role does the value $p_a = 0.45$ actually play in the design and in your inference? It turns out that, once the error probabilities α, β and the null value p_0 have been specified, in order to derive the required sample sizes and decision rules at the two stages, one must specify a numerical target value, p_a that is larger than p_0. *Essentially, p_a is a computational device rather than an actual target.* Ratain and Karrison (2007) pointed out this common misinterpretation. They called the target value p_a a "straw man," since it easy to knock it down using the hypothesis testing framework.

So, in the phase II clinical trial setting, what is the point of specifying a numerical value for the target response probability p_a? Calling p_a a "target" actually is misleading. In frequentist hypothesis testing, the actual purpose of specifying p_a is to allow one to apply a commonly used algorithm or formula for computing sample sizes. The numerical value p_a is not really a "targeted value" that you want p_E to achieve. It is just a number that you need to specify in order to compute sample sizes. Specifically, given specified numerical values of p_0, Type I error probability α, and Type II error probability β, the fixed alternative, or "target" p_a is used to derive a design that satisfies the formula

$$Pr(\text{Accept } H_0 \mid p_E = p_a) \leq \beta.$$

In words, this says that numerical values of the Simon two-stage design's sample size and decision cutoff parameters, n_1, r_1, n_2, and r_2, are derived to ensure that, if the response probability with E equals the specified alternative value p_a, then the design will incorrectly accept the null hypothesis with probability no larger than β. That is, the probability that the test will give a false negative conclusion is controlled to be no larger than whatever value of β one specifies. Equivalently, the *power* of the test is

$$Pr(\text{Reject } H_0 \mid p_E = p_a) \geq 1 - \beta.$$

To see why p_a actually is a "straw man," suppose that someone derives the design described above, but then decides that they do not want to do an 82-patient trial, and would like to do a trial with a smaller sample size. They can accomplish this, using the same numerical values of $p_0 = 0.30$, $\alpha = 0.10$ and $\beta = 0.10$, by simply specifying the larger alternative value $p_a = 0.50$. This gives a different "Simon optimal design" with $n_1 = 22$ and $n_2 = 24$, so $n_1 + n_2 = 46$, a much smaller trial. Another way to do this is to specify a larger Type II error probability, say $\beta = 0.20$ instead of 0.10. One could take a "The glass is half full" approach by stating, equivalently, "The test has power 0.80," since the power is $1 - \beta = 1 - 0.20$. This would give $n_1 = 20$ and $n_2 = 35$, so $n_1 + n_2 = 55$. Since it takes just a few seconds to run the computer program that derives an optimal two-stage design for given $(p_0, p_a, \alpha, \beta)$, people do this sort of thing all the time. That is, they adjust the value of p_a to be larger, or make α or β larger, in order to obtain a smaller sample size. If you like, you can specify a desired value of $N = n_1 + n_2$ and quickly run the computer program several times using different values of $(p_0, p_a, \alpha, \beta)$ until you obtain an "optimal design" with that given N. Then, conveniently, if you do not mention that you did this, you might pretend that the derived values $(p_0, p_a, \alpha, \beta)$ were what you wanted in the first place. This game often is played by people with a small amount of technical training, pretending to be scientists, who know just enough to be dangerous. By the way, the numerical values of α or β necessarily are arbitrary, but the value of p_0 is not, or at least should not be arbitrary.

The problem of misinterpreting p_a or some other alternative parameter, goes far beyond the Simon design. It arises in any setting where a decision is based on a test of hypotheses, with a power-sample size computation based on a fixed alternative value like p_a. For example, the same sort of misinterpretation can occur in a two-arm randomized trial to compare the survival time distributions of treatments E and S in terms of the hazard ratio $r = h_E/h_S$. The null hazard ratio usually is $r_0 = 1$, which says that the two treatments have the same death rate (hazard). Given specified Type I error probability α and power $1 - \beta$, since smaller values of $r < r_0 = 1$ correspond to E having a lower death rate than S, $h_E < h_S$, an alternative value $r_a < 1$, such as $r_a = 0.70$, is specified to derive a sample size for a clinical trial design. Assuming an accrual rate of 10 patients per time unit, a one-sided log-rank test of $H_0 : r = 1$ versus $H_a : r < 1$ with Type I error probability 0.05 and power 0.90 at alternative $r_a = 0.70$ would require 406 patients. If the more optimistic alternative $r_a = 0.50$ is specified, that E has half the death rate of S, then the required sample size is only 167. This may seem like a great way complete a study with a smaller sample size, which will save time and money. But this actually is convenient way to shoot yourself in the foot. It is very common practice to repeat the computations, as done above, for several different values of r_a in order to obtain a sample size that is practical and realistic. For example, if a 406 patient trial would take 4 years to complete, but a 167 patient trial will take only 2 years, then specifying $r_a = 0.50$ rather than $r_a = 0.70$ has very nontrivial consequences. In this test comparing hazards of death in terms of r, the particular numerical value of r_a often is nothing more than a Straw Man, so interpreting a "significant" test as evidence that $r \leq r_a$ makes little sense. Another simple point, often ignored, is that there is no free lunch with regard to sample size. All other things being equal, a smaller sample size always gives less reliable inferences. No matter how one manipulates this sort of computation, the final inferences must be based on the observed data, and a trial with 167 patients will provide much less reliable inferences than a trial with 406 patients.

In general, if the actual goal is to achieve a value at least as large as the targeted value p_a for a response probability in a single-arm trial of E based on response, or of at least as small as a targeted hazard ratio $r_a < 1$ in a randomized trial of E versus S based on survival time, *then the hypothesis testing framework may be very misleading.* In the above phase II Simon design example, the test of hypotheses may lead to the conclusion that E is "promising" even if the empirical response rate 0.366 is well below the target 0.45. As I will explain below, the data that reject H_0 actually may provide fairly strong evidence that E does *not* have a true response probability $p_E \geq 0.45$.

The Awful Truth About Hypothesis Testing

1. In a frequentist hypothesis test based on a sample of binary outcomes, the power of the test of $H_0 : p_E = p_0$ at a fixed target value $p_a > p_o$ is the probability that the test rejects H_0 if in fact $p_E = p_a$.
2. The power of the frequentist test is *not* the Bayesian posterior probability $\Pr(p_E > p_a \mid data)$, which treats p_E as a random quantity.
3. The points 1 and 2 are true much more generally, and apply to any frequentist test of hypotheses.
4. A smaller sample size gives less reliable inferences.
5. If you have been misinterpreting tests of hypothesis, you should not feel lonely.

3.4 Not Testing Hypotheses

If you are unhappy with the revelations given above, don't blame me. I'm just a messenger. Besides, as promised, here is a reasonable alternative approach. Once again, it is a simple Bayesian computation. If you really want to decide whether E achieves your targeted response probability $p_a = 0.30 + 0.15 = 0.45$ or larger, here is a way to evaluate the strength of the evidence that you actually have observed. *First of all, forget about hypothesis testing.* Instead, assume that p_E is a random quantity, which is probably what you believe anyway since you have uncertainty about p_E, and do a Bayesian analysis. This will provide a way to use posterior probabilities to quantify how the observed data have changed your uncertainty.

Start by assuming a non-informative beta(0.30, 0.70) prior for p_E. This says that you expect p_E to be 0.30, but your prior information is equivalent to an effective sample size ESS = 1, so you do not know much about p_E. The posterior quantity that you really are interested in is

$$\Pr(p_E \geq 0.45 \mid 30 \text{ responders in 82 patients treated with E}) = 0.057.$$

In words, this says that, given the observed data that 30 of 82 patients responded to treatment, there is about a 6% chance that the response probability with E is at least 0.45. So, if you take the target value $p_a = 0.45$ seriously, then from a Bayesian viewpoint it is unlikely that the response rate with E is at least 45%, and the new treatment is very disappointing. The same data that frequentist hypothesis testing would have led you to the conclusion "E is promising" actually leads to the opposite conclusion if you do a simple Bayesian analysis. To estimate p_E, the posterior mean is 30.3/(30.3+52.7)=0.36, and a 95% posterior credible interval is [0.26, 0.47]. These Bayesian probability computations are ridiculously easy. The Splus computer code is $1 - pbeta(0.45, 30.3, 52.7)$ for the posterior probability, and the code for the 95% credible interval is $qbeta(0.025, 30.3, 52.7)$ and $qbeta(0.025, 30.3, 52.7)$. Doing posterior computations with beta distributions based on this sort of count data is not rocket science. As an added bonus, one of the really nice things about a Bayesian analysis, in general, is that *you can draw a picture of the posterior distribution of any parameter of interest*, to illustrate each of your computed numerical values, as in Fig. 3.1.

Now you have a readily interpretable way to quantify strength of evidence. If you really want to take the target parameter value seriously, instead of doing a frequentist test of hypotheses, use the data that you have observed to compute posterior probabilities of events that you consider meaningful. If θ is the model parameter of interest, such as response probability, severe toxicity probability, expected length of hospital stay following surgery, or mean survival time, and a fixed value θ^* is a meaningful target that you or someone else has specified, you can compute $\Pr(\theta > \theta^* \mid data)$, the probability that, given the data you observed, the parameter is larger than θ^*. If you want to compute a 95% posterior credible interval $[L, U]$ for θ, just find the 2.5th and 97.5th percentiles of the posterior of θ, say $L = q_{.025}$ and $U = q_{.975}$, so that $\Pr(L < \theta < U \mid data) = 0.95$. Then you can draw a picture like Fig. 3.1.

Things get more interesting, and analyses usually become a lot more useful, when you have more complex data on multiple treatments given in combination or sequentially, multiple outcomes like (response, toxicity grade, time to disease progression, survival time), patient prognostic covariates or biomarkers, or processes observed over time. I will give examples of some of these in later chapters. There are Bayesian models and methods that accommodate all sorts of data structures, and new ones are being developed all the time. Of course, to make inferences in more complicated settings may require that you learn basic Bayesian statistics. You can take this as far as you like, from learning a few basic concepts to publishing

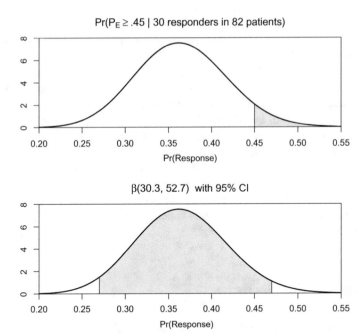

Fig. 3.1 Posterior probability 0.057 of achieving the targeted $\pi \geq 0.45$ (top panel), and 95% credible interval for π (bottom panel), based on observing 30 responses in 82 patients

cutting-edge research and becoming an internationally renowned expert. Some nice books on Bayesian models, methods, and computational techniques for data analysis are the introductory book by Bolstad and Curran (2016), the computer software-based books by Kruschke (2015) and Albert (2009), and the more comprehensive book by Gelman et al. (2013). The book by Kruschke is proof that, sometimes, you actually can judge a book by its cover. More complex Bayesian computations, for models more complicated than a simple beta-binomial, typically require the use of algorithms like Markov chain Monte Carlo (MCMC), reversible jump MCMC when your model may have two or more different possible dimensions, Gibbs sampling, or approximate Bayesian computation (ABC). You can learn about Bayesian computational methods and computer packages that use these and other algorithms from the books by Brooks et al. (2011), Gamerman and Lopes (2006), or Liang et al. (2010). There is a growing arsenal of computer packages that do Bayesian computation, such as JAGS or STAN. Or, if you are a physician scientist, you can just collaborate with your local Bayesian statistician, who will do the Bayesian modeling and data analysis for you.

By the way, a typical frequentist criticism of any Bayesian analysis is that the results depend on the assumed prior. To address this, just write down a few different but reasonable priors, repeat all the computations for each prior, and make a table with priors and corresponding computed posterior quantities. This sort of prior-to-posterior sensitivity analysis is a standard procedure in Bayesian inference, and sometimes it can be quite useful. For example, in the computation given above, if instead of a beta(0.30, 0.70) prior you assume a beta(1, 1) prior, which is a uniform distribution on p_E that has mean 0.50 and effective sample size $1 + 1 = 2$, then the above posterior probability changes from 0.067 to 0.064, so the posterior is insensitive to this change in prior. Even if you assume a more informative, more optimistic beta(4.5, 5.5) prior on p_E, which has mean 0.45 and ESS $= 10$, then the posterior probability still is only 0.071. This says that the computation, and your posterior inference, is insensitive to which of these three priors you assume, even though the third prior is somewhat informative. Of course, if you were to assume an extremely informative prior on p_E, like an optimistic beta(450, 550), which has mean 0.45 and ESS $= 450 + 550 = 1000$, then this prior will dominate the data. But assuming this sort of highly informative prior is silly, since it says that you already know so much about p_E that there really is no need to conduct an 82-patient phase II trial in the first place.

3.5   Random Standards

But wait, there's more. In the above Bayesian computation, I focused on uncertainty about p_E, but took the fixed null value p_0 as if it was known with certainty. In hypothesis testing, to make things simple and easy, people usually assume that they know p_0 or some other null parameter value like median survival time μ_0 with standard therapy. But most of the time it just ain't true. In actual practice, a numerical value of the null probability p_0 or μ_0 does not fall from the sky. If one is honest about things, the numerical value of p_0 actually corresponds to the response probability that has been seen historically with some standard treatment, S, for the disease that one is trying to improve upon. It may be the case that S represents several "standard" treatments, but let's stick to the "one standard" case for simplicity. You might think of S as being the best available standard therapy. The assumption that p_0 is a fixed constant may be a reasonable approximation to reality if there is a great deal of experience with S. In settings where the data on S are limited, however, it may not be appropriate. A more honest version of the computation given above would use such data to account for the fact that, in practice, a fixed null value like $p_0 = 0.30$ is not known with certainty. What you actually have is a statistical estimator of p_0 which, like any statistic, has inherent variability.

For example, suppose that historical data on S consist of 24 responses out of 80 patients. Under a Bayesian model where one assumes that p_0 corresponding to S is a random parameter with a non-informative beta(0.50, 0.50) prior, which has mean 0.50 and effective sample size 1, given these data, the posterior of p_0 would be beta(24.5, 56.5). Denote the historical binomial data by R_S = number of responses with S and n_S = sample size with S. Similarly, denote the binomial data from the phase II trial of E by R_E = number of responses with E and n_E = sample size with E. If one observes $R_E = 30$ out of $n_E = 80$, then a posteriori $p_E \sim$ beta(30.5, 52.5). Now the posterior probability computation given above may be elaborated slightly

Table 3.1 Posteriors of p_0 and corresponding inferences comparing p_E to p_0, starting with a beta(0.50, 0.50) prior, for observed data $R_E = 30$ responses out of $n_E = 82$ patients. Different posterior distributions of p_0 are determined by four different historical datasets, (R_S, n_S), all having sample proportion $R_S/n_S = 0.30$

(R_S, n_S)	Posterior of p_0	Posterior quantities	
		95% ci for p_0 (width)	Pr($p_E > p_0 + 0.15$)
(3, 10)	beta(3.5, 7.5)	[0.09, 0.61] (0.52)	0.26
(24, 80)	beta(24.5, 56.5)	[0.21, 0.41] (0.20)	0.12
(60, 200)	beta(60.5, 140.5)	[0.24, 0.37] (0.13)	0.09
(300, 1000)	beta(300.5, 700.5)	[0.27, 0.33] (0.06)	0.068

to account for both p_E and p_0 being random. It becomes

$$\Pr(p_E > p_0 + 0.15 \mid R_S/n_S = 24/80, R_E/n_E = 30/82) = 0.12. \qquad (3.1)$$

So, accounting for uncertainty about p_0 as well as p_E, if one takes the desired improvement of 0.15 seriously, this leads to a larger posterior probability than the value 0.067 obtained by ignoring uncertainty about p_0. But the substantive conclusion is the same, since there is only a 12% chance that E provides a 0.15 improvement in response probability over S.

To see why accounting for actual uncertainty about p_0 is important, consider the fact that the four historical sample proportions 3/10, 24/80, 60/200, and 300/1000 all equal 0.30, but are based on very different sample sizes. To quantify the basic statistical fact that the reliability of estimation increases with sample size, consider the following Bayesian analysis. If we assume, as above, that p_0 follows an uninformative beta(0.50, 0.50) prior, then the respective beta posteriors of p_0 for these four different historical samples are given in Table 3.1. The table also gives the give respective 95% posterior credible intervals for p_0, and their widths, as well as the posterior probability computed in (3.1) but repeated for each of these four different historical samples having the same observed proportion 0.30 for p_0. The value 0.057 obtained if one assumes that $p_0 = 0.30$ exactly may be thought of as corresponding to a huge historical sample, of 100,000 patients or more, treated with S.

3.6   A Fake Null Hypothesis

Before leaving the Simon design, it is worth discussing a very simple way to misuse it, or more generally any design based on a test of hypotheses where a numerical null value of a parameter is specified. Since the Simon design is very easy to construct, someone may specify a fixed value of p_0 that is substantially below the response probability that is obtained with standard therapy. This may be done inadvertently because the standard treatment response rate is not well understood, or simply as a computational device to obtain a smaller sample size so that the trial will be completed more quickly. This also may be due to confusion between a response rate that is of interest based on evidence that a drug is "active," and a response rate that is of interest based on evidence that the drug is better than standard therapy.

But there is another possible motivation for making p_0 very small. We currently are in an environment where there are many small pharmaceutical or biotechnology companies, each with its own new agent, vying to be bought by a large pharmaceutical company if they can show that their agent is "promising." From a single company's viewpoint, the use of any futility stopping rule is very undesirable. Using a futility stopping rule only makes sense from the clinician's viewpoint that, once the data show that an agent has an unacceptably low response rate, it is better to treat new patients with a different agent that still has a chance of providing a promising rate. Why would a patient, whose life is at stake, want to be treated with an agent that the current data have shown is unlikely to be effective against their disease? But from the company's viewpoint, if they have only one agent, and they have staked everything on its success, the last thing that they want is to have their phase II trial stopped early due to futility. Someone working for such a company may view the Simon phase II design as merely a device to get from phase I to phase III as quickly as possible. Since they do not want to stop the phase II trial at all, one way to help them achieve their goal is to use a Simon design, but specify a very small value of p_0. This will make the design very unlikely to stop the trial early, at the end of stage 1, and thus more likely to declare their agent promising, possibly due to the play of chance. After all, the empirical response rate required to declare an agent "promising" is smaller than the target value p_a.

Here is an example of how this may be done. Suppose that a protocol is being designed for a phase II trial to evaluate a new agent, E, in poor prognosis patients where it is claimed that there is "an unmet need" for a successful treatment. One may specify an optimal two-stage Simon design with null response probability $p_0 = 0.05$, targeted alternative $p_a = 0.20$, and $\alpha = \beta = 0.10$, equivalently, Type I error probability 0.10 and power 0.90. Using freely available software, it takes a few seconds to compute the design's sample size and decision cutoff parameters, $n_1 = 12$, $r_1 = 0$, $n_2 = 25$, and $r_2 = 3$. This design says to treat 12 patients in stage 1 and stop and accept H_0 if 0/12 responses are observed; otherwise treat 25 more patients in stage 2 and declare the agent "promising" if ≥ 4 responses are observed in the total of $12 + 25 = 37$ patients, i.e., if the empirical response rate is $\geq 11\%$. Given all this statistical jargon and notation, all these numbers, and the fact that the Simon design is used very commonly, this may seem like a good, scientifically sound design.

But here's the catch. What if historical data show that the response probability with standard therapy actually is 0.35, rather than 0.05? That is, what if the response rate of the currently best available treatment for the disease is known, but simply was ignored when specifying the null value $p_0 = 0.05$ used to construct the design? In this case, the above design would declare the agent promising based on an empirical response probability of about 0.11, which is less than 1/3 the probability 0.35 achieved with best available standard therapy.

What if the trial organizers actually know that best available standard therapy provides response probability 0.35, which is much larger than both the null value $p_0 = 0.05$ and the alternative value $p_a = 0.20$ that the PI used to construct the design? They might rationalize this by pointing out that this is a standard Simon design that is used quite commonly for many phase II trials. In saying this, they have characterized the design qualitatively, while ignoring its numerical parameters and the historical response rate. What would happen if, for example, the true response rate of the new agent actually is 0.10, which is less that 1/3 what has been obtained with standard therapy? In this case, the probability that the maximum of 37 patients will be treated is $Pr(R \geq 1 | n = 12, \pi_E = 0.10) = 0.72$. Put another way, there is only a 28% chance that a trial with a new agent that has less than 1/3 the response rate of standard therapy will be stopped early for futility. The expected number of responders in the trial would be

$$Pr(\text{Response}) \times E(\text{Number of patients treated}) = 0.10 \times (12 \times 0.28 + 37 \times 0.72) = 3.$$

If, instead, the 37 patients were treated with standard therapy, the expected number of responses would be $0.35 \times 37 = 13$. So, on average, 10 fewer patients will achieve a response if patients are enrolled in the trial of E, rather than simply treating them with standard therapy. The point of this computation is very simple. The point is that the numerical value of p_0 used to construct a design of this sort matters a lot, since it determines the early stopping rule at the end of stage 1. Setting $p_0 = 0.05$ is not just a computational device to obtain a conveniently small sample size or reduce the chance the trial will be stopped early due to futility. If whether or not a patient responds actually matters, the value of p_0 can have a profound effect on patients' lives. Consequently, intentionally describing a clinical trial design in purely qualitative terms in order to obfuscate its properties is a form of scientific fraud.

Suppose that, instead, one constructs a more honest Simon design based on the more accurate $p_0 = 0.35$ that actually corresponds to standard therapy, with targeted alternative $p_a = 0.50$, to achieve the same improvement of 0.15 over p_0. This design would require $n_1 = 34$, $r_1 = 12$, $n_2 = 47$, and $r_2 = 33$, so it requires a much larger maximum sample size of 81. If in fact the new agent has a response probability of only 0.10, then the binomial probability that this design will stop early is $Pr(R \leq 12 | n = 34, \pi = 0.10) > 0.9999$. Of course, this design treats 34 patients in stage 1, which is nearly as many as the 37 that the first design treated in two stages. So it seems that, in terms of benefit to patients in the trial, we are back where we started. How could this much more honest design treat as many patients with an ineffective treatment as the dishonest design? The answer, while perhaps not obvious, is simple. *Since a Simon design has only two stages, there is only one chance to stop the trial early if the experimental treatment is ineffective.*

Here is a different type of phase II design that fixes all of these problems. It has a high probability of protecting patients from ineffective treatments, and it also has a high probability of identifying a truly effective new treatment. So it provides a win–win. How, you might ask, is this magical design constructed? This can be done by applying a general phase II design strategy given by Thall et al. (1995). It is Bayesian, it does not test hypotheses, it has a stopping rule that monitors the data much more frequently than a two-stage design, and its parameters are calibrated by simulating the trial on The Computer before the trial is conducted. The design assumes appropriate beta priors for the response probabilities, and it includes a sequence of interim futility stopping rules, rather than only one. The design can be derived by using a menu-driven program, called *multc99*, that implements this design and lots of more complex phase II designs with multiple outcomes, and is freely available at the website https://biostatistics.mdanderson.org/SoftwareDownload.

An example of how to pull such a cute bunny rabbit out of this particular hat is as follows. Suppose that, from a Bayesian viewpoint, based on historical data on 500 patients where 175 responded, the random standard therapy response probability π_S is assumed to follow a beta(175, 325) prior, which has mean 0.35 and effective sample size ESS = 500. The parameter π_S in the Bayesian design plays the role of the hypothesis test's fixed value p_0, but π_S is random and has a prior distribution based on the historical data. Next, assume that π_E follows a beta(0.35, 0.65) prior, which has the same mean 0.35 but ESS = 1, reflecting almost no prior knowledge about π_E. Like the honest version of the Simon design with $p_0 = 0.35$, this design will treat a maximum of 81 patients, but, *after each cohort of 10 patients*, it applies the futility rule that says to stop accrual to the trial if

$$\Pr(\pi_E > \pi_S + 0.15 \mid data) < 0.025.$$

In words, this probability inequality says that, given the observed data, it is unlikely that E will achieve the targeted improvement of 0.15 over S in response probability. This rule implies that the trial should be stopped early if

[#responders]/[# patients evaluated] is $\leq 2/10, 5/20, 9/30, 13/40, 17/50, 22/60, 26/70,$ or $30/80$.

Here are some operating characteristics of this design:

1. If the true response probability of the experimental agent is $p_E^{true} = 0.35$, the same as the mean historical standard rate, then this design will stop the trial early with probability 0.83 and median sample size 40, and the probability that it will stop very quickly, after 10 patients, is 0.26.
2. If the true response probability of the experimental agent is $p_E^{true} = 0.50$, which gives the targeted improvement of 0.15 over the mean 0.35 of the standard proba-bility π_S, then this design will incorrectly stop the trial early with probability only 0.10 and median sample size 81. Equivalently, if the new agent is really promising, then the design will correctly conclude this with probability $1 - 0.10 = 0.90$. By the way, this is where the cutoff 0.025 in the stopping rule came from. I used the computer program *multc99* to calibrate the cutoff to make sure that the probability of incorrectly stopping is only 0.10 when $p_E^{true} = 0.50$.
3. *If E is very ineffective with $p_E^{true} = 0.10$, there is a 93% chance that the design will stop accrual after 10 patients.*

Point 3 says that, if the experimental agent is really poor, then with more frequent interim monitoring under this design the trial is very likely to be stopped a lot sooner than with a two-stage design.

For estimation at the end of the trial, if, for example, 40/81 responses are observed, then p_E will have a beta(40.35, 41.65) posterior, so a posterior 95% credible interval for π_E would be [0.38, 0.60], and $\Pr(p_E > 0.50 \mid 40/81) = 0.49$. So, if your goal is a response rate of 50% or larger, these results are equivocal. A similar empirical rate seen with 19/37 responses in a much smaller 37 patient trial would yield the posterior 95% credible interval [0.35, 0.66], which is 40% wider, with $\Pr(p_E > 0.50 \mid 19/37) = 0.55$, which is still equivocal. As always, a smaller sample size gives less reliable estimates, so in terms of inference a smaller phase II trial really is not more desirable. If, instead, you conduct an 81-patient trial, do not stop early, and you observe 50/81 (62%) responses, then $\Pr(p_E > 0.50 \mid 50/81) = 0.98$, so this higher response rate indicates that E actually is promising. By the way, with this Bayesian design there are no hypotheses being tested, so there is no straw man to knock down, and no p-value.

Takeaway Messages for Designing a Phase II Trial

If you wish to avoid the various problems with a two-stage hypothesis test-based phase II design, a fairly effective solution may be found if you do the following:

1. If you use hypothesis testing, make sure that you understand exactly what it is doing.
2. Take a Bayesian approach whenever it is feasible. For binary outcomes, this is not rocket science, and the computer software to implement a design like the one described above is free.
3. Specify a reasonably large maximum sample size, to get reasonably reliable inferences.
4. Define early stopping rules for futility that are applied with a reasonable frequency during the trial.
5. If it is ethical and feasible, instead of doing a single-arm trial of E and comparing it to historical data on S, design a trial that randomizes patients between E and S to obtain unbiased comparisons.

3.7 Monitoring Toxicity and Response

For most experimental agents, the maximum tolerable dose (MTD) that is used in phase II seldom is chosen reliably in the preceding phase I trial. I will discuss this in detail, with lots of examples, in Chap. 9. Consequently, to protect patient safety in phase II, it is important to monitor toxicity as well as response. Unavoidably, a design that does this is more complex, and it is more work to construct. But that is what biostatisticians and computer programs are for. Here is an example of how to do this, applying the Bayesian multiple outcome monitoring design of Thall et al. (1995),

Table 3.2 Historical data on response and toxicity with standard therapy for 500 patients. Each cell contains the number of patients, with the corresponding empirical probability given in parentheses

	Toxicity	No toxicity	Total
Response	100 (0.20)	75 (0.15)	175 (0.35)
No response	25 (0.05)	300 (0.60)	325 (0.65)
Total	125 (0.25)	375 (0.75)	500

and Thall and Sung (1998). All of the computations given below can be carried out in a few minutes using the computer program *multc99*.

Since we now are keeping track of both R=response and T=toxicity, there are $2 \times 2 = 4$ possible elementary outcomes. Historical data on the numbers of patients with each outcome when treated with S are given in Table 3.2. To construct design, first we must think about the probabilities of the four elementary events, with the experimental treatment E and with the standard S. Denoting the $\bar{R} = $ [No response] and $\bar{T} = $ [No toxicity], the four possible elementary outcomes are $A_1 = $ [R and T], $A_2 = $ [R and $\bar{T}$], $A_3 = $ [$\bar{R}$ and $\bar{T}$], and $A_4 = $ [$\bar{R}$ and T]. Notice that the elementary event A_1 is good because response is achieved, and it is bad because toxicity occurs. Denote the vector of their probabilities with S by

$$\pi_S = (\pi_{S,1}, \pi_{S,2}, \pi_{S,3}, \pi_{S,4})$$

and similarly their probabilities with E by

$$\pi_E = (\pi_{E,1}, \pi_{E,2}, \pi_{E,3}, \pi_{E,4}).$$

The marginal probabilities of R or T with S are

$$\pi_{S,R} = \pi_{S,1} + \pi_{S,2} \quad \text{and} \quad \pi_{S,T} = \pi_{S,1} + \pi_{S,4}.$$

The probability structure is the same for E, with

$$\pi_{E,R} = \pi_{E,1} + \pi_{E,2} \quad \text{and} \quad \pi_{E,T} = \pi_{E,1} + \pi_{E,4}.$$

To keep track of all this and construct a design, I will use Dirichlet distributions with four outcomes, one for π_S and one for π_E. The Dirichlet distribution generalizes the beta distribution by allowing more than two elementary events. That is, a beta usually keeps track of success and its complement, failure. In this example, there are four elementary outcomes, rather than two. For the design, the following practical approach reflects that fact that there is historical information on how S behaves but little or no information about the clinical behavior of E. To take this approach for this example, I will use the historical data to assume that $\pi_S \sim Dir(100, 75, 300, 25)$. This Dirichlet prior has effective sample size ESS $= 100 + 75 + 300 + 25 = 500$ and mean vector $\mu_S = (100, 75, 300, 25)/500 = (0.20, 0.15, 0.60, 0.05)$. A nice property of the Dirichlet distribution is that when elementary events are combined their probability follows a beta distribution with parameters obtained just by summing. In this example, $\pi_{S,R} \sim$ beta(175, 325), which has mean $\mu_{S,R} = 175/500 = 0.35$, and similarly $\pi_{S,T} \sim$ beta(125, 375), which has mean $\mu_{S,T} = 125/500 = 0.25$. So, this is a detailed probability model to keep track of the fact that, historically, S had a 35% response rate and a 25% toxicity rate.

I will assume that π_E follows a prior with the same mean outcome probability vector, but ESS $= 1$, so $\pi_E \sim Dir(0.20, 0.15, 0.60, 0.05)$. Since its ESS $= 1$, the prior on π_E is non-informative, which will allow the data obtained during the trial to dominate the adaptive decisions. The two actions of the design are to
1. Stop for futility if E is unlikely achieve a 0.15 improvement over S in response probability,
2. Stop for safety if E is likely have a higher toxicity probability than S.

These two aims are formalized by the following rules, that stop the trial if

$$\Pr(\pi_{E,R} > \pi_{S,R} + 0.15 \mid data) < 0.02 \qquad \text{(Futility Rule)},$$

$$\text{or}$$

$$\Pr(\pi_{E,T} > \pi_{S,T} \mid data) > 0.98 \qquad \text{(Safety Rule)}.$$

Since it is important to apply monitoring rules with a reasonable frequency in early phase trials, in this design these two rules are applied after each cohort of 10 patients. Together, these rules protect patients from an experimental treatment that is either unsafe or inefficacious. The two cutoffs 0.02 and 0.98 used above were obtained by using computer simulations, to calibrate them so that the design will have good OCs. The first probability inequality implies that accrual to the trial will be stopped if

[# responders]/[# patients evaluated] is $\leq 1/10, 5/20, 9/30, 13/40, 17/50, 21/60, 26/70,$ or $30/80$.

The second probability inequality says to stop accrual if

[# patients with toxicity]/[# patients evaluated] is
$\geq 6/10, 10/20, 14/30, 17/40, 20/50, 23/60, 26/70,$ or $30/80$.

The OCs of this design with these two sets of rules applied together are summarized in Table 3.3. Each scenario was simulated 10,000 times using *multc99*. The achieved sample size quartiles are denoted by (N_{25}, N_{50}, N_{75}). Only Scenario 2 is desirable, since the 0.15 improvement in response probability is achieved while toxicity is the same as with S. In Scenario 1, E has the same outcome probabilities as μ_S so no improvement is achieved. In Scenario 3, E is too toxic. In Scenario 4, E provides a 0.20 improvement over $\mu_{S,R} = 0.25$ but again is too toxic. Since the stopping probabilities are large in the undesirable Scenarios 1, 3, and 4, with median sample sizes varying from 30 to 50, and the stopping probability is the small value 0.11 in the desirable Scenario 2 with median sample size 81, the design has good operating characteristics.

Table 3.3 Operating characteristics of the phase II design to monitor both response and toxicity. The four elementary outcomes are $A_1 = [R$ and $T]$, $A_2 = [R$ and $\bar{T}]$, $A_3 = [\bar{R}$ and $\bar{T}]$, and $A_4 = [\bar{R}$ and $T]$. For numerical monitoring criteria, $\pi_{E,R}^{true} = \Pr(A_1)^{true} + \Pr(A_2)^{true} = 0.35$ is considered too low, but 0.50 is a desirably high target, while $\pi_{E,T}^{true} = \Pr(A_1)^{true} + \Pr(A_4)^{true} = 0.45$ is considered too high, but 0.25 is acceptable. Only scenario 2 has both desirable response and desirable toxicity probabilities, so it should have a low early stopping (false negative decision) probability. In each of scenarios 1, 3, and 4, the agent is inefficacious, unsafe, or both, so it is desirable to have a high early stopping probability

Scenario	π_E^{true}	$\pi_{E,R}^{true}$	$\pi_{E,T}^{true}$	P(Stop)	Sample size quartiles $(N_{25},\ N_{50},\ N_{75})$
1	(0.20, 0.15, 0.60, 0.05)	0.35	0.25	0.83	(20, 40, 70)
2	(0.20, 0.30, 0.45, 0.05)	0.50	0.25	0.11	(81, 81, 81)
3	(0.20, 0.15, 0.40, 0.25)	0.35	0.45	0.99	(20, 30, 50)
4	(0.40, 0.15, 0.40, 0.05)	0.55	0.45	0.96	(10, 50, 60)

Takeaway Messages

1. Since safety is never a secondary concern in a clinical trial, and doses are chosen unreliably in phase I, it often is very important to monitor toxicity in phase II.
2. Construct and evaluate the design under enough scenarios to characterize the interesting and important cases, including a null scenario, one where E is inefficacious, one where E is too toxic, one desirable case where both the efficacy and safety of E are acceptable, and possibly some mixed cases, such as E being efficacious but too toxic, or inefficacious but safe.
3. Constructing a phase II design that monitors both response and toxicity is much more work, but it protects patients from experimental agents that are either unsafe or inefficacious.
4. The operating characteristics of a design with two monitoring rules should be computed by simulating the design with both rules applied simultaneously, and not by incorrectly evaluating the design separately with only one rule running at a time.

Chapter 4
Science and Belief

People believe what they want to believe.
Tab Hunter

Contents

Abstract This chapter discusses the relationship between belief and statistical inference. It begins with a brief history of modern statistics, explains some important
elementary statistical ideas, and discusses elements of clinical trials. A discussion is
given of how the empirical approach used by statisticians and scientists to establish
what one believes may be at odds with how most people actually think and behave.
This is illustrated by several examples, including how one might go about determining whether dogs are smarter than cats, belief and religious wars, a story from the
early fifteenth century about how one might decide how many teeth are in a horse's
mouth, and how a prominent laboratory researcher once threw a temper tantrum in
my office. A discussion and several examples are given of cherry-picking, which is
the common practice of selecting and reporting a rare event, and how this misrepresents reality. The relationship of this practice to gambling is explained, including
examples of how to compute the expected gain of a bet. Illustrations are given of
the relationships between these ideas and medical statistics, news reports, and public
policy.

© Springer Nature Switzerland AG 2020   59
P. F. Thall, *Statistical Remedies for Medical Researchers*, Springer Series
in Pharmaceutical Statistics, https://doi.org/10.1007/978-3-030-43714-5_4

4.1 Theory Versus Practice

Statistical science is the formal basis for scientific method. As it is done today, a lot of statistics is pretty new stuff, but any of the basic ideas underlying modern statistics were established in the early twentieth century by Ronald Fisher, who is widely considered to be the father of modern statistics. He established design of experiments as an area of research, and invented many ideas, such as maximum likelihood estimation and analysis of variance, that still play fundamental roles in modern statistical practice. Fisher's work was motivated primarily by the desire to solve practical problems in biology, genetics, and agriculture. But in Fisher's time, methods for doing numerical computations were extremely limited. The recent development and wide availability of inexpensive supercomputers and clever computational algorithms has made it possible to invent and implement new statistical models and methods for an immense array of different data structures and scientific problems. Many of the statistical methods that currently are available did not exist 50 years ago. Almost none of today's sophisticated Bayesian methods could have been implemented back then, because there was no way to compute posterior distributions. This is especially true of Bayesian nonparametric statistics, as reviewed, for example, by Müller and Mitra (2013). This is a relatively new family of extremely flexible statistical models that often can be applied to solve problems for which conventional statistical models are inadequate. These days, a reasonably well trained college student with a laptop computer probably can do more statistical data analyses or simulations in a week than a typical 1960s PhD statistician could do in a decade.

But the scientific process of designing an experiment to learn about a particular phenomenon, conducting the experiment and collecting data, and making statistical inferences about the phenomenon based on the observed data, often is difficult to carry out properly. Statistical science seems to be gaining popularity, mainly because people are starting to recognize that modern statistics is a powerful tool to help achieve their goals, such as improving medical practice, making money in the stock market, or winning baseball games. But not many people are interested in the details of statistical models or methods, which often are complicated and difficult to apply properly. One of the great ironies of the modern age is that, as science, technology, and statistics are making immense leaps forward, a growing number of people are rejecting science and just expressing their opinions based on beliefs that have no basis in observable reality. Remember the fundamentalists, narcissists, hedonists, pathological liars, and politicians? Modern computing and the Internet have facilitated fast and accurate extraction of data from ever-widening sources, but they also have provided an easily accessible avenue for mass mediating factoids, distortions, statements with varying degrees of "truthiness," and outright lies.

A statistic is anything that can be computed from data using a well-defined formula or algorithm. I have discussed several examples of statistics above. The sample proportion R/n of responses, the sample mean $\overline{age}$, and the sample standard deviation s all are statistics. By far the most common statistic is a percentage, which quantifies the frequency of a given event. The response rate $100 \times R/n\%$ is a percentage. For example, about 30% of all motor vehicle traffic fatalities in the United States are alcohol-related. If you are diagnosed with acute myelogenous leukemia, the chance that you will survive at least 5 years is about 27%. A percentage often is used to predict how likely an event is to occur in the future. People compute percentages from data all the time.

Many people use percentages to quantify the strength of an opinion. Someone might say "I am 99% sure that dogs are smarter than cats," without referring to data at all. To address this important issue empirically, devising an intelligence test that is valid for both species seems like a difficult problem. One would need to obtain informed consent, account for within-species variability, and deal with dropouts. For example, a canine sample with a disproportionate number of Border Collies would lead to a biased comparison, and the feline noncompliance rate likely would be quite high. But what does this have to do with medical research? Plenty. Just get a hematologist who uses chemotherapy to treat acute leukemia in the same room with another hematologist who uses allogeneic stem cell transplantation. Ask them a few questions about how well their competing treatment modalities do in keeping people with leukemia alive, and then get out of the way, since they might end up fighting like dogs and cats.

The canine versus feline intelligence controversy, or the chemotherapy versus allosct argument, may be replaced by arguments about conflicting religious beliefs, such as whether God exists, if so how God wants people to behave, or how many angels can dance on the head of a pin. Since one cannot obtain data on such ideas, this might provide the basis for people killing each other over a religious disagreement that cannot be resolved empirically, as famously done in the Crusades, the Thirty Years' War fought in Europe between Protestants and Catholics during the seventeenth century, and modern-day Islamic terrorism. While these examples may seem to be unrelated to medical research, consider what sort of medical opinions physicians may have that are not based on actual data, and the actions that they may take based on such opinions. Tab Hunter may have been a movie actor without any knowledge of medicine or science, but he understood human behavior.

In practice, statistical inferences often lead to conclusions that contradict what people believe, which can be very unsettling. I will give lots of examples of unexpected conclusions in the following chapters. This is one of the reasons why many people hate statistics and statisticians. Who wants to have one of their cherished beliefs contradicted by a statistician? Throughout my career as a biostatistician, people have reacted to me, variously, with awe, respect, indifference, fear, jealousy, or hatred, and treated me accordingly. Statistical knowledge can be a great source of power, but you never know how people will react to some clever statistical design or data analysis.

A prominent example of how statistical practice often is at odds with the way that people actually think and behave is randomization. I will discuss this in Chap. 6. The strange truth is that randomization is the best way to obtain a fair comparison between treatments, and it actually is one of the most powerful tools that statisticians have provided to medical researchers. Still, the idea of flipping a coin to choose each patient's treatment in a clinical trial may seem very strange. Many physicians feel that explaining this to a patient during the informed consent process is an implicit admission that they do not know what they are doing. The basis for a patient entrusting their health and well-being to a physician is their belief in the physician's knowledge and expertise. If a physician has used a standard treatment S for years and statistical analyses of data from a new clinical trial show that an experimental treatment E is likely to be better than S, this may be hard for the physician to accept. Designing experiments and making inferences from actual data can get you into a lot of trouble, since most people do not like to have their beliefs contradicted.

There is a famous parable, perhaps apocryphal, about an argument among learned men that occurred sometime in the early fifteenth century. The argument was about the number of teeth in a horse's mouth, and it went on for 13 days. On the 14th day, a young monk suggested that the argument might be resolved by opening the horse's mouth and counting its teeth. The learned men, deeply insulted by this young upstart's preposterous suggestion and lack of respect for his elders, beat him severely. The bad news is that, in terms of how people come to adopt beliefs, things have not progressed much in the past 600 years. This sort of behavior persists in most societies, including the scientific and medical research communities. Because statistics is the basis for scientific method, and relies on empirical or observable evidence, medical statisticians often are forced to play the role of the young monk, and risk suffering the consequences if their analysis of the data that they are given produces inferences that do not agree with strongly held beliefs.

Many years ago, as I sat at my desk in my office, I watched a prominent laboratory researcher working in neurological science, who had both an M.D. and a PhD, throw a temper tantrum. He yelled loudly, pounded his fist on my desk, and jumped up and down. His feet actually left the ground. He was a large man, well over six feet tall, so it was quite an amazing sight. This researcher's unusual behavior was triggered by the results of a fitted regression model that I had just shown him, based on a small dataset on patients with a class of rapidly fatal solid tumors that he had provided me. The fitted model indicated that a special biomarker he had been studying in the laboratory had no relationship whatsoever with the survival times of the patients in his dataset. For each of the covariates that he had supplied, including his biomarker, when included in a survival time regression model, the p-value for the test of whether its coefficient was equal to 0 was much larger than 0.05. As you likely know, 0.05 is the Sacred Cutoff for deciding, by convention, whether the all-important "statistical significance" has been met or not for each variable in the regression model. Back then, I had not yet begun to apply Bayesian statistics routinely, and I used conventional frequentist regression models. But methodology aside, because his sample size was small, any inferences could be of only limited reliability. The sample was too small to do even a goodness-of-fit analysis reliably. Once he calmed down, I explained to him that, because the sample size was small, the power of any test based on the data that he had provided was quite low. I also explained that one or more of his covariates, including his special biomarker, might truly be associated with survival time, but that this might not have been detected statistically simply because his sample was too small. I suggested that a larger sample might enable one to detect an effect of his biomarker, if it actually existed. This seemed to palliate him, and he left my office. But he never provided another dataset for me to analyze.

I learned a valuable lesson that day. For many medical researchers, they regard a biostatistician's main job as being to compute sample sizes and p-values, since their careers often hinge on obtaining p-values that are smaller than 0.05. The reason that this researcher was so upset was that the empirical evidence, as formalized by my statistical analysis of his data, contradicted his prior belief about the biomarker, and its putative relationship with the disease, obtained from his preclinical experiments. In Chap. 5, I will discuss how the misunderstanding, misinterpretation, and misuse of p-values, and more generally the widespread lack of understanding of scientific method, have led to a war in the Scientific Community. This researcher was a casualty of that war, which still rages to this day.

Another laboratory researcher, who is quite well known internationally and shall go unnamed, once provided me with preclinical data consisting of a sample of (dose, % cells killed) pairs, and asked me to assess the relationship between these two variables. I did a regression analysis, including a plotted scattergram of the raw data, and goodness-of-fit assessments of several possible regression models. In the end, I came up with a fitted model in which % cells killed was a nonlinear function of dose, and plotted the estimated % cells killed curve over the scattergram of raw data points, along with a 95% confidence band. I provided this to the Internationally Famous Researcher, who told me that he had never seen this sort of thing done by a statistician before. He was a bit surprised by the nonlinearity of the fitted curve. When I asked him how he had determined relationships between dose and % cells killed in the past, he told me that he had always just plotted a scattergram of the data and drawn a straight line through the scattergram by hand. This primitive type of regression analysis might be called "Going Straight."

4.2 Technology and Cherry-Picking

There is a saying that you can see farther if you stand on the shoulders of a giant. In the world of medical research, these days the giants are new technologies that make it easy to see things that were hard to imagine just a few years ago. Research scientists have access to high-speed computers and specialized machines that do gene sequencing, identify surface markers on cancer cells sampled millions at a time, quantify the expression levels of thousands of proteins in the blood, create dynamic maps of brain activity, or monitor a person's heartbeat, blood pressure, and chemicals in their bloodstream continuously for weeks or months. The huge volume and complexity of data produced by new technologies have forced medical researchers to rethink how to handle their datasets. Figuring out how to store and access the immense amounts of data generated in an efficient way has made modern data management a new science unto itself. As mentioned earlier, some people like to call this "Big Data," mainly because it sounds impressive. But just because you know how to generate and store lots of data doesn't necessarily mean that the data are of any value, or that you know how to analyze it usefully. Before analyzing any dataset, large or small, two key questions are how the data were generated, and what you want to find out. This comes down to specifying inferential goals since, inevitably, statistical thinking must be done if the data are to be of any use. So, the process involves the people who generated the data, the people who stored the data, area experts who can explain their data and specify questions that they would like to answer, statisticians who can model the data structure and figure out how to analyze the data and make inferences that address the questions, and sometimes computer programmers who can write new code needed to do the statistical computations.

To provide a bit of perspective, here is yet another a cautionary tale. I once was asked to collaborate with a postdoctoral researcher who did specialized laboratory-based experiments that used very complex, very expensive machines to generate data. He sent me a file containing a dataset, but provided no explanation whatsoever of what the variables in the dataset meant or represented, or of how the experiments that had generated the data had been conducted. When I asked him how the variables were defined, and what he wanted to find out in terms of scientific questions or goals, he just said that he wanted the data analyzed. After a few more futile attempts to obtain any sort of coherent response or explanation from him, I terminated our relationship. Unfortunately, another statistician actually analyzed this dataset, although I can't imagine how, and a paper based on this was written and published in a scientific journal. A key point embedded in this sad story is that consideration of how data are generated, and what each variable in a dataset means, are critically important. A poorly designed experiment can produce poor quality data that, if analyzed using standard statistical methods, can give results that are at best useless, and at worst extremely misleading. A common phrase describing this phenomenon is "Garbage In, Garbage Out." Unfortunately, a lot of the garbage that comes out ends up in published papers.

There are lots of ways to produce garbage data. Here is a simple example. Suppose that an experiment involves treating cancer cell cultures using a range of different doses of a new agent, one dose for each cell culture, and recording the percent cells killed in each culture over a given observation period. One might construct a scattergram of the (dose, % cells killed) data with % cells killed on the vertical axis and dose on the horizontal axis. One might use statistical regression analysis, otherwise known as "fitting equations to data," to fit a straight or curved line to the data, and then plot the estimated line along with the scattergram, like I did for the Internationally Famous Researcher. Now, suppose that there actually is no effect whatsoever of the agent on the cancer cells, and about 50% of the cells in each culture die while the other 50% survive, on average, purely due to the play of chance. In this case, the scattergram is unlikely to show any pattern at all, since the agent is irrelevant, and dose has no effect on % cells killed.

Now, here is an interesting way to make a game of the laboratory research process. It is very easy to mimic this experiment by writing a computer program that generates completely random percentages, that is, random numbers that are distributed uniformly between 0 and 100. Suppose that you simply repeat this computer experiment for a given set of doses and a given sample size, drawing a scattergram of (dose, simulated % cells killed), and plotting a fitted straight line each time. Then, *purely due to the play of chance*, eventually, a linear pattern will be seen and, using standard statistical methods that ignore the repetitions, the upward slope of the fitted line will be nominally "significant." That is, eventually, a straight line fit to one of the simulated datasets will show a steep upward tilt. An example of this is to use a computer program to generate 10 samples, each representing a percentage of cells killed, with the percentage a random number generated to be uniformly distributed between 0 and 100. In each sample, the expected percentage is 50%, but a simple probability computation, from the theory or order statistics, shows that *the maximum of the 10 percentages is expected to be 91%.* Intuitively, this is because you can always order a sample of numbers from smallest to largest, and by definition, the sample maximum is larger than all of the other numbers in the sample. The general formula for the expected value of the maximum of n independent random variables uniformly distributed between 0 and 1 is $n/(n + 1)$, so above I just plugged in $n = 10$ and multiplied by 100 to express the mean as a percentage. If you do 20 repetitions, you can expect to get a maximum of 95%, and so on.

If you repeat a laboratory experiment with binomial data several times, compute the sample proportion for each repetition, and select the maximum proportion, you will see this effect. Figure 4.1 gives the distributions of the maximum sample proportion, denoted by $p_{[k]}$, from $k = 1, 3,$ or 6 independent repetitions of an experiment in which X responses are observed in each of $n = 60$ subjects, in the case where the true response probability for each subject is 0.50. So $X \sim$ binomial(60, 0.50) and the sample proportion $X/60$ varies around the mean 0.50 for each experiment. But for k repetitions of the experiment, if you cherry-pick the maximum sample proportion $p_{[k]}$, its distribution has a mean that increases with k. Figure 4.1 shows that the mean of $p_{[6]}$, the maximum proportion from six repetitions of the experiment, is 0.58. If you are willing to repeat the experiment more times, the mean of $p_{[10]}$ is 0.60 and the mean of $p_{[20]}$ is 0.62. Cherry-picking the maximum always overstates the true mean value.

Similarly, if a laboratory researcher keeps repeating an experiment until a nominally "significant" result is obtained, but then reports only this final experiment without explaining the repetitive process that preceded it, essentially, random noise is being misrepresented as an actual effect of dose on % cancer cells killed. If this cherry-picking leads to a clinical trial being conducted in which the agent is used to treat patients with the particular cancer, then this is a disaster. In some cases, the laboratory researcher may not even understand what he or she has done wrong. If the researcher is asked to replicate the reported results in a second experiment, then he/she may simply repeat the process, and cherry-pick yet another "significant" result. This is a version of the idea that, given an unlimited amount of time, a gang of monkeys playing with typewriters (or nowadays maybe iPads) eventually will reproduce all of the works of Shakespeare. It also helps explain why so many results reported in the published scientific literature cannot be reproduced.

For example, suppose that someone keeps repeating the experiment until a value $\geq 90\%$ is obtained, stops, and writes a paper describing only this last iteration of the repetitive process. The number of repetitions required to do this is a random

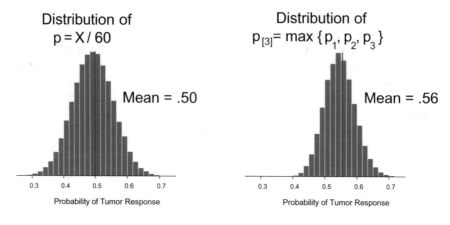

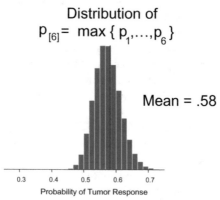

Fig. 4.1 Distributions of the maximum sample proportion from $k = 1$, 3, or 6 repetitions of a binomial experiment with sample size 60, where the true response probability is 0.50

variable that follows a *geometric distribution* with parameter $p = 0.10$, given by $\Pr(R = r) = 0.90^{r-1} \times 0.10$ for $r = 1, 2, \ldots$. The expected number of repetitions needed to achieve the above goal is $E(R) = 1/p = 1/0.10 = 10$. This is a simple version of repeating any sort of experiment until a desired outcome is obtained purely due to the play of chance.

Here is how to do the same sort of thing using a vector of genomic biomarkers, to identify a "treatment-sensitive genomic signature." Suppose that you measure five biomarkers, with each scored as being either present or absent. If a biomarker is quantitative, then you can score it as "present" if its value is above some fixed threshold. The vector of biomarker indicators can be written as $\mathbf{X} = (X_1, X_2, X_3, X_4, X_5)$ where each entry is $X_j = 1$ if the biomarker is present, 0 if absent. Suppose that you wish to compare the effects of two experimental conditions, say A and B, on survival while accounting for $\mathbf{X}$. Since $\mathbf{X}$ takes on 32 possible values, you might summarize it by just using the sum $Y = X_1 + \cdots + X_5$, and call this "overall biomarker expression."

Another example of selective pattern identification only requires continuous, casual observation. If you lie on your back outside on the grass and watch the clouds pass overhead, eventually you will see patterns that resemble familiar animals or objects, like a duck or a bunny rabbit. Essentially, this is because the human brain is, among other things, a pattern recognition machine. This may be called "The Ducks and Bunnies Effect."

I once was asked to talk to a young laboratory researcher who was involved in a large medical project for which I served as the biostatistician. The idea was that I would help him design some of his experiments and analyze his data. When I asked him about his experimental process, he quickly informed me that he did not believe that statistics was of any use whatsoever, and that he was only talking with me because the Principal Investigator of this large project had instructed him to meet with me. Since stupidity and arrogance can be a deadly combination, I asked this young genius what he would conclude if he performed the same laboratory experiment twice and got completely different outcomes. He responded that it would mean he did it wrong the first time but did it right the second time.

There are many other reasons why so much of the published scientific literature cannot be reproduced when other researchers repeat the experiment. This recently has become a source of embarrassment in the scientific community, and a huge controversy has emerged over the question of whether research results are reproducible, and what to do about it. As you might expect, the scientific community has dealt with this mainly by publishing an immense number of scientific papers about reproducibility. Some interesting papers in this vast literature are Ioannidis et al. (2009), Simmons et al. (2011), and Leek and Peng (2015). Publishing papers on the reproducibility of scientific results appearing in published papers has become quite popular. But clinical trials are expensive and time-consuming, so it is a lot harder to repeat a clinical trial many times and cherry-pick the results, compared to repeating a laboratory experiment. Still, there are other ways to go astray in clinical trial research, either unintentionally or by design. I will discuss many of these in the chapters that follow.

Takeaway Messages

1. If you do not understand how a dataset was generated, and how its variables are defined, it is a fool's errand to attempt to analyze it statistically and make inferences. If the data are garbage, any inferences based on it are at best useless, and at worst harmful.
2. Repeating an experiment until you get the results that you want risks misinterpreting random variation as a meaningful effect.
3. Repeating an experiment, computing a particular sample statistic that estimates a parameter of interest each time, and then reporting the maximum of the sample statistics as if the experiment had been done once always overestimates the parameter. The more times the experiment is repeated, the larger the overestimation error will be.

4.3   Is a New Treatment Any Good?

In medical research, a major problem is how to invent new treatments that actually work. Applying modern technologies has generated an ever increasing number of potential new treatments for the diseases that afflict human beings. In cancer research, there is constantly expanding knowledge about signaling pathways and surface markers that describe how cancer cells communicate with their biological environment and reproduce. Based on this, researchers now routinely design molecules with specific structures, or grow specialized blood cells, aimed at attacking particular cancers by disrupting their signaling pathways or selectively attacking the cancer cells by identifying their surface markers. This sometimes is called "translational" or "bench to bedside" research. These phrases may seem to imply that, once laboratory research has been carried out successfully using samples of cells, or mice that have been intentionally infected with a disease, how one uses the results to construct a new treatment in humans is obvious, and it is certain to provide a cure. This sort of optimism is an example of "The Pygmalion Effect," which occurs when someone falls in love with their own creation.

But confusing one's optimism with actual reality is bad scientific practice. Hope is not an optimal strategy. In fact, most new treatments do not turn out to have the anti-disease effect that their inventors envisioned, or do not work in precisely the way that they intended, and many have harmful side effects. For major goals like extending survival time or curing a disease, big improvements are rare. No matter how promising a potential new treatment may seem based on laboratory experiments with cells or mice, such preclinical data can only suggest how it actually works in humans who have the disease. A team of laboratory-based researchers may spend years studying a particular cancer's mechanism of growth, and design a new molecule aimed to disrupt that mechanism and kill the cancer. Unfortunately, it may turn out that, when it is given to people who have the disease, the new molecule doesn't work very well, or it produces adverse effects that severely damage or kill patients. There are many reasons for this sort of failure which, unfortunately, is much more common than success with new agents. If this were not true, then virtually all diseases would have been cured a long time ago.

Another pervasive problem is that many people do not understand the actual connections between short term outcomes, which are easy to score and use as a basis for a clinical trial design, and long-term patient benefit, such as survival time. For example, a new anticancer treatment may double the tumor response rate achieved with standard therapy, but have a trivial effect on patient survival time. I will discuss and illustrate this problem in Chap. 7. When comparing treatments, a very common mistake is to confuse the effects of confounding factors with actual treatment effects. This may be due to the failure to randomize patients between the treatments that one wishes to compare. It also may be due to the fact that a clinical trial design ignores patient heterogeneity, and the possibility that patients may react to a given treatment differently depending on their individual biological characteristics. Actual therapy often is an alternating sequence of treatments and observed outcomes, with the physician choosing the patient's treatment at a given stage based on the entire history of what has been done and seen so far. Acting as if "treatment" consists of what was given initially and ignoring the rest of the process is easy, but just plain wrong. These problems are caused by poor experimental design, flawed data analyses, or simply ignoring important information. All of these are statistical issues, which I will illustrate and discuss in later chapters.

The only way to find out how a potential new treatment actually affects people with a particular disease is to give it to people who have the disease and observe the outcomes. Dealing with this problem has led to clinical trials, which are medical experiments with human subjects. New treatments are called "experimental." There is a vast published statistical literature on how to design, conduct, and analyze the data from clinical trials. A lot of very smart people, mostly biostatisticians, have thought very long and very hard for a very long time about how to do clinical trials that will yield useful data.

Designing a clinical trial often is a complicated process. This is because diseases and treatment regimes used by physicians often are complicated, and the resources needed to conduct a trial, including time, money, administrators, physicians, nurses, and patients, are limited. Ideally, a clinical trial should evaluate treatments in a way that is scientifically valid, ethically acceptable, and based on the way that physicians actually behave. In most trials, this ideal is hard to achieve. Patient safety is never a secondary concern in a clinical trial. A central issue often is the trade-off between undesirable outcomes, often called "toxicity," and desirable outcomes, usually called "efficacy."

Clinical trials that are run using well-constructed designs that reflect actual clinical practice and lead to reliable inferences about what really matters to physicians and patients are rare. Most trial designs are flawed in one way or another, and many are such a mess that they produce useless data or misleading conclusions. Examples of typical design flaws will be given in Chaps. 9 and 10, each followed by a reasonable way to fix the problem. The prevalence of flawed clinical trial designs is because most physicians do not really understand scientific method or the fundamentals of experimental design. A practical problem is that a clinical trial involves physicians and nurses treating real people who have real diseases with experimental treatments that they know very little about. The challenge is to turn this inherently chaotic process into an ethically acceptable experiment that actually can be conducted and will yield useful data.

Statisticians often talk about "optimal designs." Methodological research to define and derive optimal designs can be quite useful if it leads to good designs that actually can be applied. Any claim of optimality almost invariably is misleading, however, unless it is qualified by a careful explanation of the particular criterion being used to determine what is best. In the real world, no clinical trial can ever be globally optimal, because the utilities of physicians, administrators, government agencies, pharmaceutical companies, patients enrolled in the trial, and future patients are all different. The practical goal of a clinical trial is not optimality, but rather to do a good job of treating the patients and producing data of sufficient quality that, when analyzed sensibly, may benefit future patients. When designing a clinical trial, one should never let the perfect be the enemy of the good. The two overarching questions in constructing a clinical trial design are whether it serves the medical needs of the patients enrolled in the trial and whether it will turn out to have been worthwhile once it is completed.

Interactions between physicians and statisticians are only part of a complex process involving medicine, statistics, computing, ethics, regulatory issues, finances, logistics, and politics. At the institutional level, elaborate administrative processes often must be followed for protocol review that involves one or more Institutional Review Boards. A major logistical issue is that trial conduct can be complicated by interim outcome-adaptive rules that must be applied in real time. A clinical trial protocol, no matter how detailed, is an idealized representation of how the trial actually will play out. One can never know in advance precisely how new medical treatments or regimes will act in humans, or precisely how physicians, nurses, or patients will behave if unexpected events occur. For example, it may be necessary to suspend accrual and modify a design in mid-trial if unexpected AEs occur, the accrual rate is much higher or much lower than expected, or results of another ongoing or recently completed trial substantively change the original rationale for the trial design.

An alternative approach to evaluating new experimental treatments is to not conduct a clinical trial at all. Instead, investigators may collect data from patients who have been given the new treatment by physicians during routine medical practice. The investigators obtain historical data on one or more standard treatments and analyze the data to figure out whether the new treatments are better than the standard. This is called "observational data analysis." Since patients in the dataset were not randomized between the experimental and standard treatments, observational data analysis easily can lead to biased treatment comparisons, and incorrect conclusions. This can happen if the actual effects of the treatments are confounded with effects of other factors related to clinical outcome. Such factors may be known or unknown prognostic variables that were not balanced between the two treatments. There is an immense statistical literature on methods to correct for bias in observational data analysis. I will discuss some of these in Chap. 6. Unfortunately, to analyze an observational dataset sensibly, one often must make assumptions that cannot be verified. A very common practice in cancer research is to conduct a single-arm clinical trial with all patients given the same experimental treatment, and then use standard statistical methods to compare the results to historical data on standard treatment, while ignoring the fact that patients were not randomized between the experimental and standard treatments. With this common practice, it is very easy to mistake a difference in confounding variables for an experimental versus standard effect. Unfortunately, this convention has been given a scientific halo by calling it a "Phase II Clinical Trial."

Perhaps the worst problem is that many deeply flawed statistical methods have become conventional practice, and they are not questioned. Unfortunately, clinging to conventional behaviors is human nature. This is a sociological problem in the medical research community. In what follows, I will give many specific examples of flawed conventions, and provide an alternative method for each problem.

4.4 The Las Vegas Effect

Despite claims to the contrary, what happens in Vegas may not stay in Vegas. Have you ever heard someone who just got back from Las Vegas tell a story about somebody they saw who gambled for a few hours and lost all of their money? No? Me neither. The type of story that gets told and retold is the time someone playing roulette kept betting on black and letting their winnings ride, won 10 times in a row, and walked away with a bundle of money. Telling that story, again and again, is typical human behavior. Common events aren't interesting, so people don't talk about them much. But rare events receive a lot of attention, especially if they involve some combination of money, sex, and violence. People tend to talk about them a lot precisely because they are unexpected, and therefore interesting. In statistics, this is called *selection bias*. If the proportions of stories about casino gambling outcomes matched their probabilities of occurrence, most of them would be boring. So the rare events of people winning big are overrepresented, because they are exciting and interesting. This has consequences that are very bad, unless you own a casino. Based on the stories that people hear or see advertised on TV, going to Las Vegas to gamble may seem like a good way to make money. The same thing goes for buying lottery tickets. You never see news stories about the millions of people who buy lottery tickets every week for years and years and always lose. But the news media are certain to tell the story of someone who bought a lottery ticket and won $300,000,000. This actually is a lot less likely than being struck by lightning. Every time I go to or from an airport, I see huge billboards along the highways with the current amount of money that can be won in one of the various state lotteries. But I have never seen a billboard saying that the probability of winning a lottery is about 1 in 25,000,000. What does this have to do with medical research? Plenty. Precisely the same considerations for casino gambling and lotteries can, and should, be applied when evaluating and reporting the effectiveness of new medical treatments.

Biased representations of possible events can mislead people to behave very foolishly. To provide a formal perspective, here is the idea of fair and unfair bets. Suppose that someone offers to pay you $60 if you flip a coin three times and you come up with heads every time, but you must pay $20 if you don't. That may seem fair, because you flip the coin three times, and the ratio of the $60 you may win to the $20 you may lose is 3 to 1. But let's compute some probabilities. In this simple experiment, if the coin is fair, the probability of winning is $1/2 \times 1/2 \times 1/2 = 1/8 = 0.125$, and the probability of losing, otherwise known as "not winning," is $1 - 0.125 = 0.875$. You may have noticed that I just applied *The Law of Complementation*. If you take the bet, your gain is either $60 or $-$20. But your *expected gain* from this gamble is the average of these two values, weighted by the probabilities of winning and losing:

$$\begin{aligned}
\text{[Expected gain]} &= \text{[amount gained from a win]} \times \Pr(\text{win}) \\
&\quad + \text{[amount gained from a loss]} \times \Pr(\text{lose}) \\
&= \$60 \times \Pr(\text{win}) + (-\$20) \times \Pr(\text{lose}) \\
&= \$60 \times 0.125 + (-\$20) \times 0.875 = -\$10.
\end{aligned}$$

So, if you take the bet and play the game, *your expected gain is the probability-weighted average* $-$*$10*. This is called a *sucker bet*, since the expected gain is negative. The person who offered you the bet may be called a *grifter*, although there are other, more colorful names for such people. If the grifter gets enough people to play the game, he or she will end up making a lot of money. For example, if 20 people play the game, the grifter will win some and lose some, but can expect to make 20 $\times$ $10 = $200. People who run this sort of scam on the street or in a park typically flash a roll of $20 bills so that people can see what they might win. Motivationally, this is a small scale version of a highway billboard showing a pile of cash that can be won in a casino or lottery.

What would a fair version of the coin-flipping gain game look like? By definition, *a fair bet has an expected gain of 0*. So, if you are paid $X if you win, this says that to have a fair bet, $X \times 0.125 - $20 \times 0.875 = 0. Solving this equation shows that the payoff for winning should be $X = $140 to make the bet fair. It's a very simple probability computation, but most people don't know how to do it, so they can't figure out that they have been offered a sucker bet. This sort of thing can be made a lot more elaborate, with more than two different possible outcomes, but it applies to any gamble. There are lots of casino games, and lots of lottery games, but all of their expected gains are negative. They are all sucker bets. The casino operators are like the grifter with the coin-flipping bet, but their games are much more elaborate, and they play them 24 h a day. The simple takeaway message is that you should never take a bet offered by a street hustler, walk into a casino, or buy a lottery ticket, unless you enjoy gambling and losing money.

What does any of this stuff about gambling have to do with medical statistics? Plenty. In medicine, it usually is the case that you have to play. When rare events are overrepresented, or if you do a poor job of computing the probabilities of possible outcomes, it misrepresents what you should expect to gain or lose by taking a given action. This can lead intelligent people to choose treatments that are the therapeutic equivalent of making a sucker bet.

It may seem strange, but The Evening News often is a lot like the story you hear from someone who just got back from Las Vegas, or the billboard advertising the lottery that you see along the highway. The Evening News does not run stories about the millions of people who drove to and from work without incident, ate dinner, watched TV, and went to bed. That isn't news. But a wreck on the highway that killed six people is almost certain to be reported, possibly with a filmed interview of someone who saw it happen. The problem is that the routine practice of overreporting rare events and not saying anything about common events misrepresents what actually is going on in the world. Years of watching the Evening News can give you a very distorted picture of reality. For example, mass shootings in the USA have received a great deal of news coverage lately, so very often are in the minds of Americans. Mass murders in schools are horrible, but let's look at some data and do a simple risk assessment. In the USA in 2017, about 40,000 people were killed by guns, and about 40,000 people were killed in automobile accidents. So, on average, an American was about as likely to be killed by a gun as in an automobile accident. Since both guns and cars are machines created by our technologies and thus, at least in principle, are under our control, maybe we should do away with both guns and cars. We could expect to save about 80,000 lives per year. Or maybe parents should not allow their children to either go to school or get into a car. Of course, life would involve a lot more bicycle rides, long walks, and homeschooling, but armed robbery, hunting, and mass murder all would be a lot harder.

More than 200,000 women in the USA die each year from heart attacks, which is about five times the number of women who die due to breast cancer. Maybe NFL players should wear red ribbons for five months to remind us about heart disease research. But that's roughly the whole NFL season, so maybe it's a bit too much to ask. They also might wear images on their helmets of a damaged brain all year to remind us that they all suffer from chronic traumatic encephalopathy (CTE) from getting banged on the head repeatedly over the course of their careers. Of course, being shot, killed in a car crash, dying from breast cancer, dying from a heart attack, or suffering permanent brain damage all are pretty horrible. But, after all, life is full of dangers. Amazingly, some very successful professional football players who have learned about CTE actually have quit playing the game. Maybe they did a cost-benefit analysis of money and glory versus the likely prospect of becoming unable to remember their own name.

When making a decision, if you want to increase the chance of a favorable outcome, it helps to know the odds. But some people do not want to know what a statistician or scientist has to say. In the movie *Star Wars Episode V: The Empire Strikes Back*, after being told by the robot C-3PO "Sir, the possibility of successfully navigating an asteroid field is approximately three thousand, seven hundred twenty to one!" Han Solo famously replied, "Never tell me the odds!" Heros don't want to hear about statistical estimates. But neither do cowards. I once gave a talk at a conference in which I discussed statistical methods for designing clinical trials to evaluate allogeneic stem cell transplantation regimens. I began with a description of how this sort of treatment modality actually works, including a description of chemical ablation of the bone marrow, and what the probabilities were for possible adverse events, including graft-versus-host disease, infection, and regimen-related death. When the discussant for the session got up to talk, the first thing that he said was how terribly upsetting he found my presentation. The poor man appeared to be quite distraught. Then he discussed some statistical aspects of my talk. I think he was a college professor.

This sort of thing also applies to news reports about medical breakthroughs. People who spend their entire professional lives working on preclinical experiments, in cells or mice, are likely to get very excited when a laboratory experiment with a new agent gives promising results. Suppose that an agent is tested, using various doses and schedules, either in cancer cells or in mice that have been artificially infected with a given disease. This commonly is done using xenografts of particular cancers. Suppose that these experiments are done repeatedly until a promising result is obtained, in terms of a large percentage of cancer cells killed. Of course, if you repeat an experiment enough times, eventually the random variation in percent cancer cells killed will produce a large number. Remember that the maximum of 10 percentages that are distributed uniformly between 0 and 100 is expected to be 91%? If you only report the largest number, you are cherry-picking. It is The Las Vegas Effect, but now in a laboratory instead of a casino.

In terms of what people come to believe about medical science, this sort of cherry-picked laboratory result may be communicated to a science reporter who then describes the results as a "potential breakthrough" in treatment of the particular disease. The science reporter is behaving honestly, in terms of what they have been told. What may happen next is that stringers, who are science reporters who work independently, may pick this up, rewrite the story so that it essentially repeats what they have read, and sell it to one of the numerous online science magazines. If this happens enough times, the large number of online reports may end up being a story televised on the Evening News, possibly with an on-camera interview of the laboratory researcher who initially reported the breakthrough, maybe wearing a white lab coat so that they look like a Scientist. No mention is made of the dozens of earlier experiments that yielded negative results, or the fact that the experiment yielding the positive result was done exactly once and cherry-picked. What may seem like an exciting, wonderful scientific breakthrough is mass-mediated to millions of people, when in fact it is nothing more than one step in a process that may go on for years, and in the end may never lead to an actual treatment advance. After all, if every new medical breakthrough reported on The Evening News over the years actually led to a cure, by now nobody would ever have to worry about dying from any disease.

Takeaway Messages

1. The Las Vegas Effect is very common in human society, since people tend to focus on unusual events rather than common events that occur frequently. Unfortunately, this gives a very biased, distorted representation of reality, especially in mass media such as news reports on television or the Internet.
2. Cherry-Picking is a general phrase to describe selection of experimental results that are desirable because they appear to support a preconceived idea. Like The Las Vegas Effect, Cherry-Picking often gives a very biased, distorted representation of reality, produces experimental results that typically cannot be replicated, and is a prominent example of Bad Science.
3. The practice by journal editors and referees of being more likely to publish papers that report nominally significant results produces a very biased representation of reality. It is a prominent example of Bad Science.
4. If data are garbage, due to poor experimental design, unclear or missing definitions of variables in the dataset, Cherry-Picking, or data being missing in a way that is related to the variables' values, then any statistical inferences based on the data are garbage.

Chapter 5
The Perils of P-Values

A significant p-value is like obscenity. I can't define it, but I know it when I see it.
Anonymous

Contents

Abstract The chapter gives an extensive discussion of p-values. It begins with a metaphorical example of a convention in which an arbitrary cutoff is used to dichotomize numerical information. The ritualistic use of p-values as a basis for constructing tests of hypotheses and computing sample sizes will be presented and discussed. This will be followed by a discussion of the use and misuse of p-values to establish "statistical significance" as a basis for making inferences, and practical problems with how p-values are computed. Bayes Factors will be presented as an alternative to p-values. The way that the hypothesis testing paradigm often is manipulated to obtain a desired sample size will be described, including an example of the power curve of a test as a more honest representation of the test's properties. An example will be given to show that a p-value should not be used to quantify strength of evidence. An example from the published literature will be given that illustrates how the conventional comparison of a p-value to the conventional cutoff 0.05 may

© Springer Nature Switzerland AG 2020   75
P. F. Thall, *Statistical Remedies for Medical Researchers*, Springer Series
in Pharmaceutical Statistics, https://doi.org/10.1007/978-3-030-43714-5_5

be misleading and harmful. The problem of dealing with false positive conclusions in multiple testing will be discussed. Type S error and the use of Bayesian posterior probabilities will be given as alternative methods. The chapter will close with an account of the ongoing P-value war in the scientific community.

5.1 Counting Cows

Imagine that, a long time ago in a primitive society, people decided that any number larger than 10 should be considered "large," with no mention of the specific number allowed, and that everyone had to use this convention. If you think about it, this actually could be considered a very practical rule. After all, you can keep track of any number up to 10 by counting on your fingers, but beyond that things rapidly become much more difficult. The prospect of having to count further, say by taking off your shoes and using your toes, is far too much trouble, and obviously should be avoided. Suppose that this was codified as law, so people wishing to make a distinction between, say, 12 and 100 would be tried by a jury of their peers and, if convicted, put into jail. Using this rule, a law-abiding dairy farmer might say to a cattle breeder, "I would like to buy a large number of cows." After agreeing on a price for "a large number of cows," the breeder then would assemble a group of cows and, using the standard method of matching cows to fingers, determine with certainty that "a large number of cows" had been assembled. A legal sale then could proceed, once a reasonable price for a large number of cows had been agreed upon. Someone foolish enough to risk breaking the law might use some sort of magical algorithm to count beyond 10, and declare that there were "fifteen cows." Of course, no law-abiding citizen would have any idea what "fifteen" might mean. If caught, the lawbreaker would be tried for their criminal behavior. After all, of what use could an illegal number like "fifteen" possibly be? Furthermore, why confuse things for law-abiding citizens?

This story of a society with a "rule of 10" actually is not that far from the reality of human behavior. In some primitive societies, the number of a collection of objects would be counted as one, two, or a heap. Anything beyond two was "a heap." In my cow counting example, I just changed two to 10. But what does counting cows have to do with medical research? Plenty. Here is a very strange fact about science in the modern age. Years ago, statisticians passed a law that is very much like the rule of two, or the rule of 10. Since statisticians use probabilities to describe things, this famous law is codified in terms of the most famous probability of all, 0.05. This also can be described equivalently, and legally, as "1 in 20." The well-known probability 0.05 is considered to be a Sacred Number by tens of thousands of scientists around the world. There is a very important law associated with this probability. Nearly all scientists follow this law, and those who violate it are subject to harsh penalties, including the ultimate punishment of having their papers rejected by reviewers at professional journals, or their grant proposals given poor scores by reviewers.

5.2   A Sacred Ritual

Before discussing this important statistical law, it is useful to review hypothesis testing, which includes laws and sacred icons. You probably know how hypothesis testing works, but I will describe it somewhat differently from what you may be accustomed to seeing. I already have discussed the Simon two-stage design, which is based on a one-sided test of hypotheses. Here is the hypothesis testing ritual, in general. This is one of the many painful things that college students must learn before they are allowed to graduate, and that medical researchers also must endure before they can hope to publish papers on their research. Thousands of statistics textbooks, both elementary and advanced, describe the sacred ritual of hypothesis testing, since it plays a central role in Scientific Research.

First, there is a parameter of interest, say θ, the value of which is not known. The parameter θ may be the probability of response with an experimental agent, as in the Simon design, or mean survival time, or the difference between two such parameters for treatments that one wishes to compare. It always is best to use Greek symbols to represent parameters although, legally, Arabic letters may be used. Next, there is a null hypothesis, $H_0 : \theta = \theta_0$, for some specified fixed value θ_0, an alternative hypothesis, sometimes of the "one-sided" form $H_a : \theta > \theta_0$, a test statistic T_n computed from a sample of n data values, a test rule such as "Reject H_0 if $T_n > c$" for a specified fixed cutoff c, and a fixed alternative parameter value, θ_a. The numerical value of θ_a is used to compute the sample size of the test needed to obtain a given power, which is defined as power$(\theta_a) = \Pr(T_n > c \mid \theta = \theta_a)$. This says that *the power of the test is the probability of rejecting H_0 if in fact $\theta = \theta_a$.* There also usually is some computation, often carried out by a professional statistician, to establish the distributions of T_n, for each of the two parameter values $\theta = \theta_0$ and $\theta = \theta_a$, so that the test's Type I error probability and power can be computed. Many statisticians have built entire careers working on this important problem, in a myriad of different cases.

Here is the *Sacred Two-Step Algorithm*, which must be followed to derive the numbers that are necessary to do a statistical test of hypotheses. These include what, without question, is widely regarded as the most important number of all, the *Required Sample Size*. This following version of the algorithm is given as applied to do a one-sided test.

Sacred Two-Step Algorithm for Computing the Sample Size Needed for a One-Sided Test of Hypotheses

Step 1: Determine the smallest cutoff c so that $\Pr(T_n > c \mid \theta = \theta_0) \le 0.05$.
Step 2: Given c, determine the smallest required sample size, n, which will ensure that power$(\theta_a) = \Pr(T_n > c \mid \theta = \theta_a) \ge 0.80$.

Notice that the Sacred Two-Step Algorithm uses the Sacred Probability, 0.05. This time-honored two-step computational algorithm has been followed for many decades by countless members of The Scientific Community, as well as college students taking required introductory statistics courses. Many scientists believe that a statistician's primary responsibility is to apply the Sacred Two-Step Algorithm and tell them the all-important value of n. Statistics textbooks are full of formulas for carrying out this algorithm. Numerous computer programs have been written for doing the necessary computations. The probability 0.05 is called the "Type I error probability" of the test, also known as its "size," or "false positive probability." The Type II Error of the test equals $1 - \text{power}(\theta_a)$, and also is known as the "false negative probability." False positive decisions are considered to be egregious errors, and the Sacred Probability, 0.05, because it is small, is considered to be the magic talisman that protects hypothesis testers from making this terrible error. It cannot guarantee that a dreaded false positive decision will not be made, but it limits the risk to 0.05. Some people have been known to use the wildly liberal value 0.10 in place of the sacred 0.05 to compute c, but this practice is not common, and it usually is frowned upon. Why double the chance of making a Type I error? If you are lucky, some smart statistician has proved that the test statistic T_n that you are using follows an approximately $N(0, 1)$ distribution under H_0, which greatly simplifies the Sacred Two-Step Algorithm, and allows you to call your test statistic a *Z-score*. Indeed, DeMoivre and Laplace are considered by many statisticians to be saints for proving the Central Limit Theorem for binomial test statistics. The value 0.80 almost always is used for the *power* of the test in Step 2. Some people take the radical approach of requiring power 0.90. Although less common than using 0.80, many respectable statisticians construct tests with power 0.90. Still, specifying larger power makes the required sample size n larger, which may not be a good idea since smaller sample sizes are widely considered to be more desirable, and of course requiring a larger value of n will make your experiment more expensive and time-consuming. In hypothesis testing, making n as small as possible often is considered to be a great virtue, and statisticians who derive tests that require smaller values of n are highly respected. Some people do two-sided tests, which are not only permissible by law, but actually are required by many strict statistical conservatives. With a two-sided test, the alternative hypothesis is $H_a : \theta \neq \theta_0$, and the test takes the general form "reject H_0 if either $T_n > c_U$ or $T_n < c_L$" for fixed upper cutoff c_U and lower cutoff c_L. While two-sided tests are slightly more complicated, it usually is the case that they are symmetric, with $c_L = -c_U$. For example, in the happy circumstance that T_n follows an approximate N(0, 1) distribution under H_0, and so is a Z-score, the test cutoffs are $c_L = -1.96$ and $c_U = 1.96$. These two numbers are quite famous and, while arguably not quite sacred, are beloved by millions, since they ensure that their two-sided test will be considered to be "significant" if it meets the Sacred Definition, given below. I will not discuss the ugly arguments that sometimes occur when strict one-sided hypothesis testers and two-sided hypothesis testers confront each other at professional meetings. After all, every religion has its extremists. In any case, these days, no one has to suffer through the painful process of deriving the numerical test cutoffs and n by hand. Today, due to modern technological advances, there are high-speed computers and computer programs that allow one to do all of the necessary computations as quick as a flick. People test hypotheses all the time. Some are scientists, some are college freshmen. The ritual of statistical tests of hypotheses requires you to apply the *Sacred Two-Step Algorithm*, compute T_n from the observed data, compare it to the cutoff or cutoffs, and either accept or reject H_0. A key thing about this ubiquitous scientific ritual is that it uses the data to make a dichotomous final decision.

Here is an example of the Sacred Algorithm in practice. This is a toy example, constructed purely for the purpose of illustration. Suppose that you are analyzing the data from a randomized clinical trial to compare treatments A versus B, where the outcome was survival time. To do this, you fit a statistical regression model with 20 different covariates and a treatment effect indicator, $X_A = 1$ if the patient was treated with A and 0 if the patient was treated with B. Suppose that $\beta_A X_A$ appears in the linear term of the log hazard of death in your model, so that $\exp(\beta_A)$ is the multiplicative A versus B effect on the hazard of death. You wish to test the null hypothesis $H_0 : \beta_A = 0$ versus the two-sided alternative $H_0 : \beta_A \neq 0$. Luckily, well-established statistical theory says that, given an estimator $\hat{\beta}_A$ of β_A with estimated standard deviation s_A, if H_0 is true, then $Z_A = \hat{\beta}_A/s_A$ is distributed, approximately, as a N(0, 1) random variable. So you may use Z_A as your test statistic for $H_0 : \beta_A = 0$, and call it a *Z-score* which, as noted above, is quite wonderful. All you need to do is compare your computed value of Z_A to the cutoffs ± 1.96 to perform your test of whether $\beta_A = 0$, that is, of whether or not there is any difference between A and B in terms of their effects on survival time.

Suppose that the computed numerical estimate of the effect of A versus B on the hazard of death is $\hat{\beta}_A = -0.01$, and the estimated standard deviation is $s_A = 0.004$. So, the observed value of the Z-score statistic is $Z_A^{obs} = \hat{\beta}_A/s_A = -0.01/0.004 = -2.5$. This gives two-sided p-value $2 \times \Pr(Z < -2.5) = 0.0124$ which, because it is smaller than 0.05, may be described formally as "statistically significant." So, treatment A must be significantly better than B in reducing the death rate. Clearly, this is grounds for submitting a New Drug Approval application for A to the FDA. After all, a large randomized study showed that A significantly reduced the risk of death compared to B. What more could one want!!? If you own the pharmaceutical company that manufactures A, it is time to open a bottle of Champagne and celebrate.

But let's look at this example a bit more closely. The estimated multiplicative effect on the hazard of death with A versus B is $\exp(-0.01) = 0.99$. In other words, A appears to reduce the hazard of death by 1% compared to B. To account for variability in the estimate, a 95% confidence interval for β_A has lower and upper limits $L_{\beta_A} = -0.01 - 1.96 \times 0.004$ and $U_{\beta_A} = -0.01 + 1.96 \times 0.004$, which gives the confidence limits $[-.01784, -.00216]$. Exponentiating these confidence limits to get the corresponding multiplicative effects on the hazard of death with A versus B gives the limits $[0.9823, 0.9978]$ for $\exp(\beta_A)$. Let's assume, for simplicity, that survival time is exponentially distributed, and that the mean survival time with B is $\mu_B = 5$ months. That is, the disease being treated is rapidly fatal. So, while this is a toy example, it is a deadly serious toy example because, as always, death is a permanent condition. The scientific and medical issue is whether, on average, one can improve on the 5 months expected survival time obtained with B by giving A instead. The test result says yes, we can, and moreover the effect is "significant." By definition, the death rate (hazard of death) λ_B with B is the inverse of the mean survival time, so

$\lambda_B = 1/5 = 0.20$. An approximate 95% confidence interval for the hazard of death with A is $0.20 \times 0.9823 = 0.1965$ to $0.20 \times 0.9978 = 0.2000$. Finally, converting these rates to compute the corresponding lower and upper mean numbers of months of survival with A, this translates to the approximate 95% confidence interval 5.01–5.09 months for the mean survival time μ_A with A. Since a month lasts about 30 days, this is equivalent to about a 0–3 day advantage in expected survival time with A compared to B. In the language of hypothesis testing, this is known as a "significant increase in the expected survival time with A versus B." So, if you are either a physician who treats the disease, or a patient with the disease, based on this test of hypotheses, you have scientific evidence that "A is significantly better than B." You can expect A to increase your survival by about 0–3 days. That should be enough time for a patient to write their last will and testament, say adios to their family and friends, and maybe watch the Evening News one or two times. But it's statistically significant, so they can die happy.

The point of this life and death example of statistics in action is the semantics of the word "significant." In hypothesis testing, all that this word means is that the observed test statistic is beyond the specified test cutoff or, equivalently, that the p-value is less than the Sacred Value 0.05. *Statistical significance says nothing about the magnitude of the estimated A versus B effect on survival.* Tragically, this example reflects many actual applications, where p-values are used to quantify strength of evidence, but the magnitude of the estimated effect is so small that it has no practical significance. The point is extremely simple. Because the word "significant" often is misunderstood, it can be very misleading. *What matters is the practical significance of the estimated magnitude of the effect, not the statistical significance of a test of whether the effect equals 0.* In practice, β_A is almost never 0. If it were the case that $\beta_A = 0$, that is, if the null hypothesis were true, then A would be perfectly equivalent to B in terms of its effect on survival time. What is going on in this example is that, if you obtain a large enough sample, you will end up with a p-value smaller than 0.05. More generally, the essential problem is that hypothesis testing dichotomizes statistical evidence as being significant or not. *What really matters are the estimated mean survival times with A and B, or the A versus B hazard ratio or the difference in means, along with confidence intervals.*

It also is useful to think about the other extreme. What if you have a similarly structured example, but the estimated mean survival times are 25 months with A and 5 months with B, the sample sizes are smaller, and the p-value $= 0.06$? Would you conclude that the observed improvement with A is "not significant"? This begs the deeper question of how strong the evidence is that $\beta_A > 0$, or possibly that $\beta_A > \beta_A^*$ for some desired value β_A^*. That is, what actually is meant by "strength of evidence" in science? I shall discuss this question further below, and provide some illustrations.

Here is another example. Suppose that T_n is a standardized log-rank statistic computed from survival time data obtained from a two-arm randomized clinical trial. That is, suppose that T_n is a Z-score. This says that, if H_0 is true, then T_n follows, approximately, a $N(0, 1)$ distribution, so the two-sided test cutoffs are the popular values -1.96 and $+1.96$. Now, what if you observe $T_n^{obs} = 2.5$? Shouldn't that provide stronger evidence that H_0 is false than, say, $T_n^{obs} = 2.0$? To address this knotty problem, which is quite general, the *p-value* was invented. Here is the definition of the most commonly used criterion for strength of evidence in science.

Definition of P-value: Given the observed value T_n^{obs} of the test statistic T_n, the *p-value* of the test is the probability, assuming that H_0 is true, of observing a value of T_n at least as extreme as T_n^{obs}.

To be a bit more precise, for a one-sided test of $H_o : \theta = \theta_0$ versus the alternative $H_a : \theta > \theta_0$,

$$\text{P-value} = \Pr(T_n \geq T_n^{obs} \mid H_0 \text{ is true}).$$

For a symmetric test of H_0 versus the two-sided alternative $H_a : \theta \neq \theta_0$,

$$\text{P-value} = 2 \times \Pr(T_n \geq |T_n^{obs}| \mid H_0 \text{ is true}).$$

The idea behind two-sided tests is that the difference could go either way, so to account for this one must multiply by 2. Strict two-sided hypothesis testers consider individuals who fail to multiply $\Pr(T_n \geq |T_n^{obs}| \text{ if } H_0 \text{ is true})$ by 2 to be morally corrupt. In any case, the value 0.05 is a Sacred Cutoff. A p-value ≤ 0.05 is considered sufficient evidence that H_0 is false, so it may be rejected, and in this case the test is called *significant*. If one is unfortunate enough to have a p-value > 0.05, then H_0 cannot be rejected, and the test is called *insignificant*. So, this provides a nice, neat, rational way for making scientific decisions. Just do a test of hypotheses and see whether it is significant or not. Notice that the use of the Sacred Cutoff 0.05 in this way dichotomizes scientific evidence as being either *significant* or *insignificant*. But here is the catch. *Carrying out this algorithmic ritual does not require you to think.*

In the above example, if $T_n^{obs} = 2.00$ then the two-sided p-value $= 0.0456$ which, happily, is significant. But if $T_n^{obs} = 2.60$ then the two-sided p-value $= 0.009$, which also is significant, but it is a lot smaller. Since a smaller p-value is thought to give stronger evidence that H_0 is false, it seems only fair to give extra credit for smaller p-values. This has led to the elaboration that a test with a p-value < 0.01 can be considered *highly significant*. Scientists love significant p-values, and they absolutely adore highly significant p-values. Of course, an insignificant p-value is a disaster, whether it is 0.06 or 0.60. After all, one must have standards, and 0.05 is the Sacred Cutoff.

Nobody knows for sure who invented p-values, but Sir Ronald Fisher used them a lot. That's right, Fisher was knighted, so if a statistician in the United Kingdom is smart enough, they might be elevated up there with the likes of Lancelot of the Lake, Ben Kingsley, and James Paul McCartney. Fisher is credited with specifying The Sacred Cutoff 0.05, not unlike Moses descending from Mount Sinai carrying the Ten Commandments etched by the Hand of Almighty God on stone tablets. Many people have wondered how Sir Ronald came up with the particular probability 0.05. One possibility, purely hypothetical, is that, because Sir Ronald was extremely nearsighted, absolutely brilliant, and learned to do all of his numerical computations in his head to astonishing levels of accuracy, he may have considered how his less intellectually gifted colleagues might perform numerical computations to obtain p-values. He may have decided that, if someone were to remove their shoes and reach the maximum number of fingers and toes, they might conclude that anything larger than 20 must be "a large number." In any case, we shall never know. Still, the ubiquitous use of the probability 0.05 to dichotomize scientific evidence as significant or not might be called "The Rule of the Inverted 20."

A major misinterpretation of The Sacred Cutoff arises from the fact that Fisher did not intend it to be applied to a test of hypotheses to quantify strength of evidence from one experiment. What Fisher (1926) actually wrote was

"A scientific fact should be regarded as experimentally established only if a properly designed experiment rarely fails to give this level of significance."

Note Fisher's use of the word "rarely." It appears that Fisher actually was referring to an experiment that was repeated, or maybe to many different experiments conducted over time, and not to one experiment. Otherwise, the phrase "...rarely fails to give..." would make no sense. He also described the experiment as being "properly designed" which, as will be illustrated by the examples in Chap. 10, is not a trivial requirement. Jerzy Neyman repeatedly made the point that a p-value referred to application to different statistical problems, and not to application of a procedure to hypothetical repetitions of the same problem. Sadly, almost nobody who uses p-values to quantify strength of scientific evidence these days either knows or cares what Fisher or Neyman actually wrote. The Sacred Value 0.05 has been handed down by generations of statisticians and research scientists, and it is widely accepted as an article of faith.

5.3   A Dataset with Four P-Values

The following example illustrates why computing a p-value sometimes is not as simple as it may seem. Suppose that you are given data from 12 patients treated with an experimental drug, including whether or not each patient responded. Denote R = number of responders and π = Pr(response). Suppose that you wish to do a test of the null hypotheses $H_0 : \pi = 1/2$. Assuming that the patients' outcomes were independent of each other, and that the experiment was planned to treat 12 patients, it follows that $R \sim binomial(12, \pi)$, with probability distribution function

$$\Pr(R = r \mid \pi) = \frac{12!}{r! \, (12 - r)!} \pi^r (1 - \pi)^{12-r}, \quad \text{for} \quad r = 0, 1, \ldots, 12.$$

So, by definition, the p-value of a one-sided test of H_0 versus the alternative hypothesis $H_a : \pi > 1/2$ is $\Pr(R \geq R^{obs} \mid \pi = 1/2)$, computed from the above binomial formula. If you observe $R^{obs} = 9$ responses then, based on the binomial distribution with null value $\pi = 1/2$, the p-value = $\Pr(R \geq 9 \mid \pi = 1/2) = 0.073$. Compared to The Sacred Cutoff 0.05, this test is *not significant*. So, you have tested your hypothesis H_0, and you have an answer, albeit an undesirable one.

Here comes the fun part. What if you do not know whether the plan was to treat 12 patients or, instead, the experiment was designed to keep treating patients until a total of three failures (nonresponses) were observed? Then R could have taken on any nonnegative integer value $0, 1, 2, \ldots$, and under this different sampling scheme, R follows a negative binomial distribution, with probability distribution function given by

$$\Pr(R = r \mid \theta) = \frac{(r+2)!}{r!\, 2!} \pi^r (1 - \pi)^3, \quad \text{for } r = 0, 1, 2, \ldots.$$

For the same data $R = 9$, this gives p-value $= \Pr(R \geq 9 \mid \pi = 1/2) = 0.0338$, which is *significant*. This different conclusion, which is much nicer because the test is significant, is reached because the set of possible experimental outcomes for binomial sampling and negative binomial sampling are different. For example, with negative binomial sampling that stops at three failures, the possible data might have been $R = 10$ out of 13, or 11 out of 14, and so on.

Things get even more interesting, depending on whether you intended to do a one-sided test or a two-sided test. Two-sided tests would give p-values $2 \times 0.073 = 0.146$ if you did binomial sampling with predetermined $n = 12$, or $2 \times 0.0338 = 0.0676$ if you did negative binomial sampling until three failures. So, it turns out that *it is not just the data that matter when you compute a p-value*. You also need to know

1. the experimental design that generated the data, and
2. whether the intention was to do a one-sided test or a two-sided test.

So, *p-values depend on not just what was observed, but also what may have been observed, and what the experimenter intended to do*. If the experimenter intended to stop at three failures and do a one-sided test, then $R = 9$ gives a significant p-value. If the experimenter intended to treat 12 patients and do a one-sided test, then $R = 9$ gives a nonsignificant p-value. Notice, again, that the observed data has not changed. There were nine responders and three nonresponders. But intention makes all the difference between whether one can declare the test significant or not.

This leads to the some very interesting questions. What if somebody gave you the data, and asked what the evidence was for or against the hypothesis $H_0 : \pi = 0.50$, but said that they did not know what the sampling plan was, or whether a one-sided or two-sided test was planned. They only knew that there were nine responders and three nonresponders. Based on the above computations, you can tell them that the p-value for the test of H_0 is 0.073, 0.146, 0.0338, or 0.0676, and give the rationale for each value. The person supplying the data might then, after careful thought, make the determination that (1) the plan must have been to sample until three failures, and (2) a one-sided test with alternative $H_a : \pi > 1/2$ obviously is what must have been intended. Isn't science wonderful? To paraphrase Benjamin Disraeli, or maybe Mark Twain depending on which account of history you believe,

"There are lies, damned lies, and p-values."

The four p-value example described above was given in the monograph by Berger and Wolpert (1988), which mainly was written to explain *The Likelihood Principle*,

given by Jeffreys (1961). *The Likelihood Principle* says that, given a statistical model
for observed Y and parameter π, all of the information needed to make an inference
about π is contained in the likelihood function of Y and π. If you are a Bayesian, of
course you also must specify a prior on π. There is an extensive discussion in Berger
and Wolpert (1988) by famous Bayesian and frequentist statisticians, stating their
opinions, and sometimes attacking each other with varying degrees of vehemence.
It is quite a lot of fun to read.

For this example, under The Likelihood Principle, $Y = R$ and the likelihood function is $\mathcal{L}(R, \pi) = \pi^R (1 - \pi)^{12-R}$, so the sampling mechanism and resulting p-value
are irrelevant. But a Bayesian analysis is nonambiguous. For the observed data in
this example, $\mathcal{L}(9, \pi) = \pi^9 (1 - \pi)^3$. Starting with, say, a non-informative beta(0.50,
0.50) prior for π, this would give a beta(9.50, 3.50) posterior. This implies that, for
example, $\Pr(\pi > 0.50 \mid R = 9, n = 12) = 0.96$, which may seem like pretty strong
evidence that the response rate is above 50%. But before jumping to a conclusion,
notice also that a 95% posterior credible interval for π is 0.47–0.92, which is quite
wide. So, in the end, we really do not know much about π, since the sample of 12 is
very small.

While this example is fun, it is important to bear in mind that it is a toy example
constructed to be a simple illustration. If this were real data, a lot of additional
information would be needed to make any sort of useful inference. One might ask the
investigators who provided the data to explain what the disease is, what the treatment
is, what other treatments may be available, why a sample of 12 (or a stopping rule of
three failures) was chosen, what is meant by "response," whether the patients were
homogeneous or differed in important ways related to treatment response, and where
the null probability 0.50 came from. Data come from somewhere, and one needs to
know a lot of important details to have some idea of what inferences are meaningful.
A major point is that researchers often provide statisticians with datasets to analyze,
but they do not, or cannot, provide accompanying explanations of the experiment
that generated the data or what the inferential goals may have been. So, if you are
in the hypothesis testing business and someone just gives you a dataset without an
explanation, any p-value that you compute is meaningless.

5.4   Bayes Factors

People misinterpret p-values all the time. One very common misinterpretation is that
a p-value is the probability that H_0 is true, since a smaller value leads to rejection of
H_0. This is just plain wrong. After all, a p-value is computed by first assuming that
H_0 is true.

If you actually want to assign probabilities to hypotheses, one coherent way to
do this is to use *Bayes Factors*. A basic reference is the paper by Kass and Raftery
(1995). Here is how this works. For two hypotheses H_0 and H_1, start by assigning
them prior probabilities $\Pr(H_0)$ and $\Pr(H_1)$ that sum to 1. Most people just use
$\Pr(H_0) = \Pr(H_1) = 1/2$, but you can use other prior values if you think that they are

appropriate. For observable data Y, denote the likelihood functions under the two hypotheses by $p(Y \mid \theta, H_0)$ and $p(Y \mid \theta, H_1)$, where θ is the model parameter, and $p(\theta \mid H_j)$ is the prior on the model parameter θ if H_j is true. For $j = 0$ and 1, the *marginal likelihoods* are

$$p(Y \mid H_j) = \int_\theta p(Y \mid \theta, H_j) p(\theta \mid H_j) d\theta.$$

This also is called the *prior predictive distribution of Y under H_j*. Bayes' Law says that, for each $j = 0, 1$, the posterior probability of H_j given the observed data Y is

$$p(H_j \mid Y) = \frac{p(Y \mid H_j)\, p(H_j)}{p(Y \mid H_0)p(H_0) + p(Y \mid H_1)\, p(H_1)}.$$

Computing ratios gives the equation

$$\frac{p(H_1 \mid Y)}{p(H_0 \mid Y)} = \frac{p(Y \mid H_1)}{p(Y \mid H_0)} \times \frac{p(H_1)}{p(H_0)}.$$

In words, this equation can be described as

[Posterior Odds of H_1 *versus* H_0] = [Bayes Factor] × [Prior Odds of H_1 *versus* H_0].

If the priors of the hypotheses are set to $p(H_0) = p(H_1) = 1/2$, then the Bayes Factor (BF) equals the posterior odds of the hypotheses.

Sometimes, a BF is defined by using two alternative models instead of two hypotheses, but the computations work the same way. People have written down all sorts of numerical criteria for how large or small a BF has to be to favor one hypothesis over the other, such as $3 < BF < 10$ being "moderate evidence" for H_1 over H_0, $10 < BF < 30$ being "strong evidence," and so on. This is a lot like using 0.05 and 0.01 to declare a p-value significant or highly significant, or using 10 as the largest number that is not "a large number." It comes down to the deep question of how big a number should be to be called "Big." To provide some perspective on this important scientific issue, to a Labrador Retriever, a cow is big, but a mouse is small. Specifying cutoffs like 3, 10, and 30 for interpreting Bayes Factors makes about as much sense as the Rule of Ten or the Rule of the Inverted 20.

While it sometimes is a messy numerical problem to compute the integral over θ to compute a BF for continuous distributions, it often is much simpler for discrete distributions. For example, if the data are binary outcomes with assumed likelihood $Y \sim binom(\pi, n)$ for the number of successes out of n trials, and prior $\pi \sim beta(\alpha, \beta)$, then the *predictive distribution of Y* is

$$p(Y \mid \alpha, \beta) = \int p(Y \mid \pi, n)\, p(\pi \mid \alpha, \beta) d\pi = \binom{n}{Y} \frac{B(\alpha + Y, \beta + n - Y)}{B(\alpha, \beta)},$$

$$(5.1)$$

where $B(a, b)$ denotes the beta function. That is, the predictive distribution is obtained by averaging the observed data likelihood over the prior. This can be written generally as

$$f^{pred}(Y \mid \tilde{\theta}) = \int_{\theta} f(Y \mid \theta) \, p(\theta \mid \tilde{\theta}) d\theta,$$

where $\tilde{\theta}$ denotes the fixed hyperparameters that determine the prior of θ.

The idea of Bayesian prediction starts with the assumption that the parameter is random and so has a distribution, which naturally implies that the distribution of a future observable variable can be obtained by averaging over the distribution of the parameter. Equation 5.1 is called a *beta-binomial distribution*, which is much loved because it applies to any simple setting with "Success/Failure" outcomes, and there is plenty of software that computes beta functions. Essentially, the beta binomial is a distribution for predicting the future observed value of Y, based on your assumed $beta(\alpha, \beta)$ prior.

Returning to Bayes Factors, suppose that we wish to evaluate the evidence in favor of the hypothesis $H_1 : \pi \sim beta(\alpha_1, \beta_1)$ over the hypothesis $H_0 : \pi \sim beta(\alpha_0, \beta_0)$. If you want to decide between H_1 and H_0, and observe binomial data (Y, n), the Bayes Factor in favor of H_1 over H_0 is

$$BF = \frac{B(\alpha_1 + Y, \beta_1 + n - Y)/B(\alpha_1, \beta_1)}{B(\alpha_0 + Y, \beta_0 + n - Y)/B(\alpha_0, \beta_0)}.$$

There are lots of software packages that compute numerical values of beta functions. For example, suppose one researcher optimistically hypothesizes a prior 90% response rate with $H_1 : \theta \sim beta(9, 1)$, while another is less optimistic with a prior 50% response rate, but also is less certain, with hypothesis $H_0 : \theta \sim beta(1, 1)$. That is, the Bayesian hypotheses specify priors on θ, since in Bayesian statistics parameters are random. Suppose that a sample of $n = 20$ patients has $Y = 5$ responses (25%), then the Bayes Factor in favor of H_1 over H_0 is

$$BF = \frac{B(9 + 5, 1 + 20 - 5)/B(9, 1)}{B(1 + 5, 1 + 20 - 5)/B(1, 1)} = 0.027$$

or posterior odds of about $1/0.027 \approx 37$ to 1 in favor of H_0 over H_1.

5.5   Computing Sample Sizes

A curious thing about The Sacred Ritual is that, for some reason that seems to have been lost in the mists of time, in most applications, a Type I error (false positive conclusion) is four times as important as a Type II error (false negative conclusion). That is, $1 - 0.80 = 0.20 = 4 \times 0.05$. Even if someone takes the radically ambitious approach of constructing a test with power 0.90, a false positive is still twice as important as a false negative.

Maybe the collective obsession with false positive conclusions is simply our desire to keep someone from putting one over on us. It is widely agreed that the number 0.05 came from Sir Ronald Fisher. But it is less clear where the often used power figure 0.80 came from. Of course, the practical implications of a false negative or false positive conclusion depend on what is being tested. In the world of medical research, where we are desperate for medical advances, one would think that false negative results would be a lot more important than a false positive. With a false negative, a drug or treatment that might have provided a real advance is abandoned. With a false positive, a drug that does not provide an improvement is wrongly considered to be "a significant advance." But if that drug is marketed and widely used, clinical practice data on its actual effectiveness will be accumulated. For example, suppose that you are testing whether a new drug improves survival in a particular subtype of leukemia and you wrongly conclude that it does when in fact it does not. Then the consequence is that the new drug will be marketed and used by physicians believing, incorrectly, that it is a treatment advance. As they use it to treat patients, however, and clinical practice data accumulate, they will discover that it does not provide an improvement. If, instead, the new drug actually gives longer survival but you commit a Type II error, then the consequence is people who later are given standard therapy dying sooner than they might have, if the new drug had been adopted.

But people almost never determine the Type I and Type II error probabilities of their statistical test based on the consequences of the possible decisions. Hypothesis testers tend to use size 0.05 and power 0.80 nearly all the time, because that is the convention. Why not use cutoff 0.20 for p-values and do tests with power 0.95? First of all, saying that a p-value ≤ 0.20 is "significant" violates Sir Ronald's recommendation, and no respectable hypothesis tester would ever do that.

But the question remains. Why do so many people design experiments based on hypothesis tests with power 0.80?

Consider the following example. Suppose that you are comparing two means, with null hypothesis $H_0 : \mu_1 = \mu_2$ for normally distributed outcome variables, using a two-sided test based on the usual Z-score statistic

$$ Z = \frac{\bar{X}_1 - \bar{X}_2}{\sqrt{\frac{\sigma_1^2}{n} + \frac{\sigma_2^2}{n}}} = \frac{\sqrt{n}\,\hat{\Delta}}{\sqrt{\sigma_1^2 + \sigma_2^2}}. $$

Suppose that you use alternative value $\Delta_a = \mu_2 - \mu_1 = 20$ based on the target $\mu_2 = 120$ versus $\mu_1 = 100$, assume common standard deviation $\sigma = 60$ based on historical data, and construct the test to have Type I error probability $= 0.05$ and power 0.80. The two-step formula says that you will need a sample of $2 \times 142 = 284$. Now suppose that, instead, you take the radical approach of setting the power to 0.95, so that the Type I and Type II errors are equally important in that both have probability 0.05. The required sample size then becomes $2 \times 234 = 468$. The attempt to treat the two types of errors as being equally important comes with a heavy price, since your required sample size is now a lot larger, by $468 - 284 = 184$. Obviously, increasing the power to 0.95 was a terrible mistake, since you now must treat 184 more patients. But remember that the difference $\mu_2 - \mu_1 = 120 - 100 = 20$ specified by the alternative hypothesis is only a Straw Man, and its only real purpose is to provide a way to compute a sample size that will ensure a given power. And after all, people use the power figure of 0.80 all the time, so it is a well-accepted convention. So, why not just follow the conventional approach and do the study with sample size 282 and power 0.80? If you want to keep the unconventionally high power figure 0.95, another way to get around the need for a larger sample size is to change the Straw Man to $\mu_2 - \mu_1 = 140 - 100 = 40$, which gives required sample size $2 \times 59 = 118$. Isn't this fun? All that is going on here is a game in which arbitrary numerical values of various test parameters are changed until the formula gives a desired sample size. Whatever test parameters are chosen, a scientific rationale can be back-engineered by referring to the Sacred Definitions. People do this all the time, usually motivated by the need to rationalize a sample size that actually is feasible for their trial. It may seem that someone who has constructed a statistical test of hypotheses is designing a scientific experiment where the Type I error probability, power, and target alternative hypothesis value all were predetermined. But most of the time they are just playing with statistical formulas in order to get a sample size that they want.

The following is something that many Scientific Researchers do routinely. It is not widely discussed, since that might prove to be embarrassing. One might even consider it irreligious.

The Not-So-Sacred Third Step of the Algorithm

Step 3: If you do not like the value of n that Step 2 gave because it is too large, just try out different numerical values of θ_a until you find one that gives a sample size small enough to do your study. Then pretend that you intended to use that targeted θ_a value in the first place.

While this may seem disingenuous, dishonest, and deceitful, it is what many people actually do. But most of the time the problem is not really with the investigators. It arises from the common requirement in the Sacred Ritual that seems to be imposed by statisticians, that a hypothesis test-based rationale for the sample size of a study must be provided with particular Type I error and power figures. If an account of how The Sacred Ritual was performed is not included in a medical paper submitted for publication, the journal's reviewers almost certainly will demand that it be provided. After all, it is sacred.

A simple, arguably more honest way to do this is to begin by plotting and examining the power curves of the test for each of several sample sizes in a range of feasible values, as shown in Fig. 5.1. For example, in the test of $H_0 : \mu_2 = \mu_1$ described above, if a sample size of 300 is what you can afford, then you can simply provide the power curve for $n = 300$. You might note, as shown by the power curve, that the test will have power 0.80 for alternative value $\Delta_a = \mu_2 - \mu_1 = 13.7$ and power 0.90 for $\Delta_a = 15.9$. More generally, one may determine the power for any given Δ_a by examining the power curve. Notice that the bottom of the power curve is at

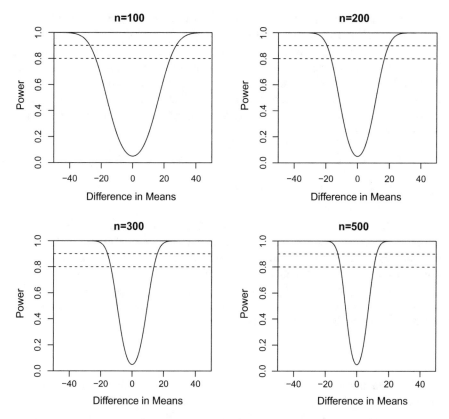

Fig. 5.1 Power curves, as functions of $\mu_1 - \mu_2$, of a two-sample symmetric two-sided 0.05-level test, based on total sample sizes 100, 200, 300, or 500

0.05, the Type I error probability of the two-sided test, since this corresponds to the null hypothesis that $\mu_2 = \mu_1$ where $\Delta_a = 0$. But, evidently, most hypothesis testers consider drawing a power curve to be too much work, since it seldom is done in practice.

Here is a radically different, irreligious alternative approach to determining the sample size of a study that avoids the power-sample size computation entirely. When a medical researcher asks me to design a clinical trial, I ask them lots of questions, including their anticipated accrual rate, maximum practical trial duration, and a range of maximum sample sizes that they can feasibly accrue. Recall the long list that I gave in the chapter that talked about the Straw Man. Once I know enough to proceed, I construct several designs, each having a different maximum sample size, and compute the design's properties, called its *operating characteristics* (OCs), by simulating the design on the computer with each sample size. Depending on the particular trial, OCs include such things as the probabilities of terminating a treatment arm early if it is truly unsafe or inferior to another treatment arm, correctly concluding that treatment

A is better than treatment B when in fact A is substantively better than B in terms of the actual difference in probability of response or mean survival time, choosing the best dose in an early phase trial, or correctly identifying a treatment–subgroup interaction, be it either good or bad. Such computer simulations are a sensitivity analysis of the design's OCs to sample size. Invariably, they show that the design's reliability increases with sample size. The investigator then uses this as a basis for choosing their sample size. I generally do this without bothering to refer to any hypotheses at all.

5.6   Not-So-Equivalent Studies

Another common error when interpreting p-values is believing that, if different studies give tests with the same p-value, then they must have provided identically strong evidence. But consider the following simple computation. Suppose that a comparative two-sample test from equal per-arm samples of size $n_1 = n_2 = n$ is based on the approximately normal test statistic as given above. Assume, for simplicity, that $\sigma_1^2 = \sigma_2^2 = 10$ and that the observed test statistic is $Z^{obs} = 1.75$. Then, if the one-sided p-value is $\Pr(Z > 1.75) = 0.04$, this implies that

$$\hat{\Delta} = \bar{X}_1 - \bar{X}_2 = 1.75(20/n)^{1/2}.$$

From this equation, $n = 10$, 100, and 1000 give corresponding estimated effects $\hat{\Delta} = \bar{X}_1 - \bar{X}_2 = 2.48$, 0.78, and 0.25. So, while the tests based on these three very different sample sizes have the identical p-value 0.04, since they all have the same $Z^{obs} = 1.75$, the estimated effect sizes differ by an order of magnitude.

Table 5.1 gives the above $\hat{\Delta}$ values, and 95% confidence intervals for $\mu_1 - \mu_2$ corresponding to these three sample sizes. The reason that the estimated effect sizes differ so greatly is that they are based on very different sample sizes, despite the fact that they all correspond to the same p-value. Suppose that, for example, in the context of the application that motivated the test, an effect size of 2.0 or larger is considered significant from a practical viewpoint. Then, based on the confidence intervals, the smallest study with $2n = 20$ seems to provide fairly strong evidence of this, while the two studies with much larger samples provide strong evidence that this is not the case. But however you interpret the confidence intervals, the idea that

Table 5.1 Three two-sample test statistics for $H_0 : \mu_1 = \mu_2$ versus the alternative $H_0 : \mu_1 > \mu_2$ that all have one-sided p-value $= 0.04$

	$2n = 20$	$2n = 200$	$2n = 2000$
$\bar{X}_1 - \bar{X}_2$	2.48	0.78	0.25
95% ci for $\mu_1 - \mu_2$	$-0.30, 5.25$	$-0.09, 1.66$	$-0.03, 0.52$

the three studies provide equally convincing evidence about the difference $\mu_1 - \mu_2$ because they all have p-value 0.04 cannot be true. Remember, all that a p-value does is quantify the probability that $Z > Z^{obs}$ under the null hypothesis that $\mu_1 = \mu_2$. In practice, whether this null hypothesis is true or not really doesn't matter much. The real issue is how different μ_1 is from μ_2. It is much more useful to estimate the difference $\mu_1 - \mu_2$, than to make a dichotomous decision about whether the data do or do not give "significant" evidence that $\mu_1 \neq \mu_2$, or that $\mu_1 > \mu_2$. The point is that the p-values of these three studies cannot possibly quantify strength of evidence, since they all equal 0.04, despite the fact that the three estimators of $\mu_1 - \mu_2$ are very different.

5.7 Acute Respiratory Distress Syndrome

P-values are not just innocuous technical devices used by researchers. A p-value actually can kill you. Here is a published multinational study that generated very simple data, in which hypothesis testing based on p-values and some simple Bayesian analyses appear to give very different inferential conclusions. It is an illustration of the different possible meanings of "Strength of Evidence" when comparing treatments statistically.

A randomized clinical trial was conducted to compare two methods for administering venovenous extracorporeal membrane oxygenation (ECMO) to patients with severe acute respiratory distress syndrome (ARDS). Details of the trial design, and data analyses, are given by Combes et al. (2018). One treatment for severe ARDS was to give Immediate venovenous ECMO (I). The other was to give continued Conventional treatment (C), but allowing crossover to delayed ECMO as rescue therapy if a patient had refractory hypoxemia. The primary endpoint was the binary indicator Y of death prior to day 60. The published final data were 44/124 (35%) deaths in the I group and 57/125 (46%) deaths in the C control group. Denote the probabilities of death in the two groups by $\pi_I = \Pr(Y = 1 \mid I)$ for Immediate ECMO I and $\pi_C = \Pr(Y = 1 \mid C)$ for the Conventional control treatment C, and denote the usual frequentist null hypothesis by $H_0 : \pi_I = \pi_C$.

The reported two-sided p-value for testing H_0 was 0.09. Based on this, Combes et al. (2018) published the following conclusion:

"Among patients with very severe ARDS, 60-day mortality was not significantly lower with ECMO than with a strategy of conventional mechanical ventilation that included ECMO as rescue therapy."

Here is what this statement was based upon.

Strength of Evidence, Quantified by the Observed P-Value

The probability of observing a value of the test statistic at least as large as the observed test statistic, multiplied by 2, is 0.09. Since this number is larger than 0.05, it is not significant.

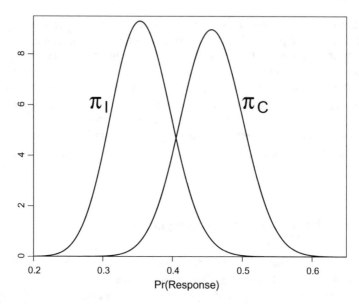

Fig. 5.2 Posterior distributions of the probabilities, π_I and π_C, of death within 60 days with I = immediate ECMO and C = conventional ECMO, based on the observed data

For those readers interested in technique, using a Z-score approximation for the two-sample binomial test gives a p-value of 0.10. Also bear in mind that the words "not significantly" refer to the failure of the p-value 0.09 to be smaller than 0.05, which is implicit in the above quote but well understood since, after all, 0.05 is The Sacred Cutoff for hypothesis testers.

Here is an alternative, Bayesian analysis of the data from this trial. One may begin by assuming π_I and π_C are random quantities that, *a priori* before the trial, were independent and both followed the commonly used non-informative beta(0.50, 0.50) prior, which has mean 0.50 and effective sample size 1. Given the observed data, the posteriors are $\pi_I \sim$ beta(44.5, 80.5) and $\pi_C \sim$ beta(57.5, 68.5). These are illustrated in Fig. 5.2. Before getting into the Bayesian analyses notice that, while they overlap, the mountain of posterior probability for π_I is to the left of the mountain of posterior probability for π_C. Since these are probabilities of death, smaller is better.

For comparative inference, a key posterior probability is

$$Pr(\pi_I < \pi_C \mid data) = 0.948.$$

Here is what this Bayesian probability equation means in words:

Strength of Evidence, Quantified by Posterior Probabilities

In patients with acute ARDS, starting with non-informative priors on the probabilities of death within 60 days with both immediate (I) ECMO compared to conventional treatment (C), given the observed data from the trial, the posterior probability that

I has a smaller 60-day death rate than C equals 0.948. Equivalently, in terms of the 60-day death rate, the posterior odds are 19 to 1 in favor of Immediate ECMO versus continued Conventional treatment.

If we change the above comparison to make it slightly more demanding for I by requiring a drop of at least 0.05 in the probability of death within 60 days by using I rather than C, then the relevant posterior probability is

$$\Pr(\pi_I < \pi_C - 0.05 \mid data) = 0.743.$$

This says that the posterior odds are 3 to 1 that immediate ECMO provides at least a 0.05 drop in the 60-day death probability compared to continued Conventional treatment. One also might do a symmetric comparison by computing the posterior probability of a small difference in either direction, like $\epsilon = 0.10$ or 0.05, between π_I and π_C. Formally, symmetric ϵ-equivalence is defined as

$$p_{equiv}(\epsilon) = \Pr(|\pi_I - \pi_C| < \epsilon \mid data).$$

This is the posterior probability that π_I and π_C differ by no more than ϵ in either direction. The values for the ECMO data are $p_{equiv}(0.10) = 0.50$ and $p_{equiv}(0.05) = 0.20$, so the trial does not provide convincing evidence that π_I and π_C are either 0.10-equivalent or 0.05-equivalent. By the way, notice that this gets at the strangeness of the null hypothesis $H_0 : \pi_I = \pi_C$ of perfect equivalence, which corresponds to $\epsilon = 0$ in the above Bayesian computation and has probability 0 under any beta distributions. If desired, however, one might cook up a Bayesian model that assigns a positive probability to H_0.

Testing hypotheses is not the same thing as making decisions. The ultimate issue here is not the conclusion of a statistical test of whether $H_0 : \pi_I = \pi_C$ is true or not. Of course, it's not true. The idea that these two parameters, which in theory can take on any value between 0 and 1, are exactly equal to each other doesn't make much sense if you look at how I and C are defined. But this is a technical issue, and really a distraction. As always, once the technical computations have been carried out, it is best to focus on what actually matters.

What Actually Matters

What actually matters is how physicians who treat patients with severe ARDS in the future will use the randomized trial data to make informed decisions between choosing to use Immediate ECMO or continued Conventional treatment.

A formal decision analysis may be carried out by first assigning numerical utilities to each of the two possible outcomes [death before day 60] and its complement [alive at day 60], to quantify how desirable they are. Denote these numerical utilities by U_{Dead} and U_{Alive}. Based on these two numerical values, the posterior mean utilities of the two actions [treat with I] = [treat with immediate ECMO] and [treat with C] = [treat with conventional Continued treatment] are computed. The posterior mean utility of using conventional treatment is the probability-weighted average

$$\overline{U}(C) = U_{Alive} \ \text{Pr}(alive \ at \ day \ 60 \ | \ treat \ with \ C, \ data)$$
$$+ \ U_{Dead} \ \text{Pr}(dead \ before \ day \ 60 \ | \ treat \ with \ C, \ data)$$

$$= \ U_{Alive} \ E(\pi_C \ | \ data) + U_{Dead} \ E(1 - \pi_C \ | \ data)$$

$$= \ U_{Alive} \ \frac{57.5}{57.5 + 68.5} + U_{Dead} \ \frac{68.5}{57.5 + 68.5}$$

$$= \ U_{Alive} \ 0.456 + U_{Dead} \ 0.544.$$

Similarly, the posterior mean utility of using immediate ECMO is

$$\overline{U}(I) \ = \ U_{Alive} \ 0.644 + U_{Dead} \ 0.356,$$

so the difference between posterior mean utilities is

$$\overline{U}(I) - \overline{U}(C) = 0.188 \ (U_{Alive} - U_{Dead}).$$

Since any acceptable numerical utilities should reflect the belief that it is better to be alive than dead at day 60, they must satisfy the obvious inequality $U_{Alive} > U_{Dead}$. It follows that, for any acceptable numerical utilities, based on the observed data, $\overline{U}(I) > \overline{U}(C)$. For example, if one uses the utilities $U_{Alive} = 100$ and $U_{Dead} = 0$, then the difference is 188. This is just a fancy version of the expected gain computation given in the discussion of gambling in Chap. 1, with $\overline{U}(I) - \overline{U}(C)$ what an emergency care physician can expect to gain by using Immediate ECMO rather than continued Conventional treatment. In this sense, a physician treating a new ARDS patient makes a gamble every time he/she decides between using Immediate ECMO or continued Conventional treatment. The utilities of a patient being alive or dead at day 60 used here replace the amount of dollars won or lost, and the treatments replace the bets. The posterior mean utilities given the observed data are computed to obtain the difference $\overline{U}(I) - \overline{U}(C)$, which quantifies what one can expect to gain by using Immediate rather than continued Conventional treatment.

Conclusion of the Utility-Based Decision Analysis

For any numerical utility that favors the ARDS patient surviving 60 days over death, based on the randomized trial data, using continued Conventional treatment rather than Immediate ECMO is a *sucker bet*.

How big a sucker someone is depends on how much they should expected to lose if they make the bet. Here is a simple way to illustrate the advantage of Immediate ECMO in a way that is easy to understand. If a physician were to treat, say, 1000 patients with severe ARDS using Immediate ECMO, then the physician should expect 355 of them to die within 60 days. But if the physician were to use continued Conventional treatment instead, the physician should expect 456 of them to die within 60 days. The difference is an additional $456 - 355 = 101$ expected deaths

Table 5.2 Posterior 80, 90, and 95% credible intervals for the expected number of deaths within 60 days out of 1000 patients treated for severe acute respiratory distress syndrome with either immediate ECMO or continued conventional treatment

Coverage probability (%)	Type of treatment	
	Immediate ECMO	Conventional
80	302–411	400–513
90	287–427	384–529
95	275–441	371–534

out of 1000 patients if the patients receive continued Conventional treatment rather than Immediate ECMO.

But these computations are based on point estimates of the 60-day death probabilities, and recall that any respectable statistical estimation should account for variability, as well as mean values. So some additional qualification of uncertainty is needed. To do this, Table 5.2 gives three posterior credible intervals for each of $\pi_I 1000$ and $\pi_C 1000$, which are the expected numbers of deaths in 1000 patients for the two treatments, computed using each of the three different posterior coverage probabilities, 0.80, 0.90, and 0.95.

Since death within 60 days obviously is a treatment failure, an emergency care physician who treats patients with acute ARDS might ask themselves two key questions.

Key Question 1: Have the above analyses of the published data convinced me that it does not matter whether I use Immediate ECMO or continued Conventional treatment, as the authors Combes et al. (2018) appear to have concluded? Or, based on the alternative Bayesian analyses, should I use Immediate ECMO to increase the chance of 60-day survival for my patients?

Key Question 2: Does the fact that the p-value 0.09 was larger than the Sacred Cutoff 0.05 make it OK for me to make a sucker bet every time I treat a new patient with severe ARDS by using continued Conventional treatment?

5.8 The Multiple Testing Problem

As people began to conduct group sequential randomized clinical trials, where a test comparing treatments is conducted repeatedly during the trial as patients are accrued, some statisticians noticed that there was a problem with the overall Type I error probability of the entire procedure. In a landmark paper, Armitage et al. (1969) explained that, if each of a series of tests performed on accumulating data in a clinical trial is conducted at nominal level 0.05, the actual Type I error of the entire multiple testing procedure is much larger than 0.05. One way of looking at this is that, if in fact H_0 is true and there is no difference between the true treatment parameters, then each computed interim test statistic, if it is standardized to be a Z-score, just varies randomly around 0 according to a normal (bell-shaped) distribution. That is, under H_0 the Z-score actually is "noise." But the more tests that you perform, the more likely it becomes that one of these noise variables will have a p-value > 0.05, due purely to the play of chance. This paper caused quite a furor, and it triggered a huge amount of research activity to develop group sequential tests with successive test cutoffs rigged so that the overall Type I error was controlled to be ≤ 0.05. Lan and DeMets (1983) suggested that one should construct a "spending function" $\alpha(t)$ that keeps track of how one "spends" a predetermined overall Type I error α at each time t that a test is performed. Various authors proposed different sorts of α-spending functions and boundaries giving cutoffs for successive group sequential tests, including O'Brien and Fleming (1979), Pocock (1977), Fleming et al. (1984), Wang and Tsiatis (1987), and many others. Current statistical software packages for designing group sequential trials provide menus of lots of different test boundaries that one may use, and a variety of ways to compute sample size and trial duration distributions depending on assumed accrual rate, number and timing of the interim and final tests, and specified overall Type I and II error rates.

Group sequential testing is a useful paradigm that is used widely, but it is not without its problems. For example, suppose that, after constructing group sequential boundaries for a trial with planned overall Type I error 0.05, during trial conduct a research physician takes an unscheduled look at the data, computes the test statistic, does an unscheduled test, and finds out that the difference is not quite significant according to the group sequential test boundary. But what if this unscheduled test has spent so much additional Type I error that, combined with all of the previous nonsignificant tests, more than the entire 0.05 has all been spent? Now, no matter what data are observed or what is done in the future, it is impossible to claim that the group sequential procedure, as actually applied, had overall Type I error 0.05. That is, essentially, every time one looks at their accumulating data, as research physicians like to do, they are punished since they have spent some of the precious Type I error probability. Of course, the physician could keep the criminally unscheduled test a secret, since if nobody else knew the unscheduled test had been performed, the crime would go undetected and unpunished. Or maybe the physician is an innocent soul who is blissfully unaware that Statistical Law requires criminal prosecution of people who conduct such unscheduled tests. Of course, ignorance of the law is no excuse for criminal behavior so, ultimately, it comes down to whether the perpetrator of such an unscheduled interim statistical test during a group sequential clinical trial gets caught. If so, and if the suspect is tried and convicted of *Illegally Looking at the Interim Data and Performing An Unscheduled Test*, then he/she likely would be sentenced to a term of imprisonment in Statistics Jail. A lenient judge might reduce the sentence to a specified period of community service, possibly teaching medical students how to compute confidence intervals and p-values.

But multiple testing can arise in many other ways. Let's return to the example, in Sect. 5.2, where we were testing whether the treatment A had a different effect than B on survival, in terms of $H_0 : \beta_A = 0$. The example showed a test of this hypothesis that was significant, with two-sided p-value 0.0124, but the magnitude of the actual

A versus B effect was trivial, with A expected to provide just a few additional days of survival. To elaborate this example by making it closer to what one encounters in practice, suppose that β_A is a treatment effect parameter in a regression model that also includes 20 patient covariates, say $X_1, X_2, \ldots, X_{20}$, such as age, disease severity, performance status, etc., and various biomarkers that may or may not be important. So, the regression model's linear component would take the form

$$\eta = \beta_1 X_1 + \cdots + \beta_{20} X_{20} + \beta_A Z,$$

where $Z = 1$ if treatment A was given and 0 if treatment B was given. For example, with an exponential or piecewise exponential survival time regression model, the baseline hazard of death is multiplied by the term $\exp(\eta)$. Thus, $\exp(\beta_j X_j)$ is the multiplicative effect of X_j on the hazard of death, for each $j = 1, \ldots, 20$, and $\exp(\beta_A)$ is the multiplicative effect of A compared to B on the hazard of death.

If we wish to test whether X_j does not or does have an effect on the death rate, equivalently, test $H_{0,j} : \beta_j = 0$ versus $H_{a,j} : \beta_j \neq 0$, for each j, then there are 20 such tests, in addition to the test of whether $\beta_A = 0$. To think about all of these tests together, it is helpful to consider first some basic facts about p-values.

1. A p-value is a statistic, since it is computed from data.

2. If the null hypothesis is true, then the p-value of the test is distributed uniformly on the interval [0, 1].

This implies that, under $H_{0,j}$, the probability of observing a p-value <0.05 for the test of β_j equals 0.05. This is just a consequence of fact 2, above, and is the reason why the probability of a false positive result from a test conducted at nominal level 0.05 equals 0.05. Similarly, if the null hypothesis is true, the probability of observing a p-value larger than 0.05 equals 0.95. Because of this correspondence between the test statistic and its p-value, and the fact that many test statistics require complicated formulas to compute, it is very common practice to just use the p-value, rather than the test statistic, to quantify the strength of evidence. If the null hypothesis is not true, and, for example, $\theta > \theta_0$, then the p-value's distribution is not $U[0, 1]$, but rather it is skewed to the right on the interval [0, 1]. This is why the power function of a test, $\text{Power}(\theta) = \Pr(\text{Reject } H_0 \mid \theta)$, of a 0.05-level test with alternative $\theta > \theta_0$ equals 0.05 if $\theta = \theta_0$, and it increases to larger values as θ moves away from θ_0 toward θ_a.

Many published scientific papers quantify strength of evidence by pairing variables with p-values, e.g., giving lists of multiple test results, such as "Compared to B, treatment A had a significant effect on survival ($p < 0.05$), age had a significant effect on survival, both A and age had insignificant effects on progression-free survival (PFS) ($p > 0.05$), bilirubin had a significant effect on survival but not PFS, and A had a significant effect on grade 3 toxicity but not grade 4 toxicity." Given this sort of vaguely worded list, one may wonder whether a given "significant effect" was beneficial or harmful for a given outcome and, because so many p-values are given, whether some nominally significant p-values were due to the play of chance rather than an actual effect.

Here are some consequences of the common practice of testing multiple hypotheses using the same dataset. Let W_{20} denote the number of "significant" tests out of the 20 tests of the hypotheses $\beta_1 = 0$, $\beta_2 = 0 \cdots$, $\beta_{20} = 0$, corresponding to the 20 covariates, say in a regression model for survival as described above. Suppose that, in reality, all $\beta_j = 0$, that is, none of the covariates actually have any effect on survival. If we assume, for simplicity, that the 20 tests are independent of each other, then W_{20} follows a binomial(0.05, 20) distribution, which has mean $0.05 \times 20 = 1$. So, *even if none of the covariates have any effect on survival, that is, if $\beta_1 = \beta_2 = \cdots = \beta_{20} = 0$, you should expect one of the 20 tests to be significant at the 0.05 level, a false positive.* Moreover, the probability that at least one of the 20 tests is significant, i.e., that you get at least one false positive test, is $\Pr(W_{20} \geq 1) = 0.26$, or about 1 in 4. This sort of thing gets a lot more complicated if separate regression models are fit, say, for survival, progression-free survival, and severe toxicity, as often is done in practice. So there would be 60 tests for covariate effects, but they no longer would independent, so W_{60} would have a more complicated distribution. But the same problems with multiple tests would persist, and become much worse since three times as many tests are being performed. In such a setting, it would be surprising if there were no significant tests.

Let's take this example a bit further, into the modern age of Big Data. As everyone knows, bigger is better, and that certainly should apply to data. Since genomics, proteomics, and bioinformatics have become quite fashionable in the medical research community, here is an example of how that line of inquiry often goes. Suppose that, through skillful application of modern bioinformatics methods, you are able to measure 100 genomic biomarkers on each subject in a sample. If you test whether each biomarker has an effect on survival, or maybe on the probability of response to a particular agent, *when in fact none of the biomarkers have any effect on patient outcome*, i.e., if $\beta_j = 0$ for each $j = 1, \ldots, 100$, then you can expect $0.05 \times 100 = 5$ of the 100 tests to be incorrectly "significant." That is, with 100 biomarkers that actually have no effect, you should expect about five false positive tests. Moreover, there is a 96% chance that you will have at least one false positive test. With 500 biomarkers, you expect about 25 false positive tests, and you are virtually certain to have at least one false positive test. If a given $\beta_j = 0$, then another way to describe the test statistic for X_j is that it is "noise," since $\hat{\beta}_j / s_j$ just varies randomly around 0 according to an approximately N(0, 1) distribution. If all the genomic covariates are noise, and someone performs enough tests, they will get significant p-values, so noise will end up being described as a "gene signature."

Many talks or papers presenting a new genomic breakthrough begin with a previously identified signature, and quickly go from there to discuss the signature's relationships to such things as patient prognostic covariates, tumor response, or survival time. Suppose, for example, that the reported signature consists of seven genes or proteins. Initially, it is mentioned that the signature was determined by examining an initial set of 500 gene/protein biomarkers and applying various microarray analyses, involving several clustering and selection algorithms, described vaguely, briefly, or not at all. That is, the talk or paper begins by identifying a "gene signature" that, implicitly, is expected to be accepted as an article of faith. So, essentially, it is a religious exercise. One may wonder whether the researchers might have been led astray by false positive tests, or some other misapplication of statistical methods. This is the "Ducks and Bunnies" effect, in the world of omics research. But a statistical scientist might wish to ask the following questions:

1. Precisely what statistical methods and algorithms actually were used to identify the signature?
2. How were the seven genes selected from the initial 500, and in particular were some genes included in the signature based on external information rather than only the data on the 500 genes?
3. Where did the set of 500 genes come from in the first place?
4. What sources of experimental variability were present, and what experimental design methods, if any, were used to account for them?
5. What computer programs were used to do the computations, and precisely how were they implemented?

If these questions are not addressed then, essentially, the statistical methodology for obtaining the signature is a black box, and any inferences based on it are worse than useless. Whether one regards this sort of thing as bioinformatics, religion, or garbage, it certainly is not science.

Readers interested in exploring the state of the art in bioinformatics research, as it actually is conducted in practice, might read the landmark paper by Baggerly and Coombes (2009). It involves a rather skillful application of forensic bioinformatics, in which the authors figured out and explained what was wrong with the methodology and the data in several published papers that claimed to have identified important microarray-based signatures of drug sensitivity. Since, at the time, the erroneous results of these papers were being used as a basis for allocating patients to treatments in several clinical trials, the Baggerly and Coombes (2009) paper raised a red flag. It triggered many articles in *The Cancer Letter* and the online popular science press. As a consequence, quite a few published papers that had made various erroneous claims about the microarray results investigated by Baggerly and Coombes (2009) were retracted, and the clinical trials were terminated. It is quite an interesting story that played out over several years, filled with intrigue, flawed data, outrageously incorrect claims, good guys, bad guys, and some amazingly clever detective work. Someone should write a book about it. With a really good screenplay, director, and cast, it might make a great movie.

I once asked Keith Baggerly how much of the published bioinformatics literature he thought had fundamental flaws, and he answered that he thought it was roughly around 50%. Dupuy and Simon (2007) reviewed 42 published microarray studies and checked whether statistical issues had been addressed reasonably well, such as doing cross-validation or excluding a test dataset from all aspects of model selection. They found at least one fundamental statistical problem in 21 of the 42 papers. Begley and Ellis (2012) described the results of their attempts, over several years, to replicate what had been presented as "breakthrough" results in the published scientific literature, with the aim to determine what might be commercialized successfully by a pharmaceutical company. After scientists at Amgen, working with the authors of these papers, attempted to replicate the published results, they were able to do so in only 6 of 53 studies. There are numerous other examples of published bioinformatics or other preclinical results that were obviously wrong or could not be reproduced. It makes you wonder what the latest sound bite about a recent medical breakthrough on *The Evening News* actually means.

All of this may seem to contradict the idea that more biomarkers should be better, since they provide more information. For example, if a biomarker identifies a cancer cell surface binding site, this may be a potential therapeutic target for a molecule designed to attach to the binding site and kill the cancer. But why devote time and resources to developing a molecule to attack a target that seems to be important on biological grounds, but in fact may not be related to patient survival at all? An important scientific question must be addressed. How can you protect yourself against making, or being victimized by, false positive conclusions, and how can you distinguish between the biomarkers that really are related to survival, at least to some degree that one may estimate, and those that are nothing more than noise? Put another way, how can you take advantage of the available Big Data on biomarkers without making lots of false positive or false negative conclusions?

Many statisticians have thought long and hard about the knotty problem of controlling the probabilities of false positive conclusions when performing multiple tests. There is a vast published literature, with numerous ways to compute so-called corrected p-values that are adjusted to account for the effects of doing multiple tests. Some references are Holm (1979), Simes (1986), and the books by Hsu (1996), Bretz et al. (2010), and many others.

Benjamini and Hochberg (1995) defined the "false discovery rate" (FDR), motivated in large part by analysis of gene expression data where an extremely large number of genes, possibly thousands, are classified as being either differentially expressed or not. The decision for each gene may be regarded as a test of hypothesis. Define the indicator $\delta_i = 1$ if it is decided that the ith gene is over-expressed and 0 if not, i.e., if the null hypothesis of no over-expression is rejected, with $D = \sum_i \delta_i$. Let r_i denote the unknown binary indicator that the ith gene is truly over-expressed. Then the false discovery rate is defined as

$$FDR = \sum_i (1 - r_i)\delta_i / D.$$

This apparently simple definition has been the focus of an immense amount of discussion and research. Müller et al. (2006) provided detailed discussions of a wide variety of Bayesian and frequentist approaches to the problem of multiple comparisons, including the FDR. Since the only unknown quantities in the FDR are the r_i's, Müller et al. (2006) suggested a straightforward Bayesian approach. If one computes the marginal posterior probabilities $\nu_i = \Pr(r_i = 1 \mid data)$ under a Bayesian model, then this allows one to compute the Bayesian estimate

$$\widehat{FDR} = \sum_i (1 - \nu_i)\delta_i/D.$$

The use of FDR, and other statistical devices, to control false positive rates has been motivated, in large part, by the fact that most published scientific results cannot be reproduced in subsequent experiments. In an extensive review of published clinical studies, Ioannidis (2005) reported that 45 of 49 (92%) highly cited clinical research studies claimed that the intervention being studied was effective but, of these 45, seven (16%) were contradicted by later studies, and seven others (16%) reported effects that were stronger than those of subsequent studies. He found that only 20 (44%) of the studies were replicated. Of six highly cited non-randomized studies, five later were contradicted, and this occurred for 9 of 39 (23%) highly cited randomized controlled trials. These are fairly troubling results, since it appears that much of the published medical literature cannot be reproduced. Here is a list of some possible reasons for this ongoing scientific disaster, many of which I have discussed elsewhere in this book.

Some Reasons for This Ongoing Scientific Disaster

1. Publication bias due to the fact that researchers are much less likely to attempt to publish papers reporting negative results.
2. Publication bias due to the fact that many journal editors are much less likely to accept papers reporting negative results, motivated in part by the desire to maintain a high "impact factor" for their journal. The irony of this behavior is that screening out papers with negative results is likely to have a destructive impact on medical practice.
3. Cherry-picking by repeating the same laboratory experiment until a positive result is obtained, in order to increase one's publication rate and chance of obtaining grant support.
4. Statistical bias in observational data arising from confounding of treatment effects with effects of patient covariates, supportive care, or institutions where studies were conducted.
5. Reporting p-values of tests of various hypotheses while ignoring the fact that multiple tests are being performed, and not accounting for this multiplicity.
6. Analyzing a dataset many different ways, including analyses of many different outcome variables and various subsets, until a positive result is obtained, and then selectively reporting or emphasizing the analyses that gave positive ("statistically significant") results.
7. Reporting conclusions based on misinterpretation of p-values.

These all may be regarded as errors made within the context of what is, at least nominally, a scientific experiment. But there is a more insidious source of misleading published medical research. Many years ago, when I was young and frisky, I presented a 3-day short course on methods for Bayesian clinical trial design to a large group of statisticians at the U.S. Food and Drug Administration. As my lectures proceeded, there was much discussion, and the course was quite well received. I was very pleased with how the short course played out. But at the end of the third day, after I had finished my final lecture, I learned *The Awful Truth*. A senior FDA statistician took me aside, and explained to me that their greatest concern as statisticians at the FDA was not misapplication of statistical methods. Rather, it was the problem of detecting falsified data used in submissions to the FDA of requests for approval of new drugs.

5.9   Type S Error

But let's return to the realm of scientific activity where the goal is to reach empirical conclusions honestly. A simple, transparent Bayesian approach to multiple comparisons was proposed by Gelman and Tuerlinckx (2000). First, for two parameters, they suggested that evaluation of the parametric difference $\theta_A - \theta_B$ should be done, using a Bayesian hierarchical model-based analysis, by stating one of the following three possible conclusions:

> 1. $\theta_A > \theta_B$ with confidence.
>
> 2. $\theta_B > \theta_A$ with confidence.
>
> 3. No claim with confidence.

They defined a "Type S error" to be making the conclusion "$\theta_A > \theta_B$ with confidence" when in fact $\theta_A < \theta_B$, and similarly for the wrong conclusion with the roles of A and B reversed. The "S" is used for the word "sign," since the idea is that evidence of a difference will go in one direction or the other, and the worst mistake that you can make is to get the sign of $\theta_A - \theta_B$ wrong. The method is based on two ideas. The first is the observation that a frequentist 95% confidence interval for $\theta_A - \theta_B$ not containing 0 leading to rejection of $H_0 : \theta_A = \theta_B$ is a test with Type I error probability 0.05. Hence, using a Bayesian credible interval in place of a confidence interval is sensible, especially since many people think that confidence intervals are credible intervals, anyway. The second idea is that $H_0 : \theta_A = \theta_B$ does not make much sense for continuous, real-valued parameters θ_A and θ_B, since they never are exactly equal. Bayesians have been saying this for many years, but, evidently, this has been ignored by most hypothesis testers.

   In the multiple comparisons setting, here is a sketch of how the Gelman and Tuerlinckx (2000) Bayesian modeling and computations for Type S error work, and some comparisons to error rates of a frequentist method. The following computations are carried out assuming a "normal-normal" Bayesian model, but the main idea is very simple and can be applied quite generally. The method begins by considering data from J independent experiments, where a summary statistic Y_j, which often is a sample mean, observed from the jth study, is assumed to follow a $N(\theta_j, \sigma^2)$ distribution. The θ_j's are assumed to be independent and identically distributed (iid), with prior $\theta_j \mid \mu, \tau \sim_{iid} N(\mu, \tau^2)$. Some computations show this implies that the joint distribution of the differences between the observed statistics and the corresponding differences between their means is

$$(Y_j - Y_k, \theta_j - \theta_k) \sim N_2(0, \Sigma), \quad \text{for all } 1 \leq i \neq j \leq J,$$

where Σ is a 2×2 matrix with $\text{var}(Y_j - Y_k) = 2(\sigma^2 + \tau^2)$, and $\text{var}(\theta_j - \theta_k) = 2\tau^2 = \text{cov}(Y_j - Y_k, \theta_j - \theta_k)$. The analysis compares frequentist confidence intervals and Bayesian credible intervals. The frequentist inference about $\theta_j - \theta_k$ is made "with confidence at a 95% level" if

Frequentist Threshold: $| Y_j - Y_k | > 1.96\sqrt{2}\sigma$.

The corresponding Bayesian threshold is computed as follows. Since, *a posteriori*, the θ_j's are independent with

$$\theta_j \mid \mathbf{Y}, \sigma, \mu, \tau \sim N(\hat{\theta}_j, V_j),$$

where

$$\hat{\theta}_j = \frac{Y_j/\sigma^2 + \mu/\tau^2}{\frac{1}{\sigma^2} + \frac{1}{\tau^2}} \quad \text{and} \quad V_j = \frac{1}{\frac{1}{\sigma^2} + \frac{1}{\tau^2}},$$

it follows that a 95% posterior credible interval for $\theta_j - \theta_k$ is

$$(\hat{\theta}_j - \hat{\theta}_k) \pm 1.96 \, (V_j + V_k)^{1/2}.$$

So, the corresponding Bayesian posterior claim about $\theta_j - \theta_k$ is made "with confidence at a 95% level" if

Bayesian Threshold: $| Y_j - Y_k | > 1.96\sqrt{2}\sigma\sqrt{1 + \sigma^2/\tau^2}$.

The multiplicative term $\sqrt{1 + \sigma^2/\tau^2}$ results in the Bayesian method having a larger limit for making a conclusion "with confidence." So, the Bayesian method may be considered more conservative.

With either the frequentist or Bayesian method, a "Type S error" occurs if a claim is made "with confidence," but it is in the wrong direction, that is, the sign of $\theta_j - \theta_k$ is wrong. The conditional probability of a Type S error is

$$\text{Pr(Type S Error} \mid \text{Claim Made with Confidence)}$$

$$= \text{Pr(sign}(\theta_j - \theta_k) \neq \text{sign}(Y_j - Y_k) \mid Y_j - Y_k > \text{cutoff)}.$$

For other data structures and models, the computations likely will be more complex, and probably will require numerical computations of posteriors using some sort of Markov chain Monte Carlo (MCMC), as described in the books by Robert and Cassella (1999) and Liang et al. (2010). But the general idea is to compute the limits using either confidence intervals or credible intervals, and the general principle is identical.

Table 5.3 Comparison-wise Type S error rates for $J(J-1)/2$ pairwise comparisons using the Bayesian method or Tukey's frequentist honest significant difference (THSD) method. Each entry is computed as the number of comparisons in which a Type S error was made divided by the number of comparisons made with confidence

τ/σ	$J = 5$		$J = 10$		$J = 15$	
	Bayes	THSD	Bayes	THSD	Bayes	THSD
0.5	0.000	0.083	0.000	0.068	0.004	0.054
1.0	0.013	0.015	0.011	0.006	0.008	0.005
2.0	0.006	0.003	0.003	0.001	0.009	0.000

In the multiple comparisons setting, there are $J(J-1)/2$ differences $\theta_j - \theta_k$ to compare. As noted above, there are many frequentist methods to control the overall comparison-wise false positive rate at 0.05, reviewed by Miller (1981) and Bretz et al. (2010). Gelman and Tuerlinckx (2000) use John Tukey's Honestly Significant Difference and Wholly Significant Difference methods, both of which are based on the *studentized range* statistic $\max_{j \neq k}(Y_j - Y_k)/s$, where s is the sample standard deviation of $Y_1, \ldots, Y_J$. Table 5.3 gives the comparison-wise Type S error rates for the Bayesian method and Tukey's HSD method for different values of τ/σ. Recall that $\sigma =$ standard deviation of observed statistic Y_j, estimated by $s = \hat{\sigma}$, while $\tau =$ prior standard deviation of $\theta_j =$ mean of Y_j. The numerical results in Table 5.3 show that the Bayesian method has much smaller experiment-wise Type S error rates than the frequentist HSD for $\tau/\sigma = 0.5$, but this pattern is reversed for $\tau/\sigma = 1.0$ or 2.0, although in these latter two cases both methods give extremely small comparison-wise Type S error rates. Thus, for practical applications, the methods will give substantively different results if τ/σ is not large, that is, if the prior between-study variance is not much larger than the estimated within-study variance.

5.10   A Simple Bayesian Alternative to P-Values

Here is something easy to understand that you can use routinely instead of p-values. It is a statistic for quantifying strength of evidence that makes sense, and it is easy to compute. You can even draw a picture of it. As you might expect, it is a posterior probability computed under a Bayesian model.

First, it is useful to examine the duality between the numerical p-value of a test of hypotheses and the probability of obtaining an effect that is large enough to be practically significant. A key point here is the sharp distinction between practical significance, which actually matters when one is choosing a patient's therapy, and statistical significance, which may or may not have practical implications. Remember the "significant" treatment effect that improved expected survival by 0–3 days?

The following example is structurally similar to the two-sample comparison of means given earlier. Consider a two-sample binomial test. I have chosen this setting because the computations are simple, but the basic idea is the same in virtually any setting where one or two parameters are the focus. Suppose that one wishes to compare the response probabilities π_A and π_B for two treatments. Say A is standard therapy, and B is experimental. A randomized clinical trial is done with equal planned sample sizes $n_A = n_B = n$, so the total study sample size is $N = 2n$. To do a test of the null hypothesis $H_0 : \pi_A = \pi_B$ versus the two-sided alternative $H_a : \pi_A \neq \pi_B$, with Type I error 0.05 and power 0.80, suppose that a null value 0.20 is assumed based on historical experience with A and power 0.80 at fixed alternative $p_B = 0.40$ is desired. So, the targeted difference is $0.40 - 0.20$. That is, you are interested in at least doubling the response rate. It is based on binomial distributions for R_A = number of responses out of 80 patients treated with A and R_B = number of responses out of 80 patients treated with B, and with sample proportions $\hat{\pi}_A = R_A/80$ and $\hat{\pi}_B = R_A/80$. I will take the usual approach of exploiting the assumption that the test statistic

$$Z = \frac{\hat{\pi}_B - \hat{\pi}_A}{\sqrt{\bar{\pi}(1 - \bar{\pi})(\frac{1}{80} + \frac{1}{80})}}$$

is approximately N(0, 1) under H_0, where $\bar{\pi} = (X_A + X_B)/(80 + 80)$ is the pooled estimator of the common value of π under H_0. This is just an extension of the DeMoivre–Laplace Theorem, for two independent binomial samples instead of one. Applying The Sacred Algorithm, the above implies that the required sample size to do the test of hypotheses is $N = 2 \times 79 = 158$. Let's make it $N = 160$ to simplify things, and also to be on the safe side with regard to Type I error probability 0.05 and power 0.80. The two-sided test will (1) reject H_0 in favor of the alternative $\pi_B > \pi_A$ if the observed test statistic is $Z^{obs} > 1.96$, or (2) reject H_0 in favor of $\pi_A > \pi_B$ if the observed test statistic is $Z^{obs} < -1.96$. Otherwise, (3) the test will accept H_0. Equivalently, if the p-value of the test is < 0.05, then H_0 will be rejected. This is standard stuff that can be found in thousands of textbooks on elementary statistics.

But this is not a standard textbook on elementary statistics. Here is something that almost nobody talks about. Nearly all elementary statistics textbooks ignore it. It hinges on the key fact, which I have stated repeatedly, that *If H_0 is rejected, this does not mean that H_a is accepted*. This simple point was central to the discussion of the Simon design. In this example, $Z^{obs} > 1.96$ does not imply that one should conclude that $p_B \geq 0.40$, despite the fact that determining whether this is true or not is the real goal of the clinical trial. The words "…reject H_0 in favor of the alternative $\pi_B > \pi_A$" do not mean that the fixed alternative $\pi_B = \pi_A + 0.20 = 0.40$, which was used to compute the test's power, is accepted.

In classical frequentist two-sided hypothesis testing with an approximately normal test statistic, all that one may conclude if $Z^{obs} > 1.96$ or $Z^{obs} < -1.96$ is that H_0 is not true.

As in the discussion of the Simon design, here the specified alternative value $p_B = 0.40$ is a Straw Man, set up for the sole purpose of doing the sample size computation. The power of the test is the probability that it rejects H_0, under the assumption that

Table 5.4 P-values and Bayesian posterior probabilities for outcomes of a two-sample binomial experiment to compare response probabilities π_A and π_B of treatments A and B. The sample sizes are 80 patients for both A and B. In all cases, the number of responses with treatment A is $R_A = 16$, so $\hat{\pi}_A = 0.20$. The p-values are for a two-sided test of the null hypothesis $\pi_A = \pi_B$, computed using a normal approximation. The Bayesian posterior probabilities are computed assuming independent beta(0.50, 0.50) priors for π_A and π_B

R_B	$\hat{\pi}_B$	$\hat{\pi}_B - \hat{\pi}_A$	Z-score	p-value	$\Pr(\pi_B > \pi_A + 0.20 \mid data)$
27	0.3375	0.1375	1.9617	0.0498	0.174
30	0.3750	0.1750	2.4454	0.0145	0.349
35	0.4375	0.2375	3.2234	0.0013	0.691
40	0.5000	0.3000	3.9780	0.00007	0.912

$\pi_B = 0.40$. *The power of the test is not the probability that* $H_a : \pi_B \geq 0.40$ *is true.* Frequentist tests do not assign probabilities to parameters or hypotheses, since they are not considered random, and hence they cannot have probabilities of being true or false. Again, frequentist tests just accept or reject a null hypothesis.

In contrast, Bayesians assign probabilities to hypotheses, and more generally to statements about parameter values, all the time. From a Bayesian viewpoint, if non-informative beta(0.20, 0.80) priors on both π_A and π_B are assumed, then a posterior probability that is of great interest is

$$\Pr(\pi_B > \pi_A + 0.20 \mid R_A, n_A, R_B, n_B).$$

This is the probability that, given the observed data, the response probability with B is at least 0.20 larger than that with A. This posterior probability takes the specified target $0.20 + 0.20 = 0.40$ seriously, rather than just using it as a device to compute sample size to obtain a hypothesis test with a given power. Table 5.4 gives the p-values and corresponding values of this important posterior probability for data where $R_A = 16$ in all cases and, as a sensitivity analysis, the strength of evidence that B is better than A is determined by one of the four observed statistics $R_B = 27, 30, 35$, or 40 responses out of 80 patients treated with B. If one uses the four corresponding posterior probabilities computed from the respective datasets as criteria to quantify the evidence that $\pi_B > \pi_A + 0.20$, then Table 5.4 suggests that at least 40/80 (50%) responses would be required to obtain at least a 90% posterior probability that this is true. The corresponding p-value is extremely small, it also is very interesting that the "significant" p-value obtained with 27/80 (34%) responses with B gives a posterior probability of only 0.173 that $\pi_B > \pi_A + 0.20$.

This example illustrates an important general fact about the difference between frequentist hypothesis testing and Bayesian inference. It shows that there usually is a large disagreement between what a p-value says about the targeted "Straw Man" in the specified alternative of a test of hypothesis, and the Bayesian posterior evidence that the specified alternative has been achieved. To obtain convincing evidence that one treatment is better than the other by a given increment $\delta > 0$ of some parameter, such as a response probability or mean survival time, i.e., that $\theta_B > \theta_A + \delta$, the empirical difference between θ_B and θ_A must be much far larger than what is required to obtain a "significant" p-value < 0.05. This may help to explain why so many "significant" tests are obtained from data where the actual difference in treatment effects is not practically meaningful.

5.11   Survival Analysis Without P-Values

Here is an example of how p-values may be avoided entirely when summarizing the results of a time-to-event regression analysis. Andersson et al. (2017) reported the results of a randomized clinical trial to compare pharmacokinetically (PK) guided versus fixed-dose intravenous (IV) busulfan as part of the pretransplant conditioning regimen in allogeneic stem cell transplantation (allosct). Allosct is used to treat severe hematologic diseases, including acute myelogenous and lymphocytic leukemia (AML and ALL), and non-Hodgkin's lymphoma. The therapeutic strategy is first to ablate (destroy) the patient's bone marrow in order to kill the cancer cells where they originate, using high dose total body radiation or combination chemotherapy, the "preparative" or "conditioning" regimen, and then infuse cells from a human leukocyte antigen matched donor into the patient. The goal is that the infused donor cells will enter and repopulate the patient's bone marrow, known as "engraftment," and restart the process of blood cell production, "hematopoiesis." This process carries many risks, including various toxicities, infection, and graft versus host disease (GVHD), where the newly engrafted donor cells attack the patient's normal cells. The rates of these adverse events have been reduced greatly over the years, and the main cause of death following allosct is disease recurrence. Numerous new therapies always are being tested with the goal to stop or reduce the rate of disease recurrence following allosct.

A total of 218 patients with either acute myelogenous leukemia (AML) or myelodysplastic syndrome (MDS) were randomized to receive either IV busulfan at fixed dose $108\,\text{mg/m}^2$ ($n = 107$) or PK-guided dosing that targeted 6000 μ_M-min $\pm$ 10% ($n = 111$). Patients were stratified by whether they were in complete remission (CR) at transplant or not, i.e., had active disease. As a regression analysis to assess the effect of dosing method and prognostic covariates on survival time, a Bayesian piecewise exponential regression model was fit to the data, as described generally by Ibrahim et al. (2001). In this model, the hazard function is multiplied by e^{η} where, denoting the binary indicator $I[A] = 1$ if the event A occurs and 0 if not, the linear term in this application was defined to be

$$\eta = \beta_1\,I[\text{PK guided}] + \beta_2\,(\text{Age} - \overline{\text{Age}}) + \beta_3\,I[\text{MDS}]$$
$$+ \beta_4\,I[\text{Poor Risk Cytogenetics}] + \beta_5\,I[\text{CR}] + \beta_6\,I[\text{FLT 3+}].$$

In this piecewise exponential model, $\exp(\eta)$ multiplies the baseline hazard function. So, a negative value of the parameter β_j for variable X_j corresponds to larger X_j

Table 5.5 Fitted Bayesian piecewise exponential regression model for survival time, for 216 leukemia patients, with 114 deaths, treated with PK-guided versus fixed-dose IV busulfan as part of their pretransplant conditioning regime in allogeneic stem cell transplantation. PBE = posterior probability of a beneficial effect for larger values of the covariate

Variable	Posterior quantities			
	Mean of β (sd)	95% credible int. for β		PBE
		L	U	
PK-guided dose	−0.27 (0.19)	−0.64	0.10	0.92
Age − $\overline{\text{Age}}$	0.02 (0.01)	0.00	0.03	0.03
MDS versus AML	−0.42 (0.25)	−0.88	0.09	0.95
Poor Risk Cyto	0.39 (0.19)	−0.01	0.75	0.02
CR	−0.91 (0.22)	−1.32	−0.47	1.00
FLT-3 +	−0.11 (0.20)	−0.28	0.51	0.29

(or $X_j = 1$ if it is a binary indicator variable) being beneficial, since larger X_j must make $\exp(\beta_j X_j)$ smaller when $\beta_j < 0$. Thus, larger X_j (or $X_j = 1$ rather than 0) must make the hazard of death smaller. Equivalently, if $\beta_j < 0$, then larger X_j, or $X_j = 1$ if it is a binary indicator variable, must make expected survival time longer. Poor risk cytogenetics was defined as any of (i) a deletion of chromosome 5 or 7, (ii) multiple chromosomal abnormalities, or (iii) trisomy of chromosome 8, all well known to be negative prognostic abnormalities. Independent non-informative N(0, 100) priors were assumed for the β_j's. The fitted model is summarized in Table 5.5.

In the table, posterior probabilities of the general form $\Pr(\beta_j < 0 \mid data)$ are labeled simply as PBE for "probability of a beneficial effect." This provides an easily interpretable way to quantify how survival increases with the covariate. For example, the PK-guided Bu dose group had a PBE = 0.92. While the posterior 95% credible interval contains 0, this says that the posterior probability that PK-guided IV Bu is superior to fixed-dose IV Bu in terms of survival time was 0.92. A great advantage of summarizing a fitted Bayesian regression model in this way is that the posterior values are easy to interpret. The reader may judge whether they consider a PBE = 0.922 to be large enough to convince them that PK-guided dosing is superior to fixed dosing. Another way to state this is that the posterior odds are about 12 to 1 in favor of PK-guided over fixed IV Bu dosing if survival time is what matters. The PBE = 0.028 for Age is equivalent to a posterior probability of a harmful effect (PHE) of older age being PHE $= 1 - $ PBE $= 1 - 0.028 = 0.972$, which reflects the well-known fact that older age increases the risk of death in allogeneic stem cell transplant patients. Similarly, on average, the MDS patients survived longer than the AML patients, having poor risk cytogenetics had large PHE $= 1 - 0.024 = 0.976$, as expected, being in CR rather than having active disease at transplant was certain to be advantageous with PBE = 1.00, and whether a patient had the FLT-3 mutation did not appear to have a meaningful effect on survival time.

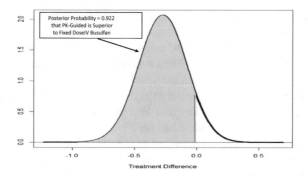

Fig. 5.3 Posterior distribution of the PK-guided versus fixed-dose effect on survival time. The shaded area illustrates the probability 0.922 that PK-guided is superior to fixed IV busulfan dosing in terms of survival of leukemia patients undergoing allogeneic stem cell transplantation

A very useful thing about Bayesian regression models is that you can plot posterior distributions of important parameters, as done for the PK-guided dosing effect in Fig. 5.3. This is a very easily interpretable graphic to illustrate what the posterior probability PBE = 0.922 for PK-guided dosing looks like in terms of the posterior distribution of the parameter that multiplies $I[PK\ guided]$ in the linear term of the hazard function. This simple plot may help the reader decide how convincing the evidence about treatment effect from the fitted Bayesian survival time regression model may be. If some numerical improvement in the log hazard effect β_1 is of interest, or has been prespecified, possibly in the context of a test of hypotheses, say $\exp(\beta_1) = 0.80$ as a 20% drop in the hazard of death by using PK-guided rather than IV Bu dosing, then one also may compute $\Pr(\exp(\beta_1) < 0.80 \mid data) = \Pr(\beta_1 < -0.097 \mid data)$. To illustrate this, draw a vertical line at -0.097 in the plot of the posterior of β_1 and shade in the area to the left of the line. If, instead, your model is, say, a lognormal, where $\beta_1 X_1$ appears in the logarithm of the mean, then larger $\beta_1 > 0$ means that, on average, larger X_1 increases survival time, so it favors a treatment like PK-guided dosing. In this case, the area under the curve of the posterior distribution of β_1 to the right of 0 would be shaded in to illustrate the PBE of the treatment. Using this sort of Bayesian regression analysis, a table like Table 5.5, and maybe a figure like Fig. 5.3, provide an easily interpretable way to summarize inferences, without ever having to compute, or even mention, p-values.

5.12 The P-Value War

If you want to understand human behavior in any society or institution, just look at the system of rewards and punishments that are in place.

Peter F. Thall, in the book 'Statistical Remedies for Medical Researchers'

Until recently, most statisticians and research scientists put up with the ubiquitous misuse and misinterpretation of p-values, as well as blind adherence to the Sacred Cutoff 0.05 to dichotomize strength of evidence as either "significant" or "not significant" in the published scientific literature. For decades, to protest the misuse of p-values was much like someone counting past 10 cows in my earlier example. But it appears that the pendulum may have begun to swing in the opposite direction. Among many scientists, p-values have fallen into disrepute, and the formerly blasphemous idea that 0.05 is not a Sacred Cutoff has been put forth by many prominent statisticians and scientists. In 2015, the editors of the journal *Basic and Applied Social Psychology* banned both p-values and confidence intervals, widely considered to be a highly significant event. This recent movement has been somewhat like people, in my earlier example, not only embracing the idea that there may be such a thing as "15 cows," but even demanding that the tyranny of "The Rule of 10" be overthrown. There was The Trojan War, The War of the Roses, the American Civil War, two World Wars, and now we are in the midst of the P-value war. If you do not believe that lives are at stake in this war, go back and look at Table 5.2.

There have been so many published papers discussing problems with the misuse of p-values that I will mention only a few. They appear to be as numerous as the stars in the night sky. If you have a few months of free time, and nothing better to do, you might try to read them all. Some are quite interesting and entertaining. Actually, there were rumblings about misuse and misinterpretation of p-values quite some time ago. Perhaps some of the most egregious examples are those that confuse statistical significance, defined as a p-value < 0.05, with a clinically significant effect. Of course, these are completely different things. In an early paper, Altman and Bland (1995) explained the problem of misinterpreting a "nonsignificant" test with p-value > 0.05 as implying that a clinically significant effect did not exist. Gelman and Stern (2006) provided interesting examples, both toy and real, to illustrate problems with dichotomizing an experiment's evidence as significant or not. Ioannidis (2005) provided a detailed technical argument that the majority of published research results are false. As one might expect, this paper received a lot of attention, and stimulated quite a lot of controversy. Unfortunately, the medical literature is full of statistical errors, many of which, as I discuss in this book, go far beyond misinterpretation of p-values.

As the controversy grew, the mighty American Statistical Association Board of Directors came out with an official *Statement on P-Values and Statistical Significance*, edited by Wasserstein and Lazar (2016). One could feel the ground shake as the *Statement* was published. It includes *Six Principles* to guide one in the proper use of p-values. The six easily might have been enumerated as 10 or 20, since they contain many wonderful statements, definitions, guidelines, clarifications, and insights. The first Principle begins with cautionary references to such blatantly evil activities as "cherry-picking," "data dredging," and "selective inference." The *Statement* is great fun to read, and very informative. If you are a non-statistician, you might memorize the *Six Principles* so that you can impress, or possibly even attempt to reform, your less informed colleagues.

Depending on how you look at it, The Board may have been preaching to the choir, but they certainly stirred the pot. Immediately upon publication of the *Six Principles*, a storm of papers, editorials, news reports, and online blogs erupted. Some authors damned p-values to Hell for all eternity. The cutoff 0.05 was declared by many to be a false god that had brought pestilence to the world of science. Some suggested that 0.05 should be replaced with 0.005, as a newer, better, more conservative god that would do a much better job of protecting us from the evils of false positive decisions. A bit of thought shows that the new god 0.005 actually is a modern, revisionist hybrid of Sir Ronald's iconic cutoff 0.05 and "The Rule of the Inverted 10." After all, exactly how many cows should be considered "a large number"? Many people wrote papers explaining, often quite passionately, the things that they felt are wrong with how p-values are misused and misinterpreted by scientists.

Flawed statistical practices, including misinterpreted p-values, errors in experimental design or data analysis, and incorrect inferences, have been a pervasive problem for a very long time in the Scientific Community. Over a half century before the *Six Principles* were published and this recent War on P-Values erupted, Frank Yates (1964) wrote

"The most commonly occurring weakness in the application of Fisherian methods is undue emphasis on tests of significance, and failure to recognize that in many types of experimental work estimates of the treatment effects, together with estimates of the errors to which they are subject, are the quantities of primary interest."

Another way to put Yates' statement is that, when analyzing data and making inferences, we are in the estimation business, not the hypothesis testing business. More recently, Goodman (2008) listed and explained 12 misconceptions about p-values that commonly are made in the scientific literature. So Goodman provided a list twice as long as ASA Board's list, and he did it 8 years earlier. He called these misconceptions a "dirty dozen," which appeared to be an oblique reference to a famous 1967 B-movie directed by Robert Aldrich. Goodman's paper provides useful guidance for what not to do with p-values. I already have noted a few of these, including the mistaken idea that the p-value of a test is the probability of H_0, which cannot be true since a p-value is computed by first assuming that H_0 is true. Another common mistake is the idea that two studies that produce the same p-value must provide identical evidence, which Table 5.1 shows is complete nonsense. A detailed technical explanation of misinterpretations of p-values is given by Greenland et al. (2016). A review of the official ASA statement and a series of interesting illustrative examples are given by McShane and Gal (2017) who, like Gelman and Stern (2006), decried the common practice of dichotomizing statistical evidence as being either significant or not. After all, statisticians use probabilities all the time, so why dichotomize evidence as being either "significant" or "not significant" when most of the time the truth is somewhere in between? All of this gets at the idea, illustrated earlier, that estimating effects and determining their practical significance is far more important than making a dichotomous "accept or reject" decision about a null hypothesis.

When Neyman and Pearson (1933) published the paper with their famous Neyman–Pearson lemma for hypothesis testing, it is doubtful that they imagined their ideas would be misused so egregiously. Nonetheless, Fisher (1958) foresaw the coming plague of statistical misapplications, and he provided the following prescient warning:

"We are quite in danger of sending highly trained and highly intelligent young men out into the world with tables of erroneous numbers under their arms, and with a dense fog in the place where their brains ought to be. In this century, of course, they will be working on guided missiles and advising the medical profession on the control of disease, and there is no limit to the extent to which they could impede every sort of national effort."

Perhaps what often goes wrong in the world of scientific research has more to do with sociology than how p-values or other aspects of statistical reasoning are misused or misinterpreted. If you want to understand human behavior in any society or institution, just look at the system of rewards and punishments that are in place. Scientists are rewarded for publishing papers, and any given scientific paper is more likely to be published if it provides "significant" results. Some journal editors and reviewers even reject papers that report the results of a laboratory or genomic experiment, or a clinical trial, for which the results are not statistically significant at p-value level 0.05. This practice actually increases the likelihood of false positive conclusions in the published scientific literature, since it is an inherently biased selection of positive over negative conclusions. This is ironic, given the pervasive obsession in the scientific community with protecting the world from false positive conclusions. Scientific careers are made or broken, and grants are funded or not, based on published research. One of the methods for evaluating a scientist is first to count his or her publications. Whether interpreted correctly or not, p-values happen to be the currency used most commonly to quantify strength of evidence.

While writing this book, I showed Table 5.4 to several of my physician-scientist collaborators, to see how they would react. These are highly educated, highly successful medical researchers with years of experience treating patients, conducting clinical trials, doing laboratory research, and publishing papers in medical journals. Invariably, they were quite impressed and surprised by the sharp disagreement between what the numerical p-values and posterior probabilities appear to imply about strength of evidence. But, also invariably, they admonished me that they must continue to use p-values in their research papers if they wish to get them published. Sometimes, I ask collaborators who are not statisticians to define "p-value." Without exception, none of them know the definition. For example, if a non-statistician is asked what they would conclude if a properly designed and properly conducted randomized clinical trial of a new treatment E compared to a conventional standard treatment S gave a two-sided p-value < 0.05 in favor of E, they typically say things like "The probability that E is not better than S is less than 5%." Formally, if θ_E and θ_S are the parameters and larger is better, as with response probabilities or mean survival time, this statement says $\Pr(\theta_E \leq \theta_S \mid data) < 0.05$, which is equivalent to $\Pr(\theta_E > \theta_S \mid data) > 0.95$. This is a Bayesian posterior probability, and certainly is not what observing a p-value < 0.05 means. One interesting, rather circular response was "Something that represents significance and that occurs when the p-value is less than 0.05." Another answer was "A p-value less than 0.05 is the observation that a particular event studied has less than a 5% probability of occurring by chance only." Another response was "The probability of inappropriately accepting the null hypothesis." Still, despite the fact that they don't know the definition, the one thing that they all know for sure is that a p-value <0.05 is what matters.

I have shown some of my medical collaborators the posterior distributions in Fig. 5.2 and Table 5.2, and the posterior credible intervals of the probabilities of 60-day death with the two ECMO methods discussed in Sect. 5.7. Some have become alarmed and suggested that I write a letter to the journal that published the paper by Combes et al. (2018), in order to present my Bayesian analysis as an alternative. The authors of this paper are experts in their field who certainly must care deeply about their ARDS patients' benefit as they struggle to keep them alive. But, by deeply entrenched convention, the two-sided p-value 0.09 was "not significant." Evidently, following the convention of making dichotomous inferences based on p-values led the authors to their conclusion, rather than using estimation of the two treatments' probabilities of death within 60 days as the primary basis for inference and therapeutic decision making. In general, one may replace p-values with other statistical criteria, such as Bayes Factors, Type S error, a decision analysis, posterior probabilities, or parameter estimates with accompanying estimates of their reliabilities. But we should not expect human behavior to change, for scientists, physicians, statisticians, or dairy farmers.

The fault, dear reader, is not in our p-values, but in our widespread ignorance of how they are defined, our incorrect use of them to quantify strength of evidence, and our frequent misuse of them to make dichotomous conclusions.

Chapter 6
Flipping Coins

Take away the cause, and the effect ceases.
Miguel de Cervantes

Contents

Abstract In this chapter, I will discuss the use of randomization as a fundamental scientific tool in experiments where one wishes to make fair comparisons. I will begin with a brief discussion of the use of randomization in agricultural experiments many years ago, and how much later it became a prominent component of comparative clinical trials in medical research. To illustrate its usefulness and importance, I will describe a famous example of how randomization was used to obtain an unbiased comparison of two very different treatments for breast cancer that changed medical practice worldwide. An explanation of why randomization provides unbiased estimators of causal effects will be given. I will provide an example showing how a covariate effect can be mistaken for a treatment effect if one does not randomize, and instead relies on a conventional regression analysis of observational data. Reviews and illustrations will be given of statistical methods to correct for bias when analyzing observational data from non-randomized studies, including stratification, inverse probability weighting, and pair matching. Rubin's (1978) Bayesian rationale for randomization will be described. A review will be given of two simulation studies of outcome-adaptive randomization, including potentially severe scientific flaws with this methodology that may not be apparent.

© Springer Nature Switzerland AG 2020
P. F. Thall, *Statistical Remedies for Medical Researchers*, Springer Series
in Pharmaceutical Statistics, https://doi.org/10.1007/978-3-030-43714-5_6

6.1   Farming and Medicine

Many years ago, statisticians convinced medical researchers that, if they wanted to obtain a fair comparison between two treatments, the best thing that they could do was run a clinical trial with each new patient's treatment chosen by flipping a coin. Much earlier, for designing agricultural experiments involving comparison, R.A. Fisher presented randomization as an essential component of experimental design, in Fisher (1925), and Fisher (1935). An account of the early history is given by Armitage (2003). It is interesting that, when Fisher and his colleagues were inventing experimental design methods at Rothamsted Experimental Station in the early twentieth century, they were motivated by problems arising in agricultural experiments. This is why many early books on experimental design talk about things like "split plot designs." In Fisher's time, a major problem facing the human race was producing enough food to feed the growing population. So a major statistical problem was how to grow crops more efficiently, with a common goal to increase yield per acre. Since the world population has grown from less than two billion in 1920 to over seven billion today, it appears that they may have succeeded.

Today, a lot of the scientific community's focus has shifted to evaluating new medical treatments. Basic statistical science that was developed for designing agricultural experiments has been used, modified, and extended in the past few decades to accommodate clinical trials and laboratory-based biological experiments. A major tool that has been retained, and elaborated in many different ways, is randomization. A basic reference on randomization in clinical trials is the book by Rosenberger and Lachin (2004). The published literature on various randomization methods in clinical trials is immense. Some discussions and applications are given by May et al. (1981), Simon et al. (1985), Christen et al. (2004), Rubinstein et al. (2005), Sharma et al. (2011), Jung (2013), and Korn et al. (2012). A useful explanation of some common misunderstandings about randomization is given by Senn (2013), who is a global expert on statistics, playing dice, and mythology.

For example, suppose that a group of physicians wish to compare treatments A and B for a particular disease, in terms of how long patients survive from the start of treatment. They might design and conduct a clinical trial to achieve this goal, requiring each physician or medical center participating in the trial to do the following. When a new patient is diagnosed with the disease and enrolled in the trial, flip a coin. If it comes up heads, treat the patient with A. If it comes up tails, treat with B. Call the two resulting treatment groups "Arm A" and "Arm B," and keep track of each patient's time to death, or how long they have been followed without dying. At the end of the trial, use the survival time data to decide whether one treatment is better than the other.

Of course, people who are eligible for a medical experiment are not the same thing as plots of land. It is necessary to explain the coin-flipping step to each patient during the informed consent process, in order to find out whether they will agree to be enrolled in the trial and randomized. The ethical rationale for randomization is twofold. First, based on current knowledge, there must be no reason for any of the participating physicians to favor one treatment over the other. This condition is known as "equipoise." This says that, given current knowledge, in terms of the patient's potential benefit it does not matter whether they receive A or B. Since equipoise is subjective, and is based on both knowledge and belief, at the same point in time one physician may have equipoise while another does not. A physician who favors one treatment over the other, hence does not have equipoise, should not participate in a randomized trial. Some physicians feel that admitting their ignorance by telling a patient that they do not know which of the two treatments is better is bad medical practice, since it may undermine the patient's trust in their medical expertise. Second, flipping a coin to choose each patient's treatment will produce data for comparing A to B fairly, which serves as a reliable basis for deciding whether one treatment is better than the other, and thus benefiting future patients. In practice, one does not actually flip a coin, since a computer will do the equivalent of this to ensure that each treatment is assigned with probability 1/2.

As crazy as flipping a coin to choose each patient's treatment may seem, it has turned out to be one of the most powerful scientific tools that statisticians ever gave to physicians. Two questions are why this produces a "fair" comparison, and why it is better than just letting each physician choose the treatment that they think is best for each patient. The answers to these questions have to do with the validity of any statistical comparison between A and B based on data from the experiment.

A famous example of the value of randomization came from the problem of evaluating two different treatments for women diagnosed with primary invasive breast cancer. The issue was how to obtain a fair comparison of expected survival time following either mastectomy with axillary dissection (MT), or breast conserving therapy (BCT) consisting of lumpectomy, axillary lymph node dissection, and local radiation. For many years, it was believed widely by oncologists that MT was the best treatment option, and it was used routinely, despite the fact that it left the patient severely disfigured. Then medical researchers began to take the statistical approach of conducting prospective randomized comparative trials of MT versus BCT, starting with the now famous Danish study reported by Blichert-Toft et al. (1992). To the surprise of many oncologists, several randomized trials showed that expected survival times with MT and BCT were very nearly identical. While this issue has a long history, a detailed meta-analysis by Jatoi and Proschan (2005) based on the results of six randomized studies concluded that, after long-term follow up, BCT and MT have comparable effects on survival time in women with primary breast cancer, although BCT has a significantly greater risk of locoregional recurrence.

Without randomization, and the courage of Blichert-Toft et al. (1992) to do the first randomized comparison, most women with primary invasive breast cancer still might be undergoing MT unnecessarily. Maybe it was because Danes are descended from Vikings, who were fearless explorers and great warriors.

6.2 How Not to Compare Treatments

Here is an example of how one can go astray when comparing treatments. Suppose that a practicing physician is deciding whether to give a patient with a particular disease either treatment A or B, with the goal to maximize the patient's survival time. The physician sees that published data show the expected survival time with B is 2.90 years, compared to 2.25 years with A, so obviously B is the better treatment, on average. Suppose that the data were obtained from clinical practice treating the disease, where many physicians believe that B may be harmful to patients with impaired kidney function, specifically with C = creatinine level ≥ 1.40 mg/dl, but that otherwise B is better than A. Consequently, the physicians who generated the data were more likely to give A to patients with $C \geq 1.40$ and more likely to give B if $C < 1.40$.

Suppose, for simplicity, that A and B actually have the same sample mean survival times within each of the two creatinine subgroups, $\hat{\mu}_{A,1} = \hat{\mu}_{B,1} = 3.0$ years for "low" creatinine patients with $C < 1.40$, and $\hat{\mu}_{A,2} = \hat{\mu}_{B,2} = 2.0$ years for "high" creatinine patients with $C \geq 1.40$. For brevity, I have indexed the low creatine subgroup with $C < 1.40$ with 1 the high creatine subgroup with $C \geq 1.40$ with 2 in the subscripts. So, within each creatinine level subgroup, A and B are perfectly equivalent. Next, suppose that the subsample sizes of the data, cross-classified by treatment and creatinine level, are as in Table 6.1, which was not provided in the published paper. In the table, $\hat{\mu}_{A,1}$ is the estimated mean survival time of patients with $C < 1.40$ who received A, and so on. Suppose that the overall sample means reported in the paper actually were computed as the following subgroup-weighted averages:

$$
\hat{\mu}_A = \frac{200}{800} \times \hat{\mu}_{A,1} + \frac{600}{800} \times \hat{\mu}_{A,2} = 0.25 \times 3.0 + 0.75 \times 2.0 = 2.25
$$

and

$$
\hat{\mu}_B = \frac{900}{1000} \times \hat{\mu}_{B,1} + \frac{100}{1000} \times \hat{\mu}_{B,2} = 0.90 \times 3.0 + 0.10 \times 2.0 = 2.90.
$$

While this may seem to make sense, something went wrong in the published paper's computations. The overall, subgroup-weighted sample mean survival times 2.25 for A and 2.90 for B were computed incorrectly. This is because the computation did not account for the fact that a much higher proportion of patients with good kidney function ($C < 1.40$) got B and a much higher proportion of patients with poor kidney function ($C \geq 1.40$) got A. So the sample of clinical practice data was biased in favor of B over A. Perhaps the people who wrote the paper and computed the overall means using the above formulas did not notice the relationship between treatment assignment and creatinine level. Perhaps they thought that computing the statistics given above was the right way to do things. But since the data were biased in favor of B over A, this made B appear to be a better treatment despite the fact that A and B were perfectly equivalent within each of the two creatinine level subgroups. The difference between the two overall sample means 2.25 and 2.90 actually was due entirely to the effect of creatinine on survival time.

While this example is very simple, a key question is what would you infer if the information in Table 6.1 were not available? That is, suppose all you know is that you have samples of size 800 for A and 1000 for B with the reported means. If the tabled creatinine levels are not available, and you do not know how practicing physicians

Table 6.1 Subgroup sample sizes and sample mean survival times, in years, from historical data on treatments A and B, cross-tabulated with C = Creatinine level, in mg/dl, being either <1.40 or ≥ 1.40

Treatment	$C < 1.40$		$C \geq 1.40$		Overall mean
	n	Mean	n	Mean	
A	200	3.0	600	2.0	2.25
B	900	3.0	100	2.0	2.90

chose between the two treatments based on creatinine level, then it may be very easy to be misled by the two means 2.25 years for A and 2.90 years for B. You may take the summary statistics at face value and conclude that B is superior to A. This sort of thing happens all the time.

There also is another, even more insidious possibility. For simplicity, in this example I isolated one key prognostic variable, known by the practicing physicians, that is strongly related to mean survival time. What if, instead of creatinine level, there is some other variable, completely unknown, that is strongly related to survival time, and by chance it is similarly unbalanced between A and B in the dataset? That is, what if the data actually have the structure of Table 6.1, but this cross-classification is unavailable, and all you have are the sample sizes and computed sample means 2.25 and 2.90? Such unknown variables sometimes are called *external confounders*. They are the source of a great deal of confusion and the cause of numerous incorrect conclusions. Remember the Innocent Bystander Effect? What if the unknown external confounder is the actual cause of the difference in mean survival times, and the treatments A and B are just innocent bystanders? Even worse, what if A actually is better than B, but the external confounder leads you to the opposite conclusion? This is a scary thought, since it suggests that anything you, or anyone else, may conclude from analyzing observational data might be just plain wrong. It turns out that there are lies, damned lies, and naively analyzed observational data. But don't despair. There is a statistical technique that can save you, at least in some cases.

6.3 Counterfactuals and Causality

This section has a lot of formulas, so if that scares you then you might just skip ahead. But then you will miss the wondrous explanation of why something as crazy as flipping a coin to choose treatments actually is a great idea.

If your goal is to compare two treatments in terms of mean survival time, or more generally the mean of any observed outcome, here is a way to do it that avoids the sort of biased comparison given above. It is based on the following thought experiment, which provides the basis for how to conduct a clinical trial to compare A and B. That is, the thought experiment will provide a rationale for how to conduct an actual experiment.

Consider how a physician would make a choice between treatments A and B for a single patient. In typical practice, possibly in the absence of confirmatory data, a physician first looks at each patient's vector of prognostic covariates, X, which may include variables such as creatinine level, age, performance status, disease severity, and so on. It is natural for the physician to think about how long the patient might live with each treatment, given X. If the imagined survival time of the patient if treatment A is given is denoted by $Y(A)$, and the imagined survival time if B is given is denoted by $Y(B)$, then the physician's treatment decision may be based on $Y(A) - Y(B)$. This difference is called the *causal effect of treatment A versus B* on the patient. Notice that $Y(A)$ and $Y(B)$ are imaginary future outcomes, called *potential outcomes*. Since the patient can only receive one of the two treatments, only one of these two potential outcomes can be observed. If the patient is treated with A and the patient's actual survival time Y is observed subsequently, then it will turn out that $Y(A) = Y$. But, while $Y(B)$ was imagined before the patient was treated, and if B had been given it could have been observed, in fact B was not given so $Y(B)$ was not observed. In this case, the potential outcome $Y(B)$ is called the *counterfactual* outcome. Actually, before a treatment was chosen and administered, all that was known was that one of $Y(A)$ or $Y(B)$ would turn out to be the actual observed survival time, depending on which treatment was given. The fact that only one of the two potential outcomes can be observed for each subject, so it is impossible to observe $Y(A) - Y(B)$, is called the *Fundamental Problem of Causal Inference*, Holland (1986).

Here is a thought experiment that solves the problem. First, make two identical copies of the patient. Treat one copy with A, and the other copy with B. So, now both $Y(A)$ and $Y(B)$ can be observed and the causal effect $Y(A) - Y(B)$ can be computed. As wonderful as this solution is, of course you can't do this experiment with a real patient. So, we have a wonderful yet impossible solution to the problem. You can't make copies of patients. While it is nice to imagine things like $Y(A)$ and $Y(B)$, this solution requires us to do something that is impossible in reality.

But here is something that may seem really strange. Thinking about the potentially observable outcomes $Y(A)$ and $Y(B)$, and the difference $Y(A) - Y(B)$ that is impossible to observe, is the key to solving the problem. This can be done by conducting a clinical trial with real physicians using real treatments A or B to treat real patients. *If you are willing to make a few reasonable assumptions, then you can generate real data that can be used to obtain an unbiased statistical estimate of the mean causal effect of A versus B.* The fact that the estimator is correct, formally that it is unbiased, can be proved by thinking about the potential outcomes $Y(A)$ and $Y(B)$. Here is how this is done.

Since we can imagine $Y(A)$ and $Y(B)$, we can imagine the expected survival times, $\mu_A = E\{Y(A)\}$ and $\mu_B = E\{Y(B)\}$. These leads to the following definition of what we want to estimate.

DEFINITION The *mean causal effect of A versus B* on survival time is $\Delta_{A,B} = E\{Y(A) - Y(B)\} = \mu_A - \mu_B$.

What follows may seem a bit technical, but it doesn't require any deep mathematics. Estimating the causal parameter $\Delta_{A,B}$ requires the following key assumptions. For convenience, define the treatment indicator $Z = 1$ if A is given and $Z = 0$ if B is given. Recall that X denotes the patient's vector of observed covariates.

Positivity Assumption

It must be possible to give each treatment being considered for every possible covariate vector, that is, $\pi(X) = \Pr(Z = 1 \mid X) > 0$ and $1 - \pi(X) = \Pr(Z = 0 \mid X) > 0$ for all values of X. After all, if it is not possible for patients with certain values of X to receive one of the treatments, then comparing the treatments is pointless for those patients.

Consistency Assumption

The observed outcome must be the potential outcome corresponding to the treatment actually received. Formally, $Y = Y(A)$ if $Z = 1$ and $Y = Y(B)$ if $Z = 0$. This says that, once you know the treatment, the potential outcome for that treatment and the observed outcome must be the same.

The next assumption requires the idea of independence. If the two variables U and V are independent, we write this as $U \perp\!\!\!\perp V$.

Ignorability Assumption

$Y(A), Y(B) \perp\!\!\!\perp Z \mid X$. This says that, given the patient's known covariates, their treatment assignment is independent of their potential outcomes. It is motivated by the idea that you can't see into the two possible futures and find out what $Y(A)$ or $Y(B)$ would turn out to be. After all, if you could, and Y is, say survival time, you would just give the patent the treatment that gives the larger future survival time. This also is called the *No Unmeasured Confounders* assumption.

Stable Unit Treatment Value Assumption (SUTVA)

This is a famous acronym, so you might commit it to memory so that you can impress your friends and relatives. In modern statistics, famous acronyms are a bit like incantations used by magicians. Just practice saying "SUTVA" a few times, and then casually drop it into a conversation at your next social event. Here are the two components of SUTVA:

1. There is only one form of each treatment. This implies that the potential outcomes are well defined.
2. The treatment assignment of any given subject does not affect (is independent of) the potential outcomes of any other subjects. This also is called *noninterference*.

Indexing subjects by $i = 1, \ldots, n$, noninterference says that $Y_i(Z_i) \perp\!\!\!\perp Z_j$ for each $j \neq i$. For example, the fact that $Z_1 = 1$ for subject 1 does not affect the potential outcome $Y_2(Z_2)$ of subject 2.

Now consider a sample of outcomes on N subjects, in which n_A subjects received treatment A and n_B subjects received B. For each subject $i = 1, \ldots, N = n_A + n_B$, the potential outcomes are $Y_i(t)$ for $t = A, B$, with $Z_i = 1$ if the subject got A and $Z_i = 0$ if the subject got B. The ith subject's observed outcome is Y_i, and X_i is the covariate vector. So the observed data are $\{(Y_i(Z_i), Z_i, X_i), i = 1, \ldots, N\}$, since

$Y_i = Y_i(Z_i)$. The pairs of potential outcomes $\{Y_i(A), Y_i(B)\}$ are independent with means $E\{Y_i(t) \mid X_i\} = \mu(t)$ for $t = A, B$. For brevity, denote the probability that subject i received treatment A by $\pi_i = \Pr(Z_i = 1 \mid X_i)$, the patient's *propensity score*. This notation suppresses the facts that $\mu(t)$ and π_i each may depend on the covariates, X_i. For example, in clinical practice, physicians routinely use a patient's characteristics, X_i, to choose a treatment.

The following is an estimator of the causal effect $\Delta_{A,B}$:

$$\hat{\Delta}^{causal}_{A,B} = \frac{1}{N} \sum_{i=1}^{N} \left\{ \frac{Y_i Z_i}{\pi_i} - \frac{Y_i(1 - Z_i)}{1 - \pi_i} \right\}. \tag{6.1}$$

Notice that the *positivity assumption* is needed to ensure that this estimator can be defined. Otherwise, $\pi_i = 0$ or $1 - \pi_i = 0$, and the above formula would require dividing by 0, which simply is not done by respectable people. But if it is impossible for some patients to receive a treatment, then there is no point in evaluating its effects on those patients in the first place.

It turns out that $\hat{\Delta}^{causal}_{A,B}$ is an unbiased estimator of the causal parameter. To see why this is true, just consider one subject. For any i, by the *consistency assumption*, since $Z_i = 0$ or 1,

$$Y_i = Y_i(A)Z_i + Y_i(B)(1 - Z_i)$$

This implies that

$$E\left\{ \frac{Y_i Z_i}{\pi_i} \right\} = E\left\{ \frac{Y_i(A)Z_i^2 + Y_i(B)(1 - Z_i)Z_i}{\pi_i} \right\} =$$
$$E\left\{ \frac{Y_i(A)Z_i + 0}{\pi_i} \right\} = E\left\{ \frac{Y_i(A)Z_i}{\pi_i} \right\}.$$

A useful fact in probability is the iterated expectation equality, that $E(U) = E\{E(U \mid V)\}$ for any random quantities U and V. Applying this, and the *ignorability assumption*, the above expression equals

$$E\left[E\left\{ \frac{Y_i(A)Z_i}{\pi_i} \mid X_i \right\} \right] = \frac{E\{Y_i(A) \mid X_i\}E(Z_i \mid X_i)}{\pi_i} = \frac{\mu(A)\pi_i}{\pi_i} = \mu_A.$$

By exactly the same reasoning, for each i,

$$E\left\{ \frac{Y_i(1 - Z_i)}{1 - \pi_i} \right\} = \mu_B.$$

This says that each of the N summands in the causal estimator $\hat{\Delta}^{causal}_{A,B}$ has expected value $\mu_A - \mu_B = \Delta_{A,B}$, so this also must be their average value. This is why $\hat{\Delta}^{causal}_{A,B}$ is called the *causal estimator* of $\Delta_{A,B}$.

Expression 6.1 also is called an "inverse probability of treatment-weighted" (IPTW) estimator. In practice, IPTW estimation can be used to correct for bias in observational data by assuming a regression model for the treatment assignment probability $\pi_i = \pi(X_i, \theta)$, and plugging in estimates $\hat{\pi}_i = \pi(X_i, \hat{\theta})$ obtained from the fitted model to compute 6.1. Essentially, to correct for treatment assignment bias, the IPTW estimator exploits the relationship between treatment assignment probability and known X_i to obtain each $\hat{\pi}_i$, then replaces Y_i with $Y_i/\hat{\pi}_i$ if $Z_i = 1$ or by $Y_i/(1 - \hat{\pi}_i)$ if $Z_i = 0$, and then averages the re-weighted outcomes. So now $\hat{\Delta}_{A,B}^{causal}$ computed using the estimates $\hat{\pi}_i$ is approximately unbiased. To help avoid problems that may arise if some $\hat{\pi}_i$ are near 0 or 1, one may use the normalized estimators

$$\hat{\mu}_A^{NIPTW} = \frac{\sum_{i=1}^n Y_i Z_i/\hat{\pi}_i}{\sum_{i=1}^n Z_i/\hat{\pi}_i} \quad \text{and} \quad \hat{\mu}_B^{NIPTW} = \frac{\sum_{i=1}^n Y_i(1 - Z_i)/(1 - \hat{\pi}_i)}{\sum_{i=1}^n (1 - Z_i)/(1 - \hat{\pi}_i)},$$

and $\hat{\Delta}_{A,B}^{NIPTW} = \hat{\mu}_A^{NIPTW} - \hat{\mu}_B^{NIPTW}$.

Another modification of the IPTW estimation formula, to deal with the problems that the assumed model $\pi(X, \theta)$ for the treatment assignment probabilities may be incorrect and the estimator may be unstable is the *Augmented IPTW* estimator, AIPTW, given by the formula

$$\hat{\Delta}_{A,B}^{AIPTW} = \frac{1}{N} \sum_{i=1}^N \left\{ \frac{Y_i Z_i}{\hat{\pi}_i} - \frac{(Z_i - \hat{\pi}_i)}{\hat{\pi}_i} m_A(X_i, \hat{\alpha}_A) \right\}$$

$$- \frac{1}{N} \sum_{i=1}^N \left\{ \frac{Y_i(1 - Z_i)}{1 - \hat{\pi}_i} + \frac{Z_i - \hat{\pi}_i}{1 - \hat{\pi}_i} m_B(X_i, \hat{\alpha}_B) \right\}$$

$$= \hat{\mu}_A^{AIPTW} - \hat{\mu}_B^{AIPTW},$$

where $m_A(X, \alpha_A)$ and $m_B(X, \alpha_B)$ are the respective means $E(Y \mid Z = 1, X)$ and $E(Y \mid Z = 0, X)$ of the outcome for treatment A and B, based on some assumed regression model. The idea behind the AIPTW formula is that the additional terms will correct for the possibility that the propensity score model may be incorrect. The AIPTW estimator often is called *Doubly Robust* since it has been proved that, if (1) the assumed propensity score model $\pi(X, \theta)$ is correct or (2) the assumed regression model for $[Y \mid Z, X]$ is correct, then $\hat{\Delta}_{A,B}^{AIPTW}$ is a consistent estimator of the causal effect. That is, if at least one of the two assumed models is correct then, $\hat{\Delta}_{A,B}^{AIPTW}$ will be a well behaved estimator of $\mu_A - \mu_B$, provided that you have a large sample. This is a very clever estimator and a nice theorem, but it is a bit like saying that, if at least one of The Easter Bunny or Santa Claus actually exists, then life will be good. If you want to do AIPTW estimation, the practical issues comes down to finding reasonably robust models for the propensity scores and observed outcomes, and estimating the variance of $\hat{\Delta}_{A,B}^{AIPTW}$ so that you can compute approximate confidence intervals, or computing them using a bootstrap method. A very thorough treatment of AIPTW estimation is given in the book by Tsiatis (2006).

At this point, it should be obvious that bias correction in causal inference can get very complicated very quickly. Like Alice in *Alice's Adventures in Wonderland* by Carroll (1898), once you go down the rabbit hole you may encounter all sorts of strange things. There is a huge published literature on causal inference. Some other useful books are those by Pearl et al. (2016), Imbens and Rubin (2015), Rosenbaum (2010), Morgan and Winthrop (2015), and Kleinberg (2013).

6.4 Why Randomize?

In addition to showing ways to correct for bias in observational data, the development given above can be used to establish a motivation for randomizing. If you randomize patients fairly between two treatments, then things get a lot simpler, since you know that $\pi_i = 1 - \pi_i = 1/2$ for all patients, so the causal estimator (6.1) is simply

$$\hat{\Delta}_{A,B}^{causal} = \frac{1}{N} \sum_{i=1}^{N} \left\{ \frac{Y_i Z_i}{1/2} - \frac{Y_i(1 - Z_i)}{1/2} \right\}.$$

Denote the vector of treatment assignment indicators for the sample of patients by $\mathbf{Z} = (Z_1, \ldots, Z_n)$. Denote the set of indices for patients who received A by $\mathcal{I}_A = \mathcal{I}_A(\mathbf{Z}) = \{i : Z_i = 1, i = 1, ..., n\}$. The subsample size for A is $|\mathcal{I}_A(\mathbf{Z})| = n_A(\mathbf{Z}) = n_A$. Write the first part of $\hat{\Delta}_{A,B}^{causal}$ as

$$\frac{1}{N} \sum_{i=1}^{N} \frac{Y_i Z_i}{1/2} = \frac{2}{N} \sum_{i \in \mathcal{I}_A} Y_i Z_i = \frac{2}{N} \sum_{i \in \mathcal{I}_A} Y_i = \frac{2n_A}{N} \frac{1}{n_A} \sum_{i \in \mathcal{I}_A} Y_i = \frac{2n_A}{N} \overline{Y}_A.$$

Since $\Pr(Z_i = 1) = 1/2$, the subsample size n_A of patients who received A is a random variable that follows a binomial distribution with parameters $1/2$ and N, so $E(n_A) = N/2$. Therefore, again using the independence of the treatment assignments $\mathbf{Z}$ and each observed potential outcome $Y_i = Y_i(Z_i)$, since n_A is determined by $\mathbf{Z}$,

$$E\left(\frac{2n_A}{N} \overline{Y}_A\right) = E\left(\frac{2n_A}{N}\right) E(\overline{Y}_A) = \frac{2(N/2)}{N} \mu_A = \mu_A.$$

This says that the left hand portion of $\hat{\Delta}_{A,B}$ has mean μ_A. Similarly, the right hand portion of $\hat{\Delta}_{A,B}^{causal}$ has mean μ_B, so $E(\hat{\Delta}_{A,B}^{causal}) = \mu_A - \mu_B = \Delta_{A,B}$, that is, it is an unbiased estimator of the causal B versus A effect. So, as promised, you don't have to take the impossible approach of making two identical copies of each patient, treating one with A, the other with B, and observing $Y_i(A)$ and $Y_i(B)$ for each $i = 1, \ldots, n$ in order to estimate the causal A versus B effect. Instead, just flip a coin to choose each patient's treatment.

This is the proof of unbiasedness in the case implied by common statement "Given independent samples $Y_{A,1}, \ldots, Y_{A,n_A}$ from population A and $Y_{B,1}, \ldots, Y_{B,n_B}$ from population B, the difference between the sample means, $\overline{Y}_{A,n_A} - \overline{Y}_{B,n_B}$, is an unbiased estimator of the difference between the population means." To be true, this requires that subjects were randomized. If you want to compare say, three treatments, A, B, and C, and you randomize patients fairly among them with probabilities 1/3 each, the same sort of derivation shows that your data will provide unbiased estimators of all three differences $\mu_A - \mu_B$, $\mu_A - \mu_C$, and $\mu_B - \mu_C$. So, randomization also is the right thing to do for multi-arm studies. Interestingly, this reasoning also works the same way even if you don't randomize fairly. All that is needed is a known randomization probability that was used throughout the study. For example, if patients were randomized to A and B in a 2:1 ratio, with fixed probabilities $\pi_i = 2/3$ of receiving A and $1 - \pi_i = 1/3$ of receiving B for all $i = 1, \ldots, n$, then the causal estimator would be

$$\hat{\Delta}_{A,B}^{causal} = \frac{1}{N} \sum_{i=1}^{N} \left\{ \frac{Y_i Z_i}{2/3} - \frac{Y_i(1 - Z_i)}{1/3} \right\}.$$

Since the general computation given above still holds, with the only difference being that $n_A \sim$ binomial(2/3, N) and $n_B \sim$ binomial(1/3, N), so the sample size n_A has mean $(2/3)N$ and n_B has mean $(1/3)N$, even with this unfair 2:1 randomization, the causal estimator still is unbiased. All that you need is to know the randomization probabilities.

Still, there is the practical consideration that, given the overall sample size N, the variance of the estimator $\hat{\Delta}_{A,B}$ is minimized if $\pi_i \equiv 1/2$. So, for any given N, unbiased (fair) randomization gives the smallest variability in the estimator $\hat{\Delta}_{A,B}$. A 2:1 randomization may be conceptualized by thinking of rolling a six-sided die and giving treatment A if a 1, 2, 3, or 4 comes up, and B if a 5 or 6 comes up. While some clinical trialists actually do 2:1 randomization, I always have considered it a bit difficult to defend. What do you say if, at the end of the trial, treatment B, which was given to 1/3 of the patients, turns out to be better than treatment A, which was given to 2/3 of the patients? "Oops"? Besides, if you randomize in a 2:1 ratio, then you don't really have equipoise, so what do you tell your patients when you are trying to get them to sign an informed consent form? "I really like A twice as much as B, but let me choose your treatment with a 2:1 randomization rather than just giving you A"? People who design and run clinical trials do all sorts of nutty things, and I think that this is one of them.

It is worthwhile to think about what is random and why. The potential outcomes $Y(A)$ and $Y(B)$ are random because, whether treatment A or B is given, the outcome is not certain. If the patient is treated with A, so formally $Z = 1$ and $Y = Y(A)$, this does not say that $Y = \mu_A$, only that the expected value of Y is μ_A. The sample sizes n_A and n_B are random because flipping a coin to decide each patient's treatment intentionally introduces randomness into the treatment assignment process. That is why n_A has a binomial $(1/2, N)$ distribution, rather than being a fixed number determined ahead of time. The expected value of n_A is $N/2$, but due to random variation in the treatment assignment process it is unlikely to exactly equal $N/2$. In practice, one may randomize in blocks to ensure interim balance, but the treatment assignments still are random using fixed probabilities. So, a strange but true fact is the following:

By introducing additional variability through randomization, an unbiased estimator of the causal effect $\Delta_{A,B}$ is obtained.

6.5 Stratifying by Subgroups

Here is how inverse probability of treatment weighting (IPTW) works if you want to stratify by subgroups. Suppose that you have G subgroups indexed by $g = 1, \ldots, G$. In the example above where creatinine level was associated with whether a patient was given treatment A or B, there were $G = 2$ subgroups defined by $[C < 140]$ versus $[C \geq 140]$. But there may be a lot more than two subgroups, so to keep track, the covariate X_i can be used as a subgroup identifier, with X_i giving a subgroup index in $\{1, \ldots, G\}$. That is, the covariate is used to identify the patient's subgroup. The conditional means are

$$\mu_g(t) = E\{Y_i(t) \mid X_i = g\} \text{ for } t = A, B, \text{ and } g = 1, \ldots, G.$$

Notice that there are $2G$ causal parameters, since now there are subgroup-specific treatment effects, and the means of the potential outcomes for each subgroup are $\mu_g(t) = E\{Y_{g,i}(t)\}$ for $t = A, B$.

The *causal effect of A versus B in subgroup g* is

$$\Delta_{g,A,B} = \mu_{g,A} - \mu_{g,B}, \quad \text{for each } g = 1, \ldots, G.$$

These may be combined to define one overall parameter. Let λ_g be the probability that a subject belongs to subgroup g, and define the *population, subgroup-weighted causal effect of A versus B*

$$\Delta_{A,B}^{causal} = \sum_{g=1}^{G} \lambda_g \Delta_{g,A,B} = \sum_{g=1}^{G} \lambda_g (\mu_{g,A} - \mu_{g,B}).$$

It turns out that the causal estimator of $\Delta_{A,B}$ is the usual subgroup-weighted estimator based on the within-subgroup sample means,

$$\hat{\Delta}_{A,B}^{subgroups} = \sum_{g=1}^{G} \hat{\lambda}_g \hat{\Delta}_{g,A,B} = \sum_{g=1}^{G} \hat{\lambda}_g (\overline{Y}_{A,g} - \overline{Y}_{B,g}), \tag{6.2}$$

where N_g is the sample size in subgroup g and each $\hat{\lambda}_g = N_g/N$ is the empirical estimator based on the observed subgroup sample size proportion, with $N = N_1 + \cdots N_G$.

To see why this is true, first note that, by re-indexing subjects using (i, g) to identify the ith patient in subgroup g, we can write the general causal estimator in the inverse probability-weighted form

$$\hat{\Delta}_{A,B}^{causal} = \frac{1}{N} \sum_{g=1}^{G} \sum_{i=1}^{N_g} \left\{ \frac{Y_{ig} Z_{ig}}{\pi_{ig}} - \frac{Y_{ig}(1 - Z_{ig})}{1 - \pi_{ig}} \right\}.$$

In subgroup g, let $n_{A,g}$ denote the number of patients who received A and $n_{B,g}$ the number of patients who received B, so $n_{A,g} + n_{B,g} = N_g$. If $Z_{ig} = 1$ then $\pi_{ig} = \pi_g = n_{A,g}/N_g$, and if $Z_{ig} = 0$ then $1 - \pi_{ig} = 1 - \pi_g = n_{B,g}/N_g$, regardless of i. So

$$\hat{\Delta}_{A,B}^{causal} = \frac{1}{N} \sum_{g=1}^{G} \sum_{i=1}^{N_g} \left\{ \frac{Y_{ig} Z_{ig}}{n_{A,g}/N_g} - \frac{Y_{ig}(1 - Z_{ig})}{n_{B,g}/N_g} \right\}$$

$$= \sum_{g=1}^{G} \frac{N_g}{N} \sum_{i=1}^{N_g} \left\{ \frac{Y_{ig} Z_{ig}}{n_{A,g}} - \frac{Y_{ig}(1 - Z_{ig})}{n_{B,g}} \right\}. \qquad (6.3)$$

If $Z_{ig} = 1$ then the observed outcome must be the potential outcome for treatment A, that is, $Y_{ig} = Y_{ig}(A)$, and if $Z_{ig} = 0$ then $Y_{ig} = Y_{ig}(B)$. So, as before in the general case, for each subgroup g,

$$\sum_{i=1}^{N_g} \frac{Y_{ig} Z_{ig}}{n_{A,g}} = \sum_{i \in I_A} \frac{Y_{ig}(A)}{n_{A,g}} = \frac{1}{n_{A,g}} \sum_{i=1}^{n_{A,g}} Y_{ig} = \overline{Y}_{A,g}.$$

Likewise, the second sum is

$$\sum_{i=1}^{N_g} \frac{Y_{ig}(1 - Z_{ig})}{n_{B,g}} = \sum_{i \in I_B} \frac{Y_{ig}(B)}{n_{B,g}} = \frac{1}{n_{B,g}} \sum_{i=1}^{n_{B,g}} Y_{ig} = \overline{Y}_{B,g}.$$

But these are just the usual sample means for each treatment in each subgroup. So the causal estimator must be

$$\hat{\Delta}_{A,B}^{causal} = \sum_{g=1}^{G} \frac{N_g}{N} (\overline{Y}_{A,g} - \overline{Y}_{B,g}) = \hat{\Delta}_{A,B}^{subgroups}.$$

That is, the usual subgroup-weighted average given in (6.3) is a special case of the causal IPTW estimator. Just make sure that you weight the subgroups correctly, using $\hat{\lambda}_g = N_g/N$.

To apply this to the example in Table 6.1, where survival depended on creatinine level, which was associated with treatment assignment, index the two subgroups by

$g = 1$ if $X_i < 1.40$ and $g = 2$ if $X_i \geq 1.40$. Then the subgroup-weighted causal A versus B estimator is

$$\hat{\Delta}_{A,B}^{causal} = \frac{N_1}{N}(\overline{Y}_{A,1} - \overline{Y}_{B,1}) + \frac{N_2}{N}(\overline{Y}_{A,2} - \overline{Y}_{B,2}).$$

From the table, the subgroup sample sizes were $N_1 = 200 + 900 = 1100$ and $N_2 = 600 + 100 = 700$ so the subgroup proportions are $\hat{\lambda}_1 = N_1/N = 1100/1800 = 0.61$, and $\hat{\lambda}_2 = 700/1800 = 0.39$. But, recall that in the example the within-subgroup sample means were $\overline{Y}_{A,1} = \overline{Y}_{B,1} = 2.0$ and $\overline{Y}_{A,2} = \overline{Y}_{B,2} = 3.0$. Because the within-subgroup sample means for A and B are identical, the subgroup-weighted causal estimator (6.2) must equal 0, regardless of λ_1 and λ_2.

But things can be more complicated, and the values of λ_1 and λ_2 usually matter a lot. The causal estimators show that $\hat{\Delta}_{A,B} = 0$ for the example in Table 6.1. A big question is why the computations in the table give the different estimators $\hat{\mu}_A = 2.25$ and $\hat{\mu}_B = 2.90$, which must be incorrect. The answer actually is very simple. *These estimators were computed incorrectly using within-subgroup weights for averaging $\hat{\mu}_{t,1}$ and $\hat{\mu}_{t,2}$ that were computed for each treatment $t = A$ and $t = B$ separately, which is just plain wrong.* People actually make this mistake quite often, since it is not obvious why it is wrong. But, now that you know how to compute a causal estimator, either overall or accounting for subgroups, you won't make this dumb mistake.

As a numerical illustration to see how the estimator that uses within-subgroup weights gets it wrong and the overall subgroup-weighted estimator gets it right, let's change the example by supposing that the sample means are the different values $\overline{Y}_{A,1} = 2, \overline{Y}_{B,1} = 4, \overline{Y}_{A,2} = 5$, and $\overline{Y}_{B,2} = 3$. Then the incorrect estimator is

$$\begin{aligned}\hat{\Delta}_{A,B}^{incorrect} &= (0.25\,\overline{Y}_{A,1} + 0.75\,\overline{Y}_{A,2}) - (0.90\,\overline{Y}_{B,1} + 0.10\,\overline{Y}_{B,2}) \\ &= (0.25\,\overline{Y}_{A,1} - 0.90\,\overline{Y}_{B,1}) + (0.75\,\overline{Y}_{A,2} - 0.10\,\overline{Y}_{B,2}) \\ &= (0.25 \times 2 - 0.90 \times 4) + (0.75 \times 5 - 0.10 \times 3) = 0.35,\end{aligned}$$

while the correct causal estimator is

$$\begin{aligned}\hat{\Delta}_{A,B}^{causal} &= \hat{\lambda}_1(\overline{Y}_{A,1} - \overline{Y}_{B,1}) + \hat{\lambda}_2(\overline{Y}_{A,2} - \overline{Y}_{B,2}) \\ &= \frac{N_1}{N}(\overline{Y}_{A,1} - \overline{Y}_{B,1}) + \frac{N_2}{N}(\overline{Y}_{A,2} - \overline{Y}_{B,2}) \\ &= 0.61 \times (2 - 4) + 0.39 \times (5 - 3) = -0.44.\end{aligned}$$

So, if the weights are computed incorrectly in this example, then the estimator $\hat{\Delta}_{A,B}^{incorrect} > 0$, which says that A is better than B. But the correct estimator $\hat{\Delta}_{A,B}^{causal} < 0$, which says that, aside from variability, B is better than A. If you think that all of the above formulas are a pain-in-the neck because they are so complicated, you might also think about whether you want to get things backwards, by saying that

A is better than B when the opposite is true. This is the "Type S error" defined by Gelman and Tuerlinckx (2000).

By the way, before computing any estimators, it is useful to do a little detective work by asking the question of whether the sample sizes could have come from a randomized trial. In Table 6.1, if the 1100 patients with creatine level $C < 1.40$ had been randomized fairly between A and B, then the subsample size $n_{A,C<1.40}$ would have followed a $binom(1000, 0.50)$ distribution. This implies that obtaining a sample size imbalance as large as 200 for A and 900 for B would occur with probability less than 10^{-65}. You get the same sort of answer for unbalanced randomization with probabilities, say, 2/3 and 1/3. From a Bayesian viewpoint, if $p = \text{Pr}(\text{randomized to } A)$ has a beta(0.50, 0.50) prior, then the posterior mean would be 0.18, which is the same as the empirical mean 200/1800 due to the large sample size. This suggests there might have been a 4:1 randomization probability for A:B in this subgroup, which is pretty strange. The same reasoning implies that the randomization probability may have been 6:1 for A:B in the $C \geq 1.40$ subgroup. If patients were randomized, then it must have been done using something close to these very strange and completely different proportions in the two creatinine subgroups. So it's a pretty good bet that randomization was not used at all, and treatments were assigned in some other fashion.

Takeaway Messages About Randomization

1. If you want to compare treatments A and B pre-clinically, perform a randomized experiment.
2. If you want to compare treatments A and B in humans and it is ethical, perform a randomized clinical trial.
3. If you are given observational data, when possible do some initial binomial computations to check the probability that the data came from a randomized trial.
4. Weighting subgroup-specific estimators incorrectly can lead to a conclusion that is the opposite of what the data actually are saying.
5. Watch out for incorrect estimators in the published literature that were obtained by using the incorrect subgroup weights when analyzing observational data. Just because a paper was refereed and published does not make it correct.
6. If someone analyzes observational data as if it were obtained from a randomized trial, do not trust their conclusions. Contact The Causality Police immediately and have them arrested.

6.6 Inverse Probability-Weighted Survival Analysis

If you have observational data, where patients were not randomized, the treatment of patient i was chosen by their physician in some way that may depend on baseline covariates X_i, and possibly time-varying covariates $Z_i(t)$, then the causal argument given above will not work. So the problems with making comparative inferences

based on observational, non-randomized data are that the treatment assignment probabilities $\{\pi_1, \ldots, \pi_n\}$ may differ from patient to patient, they are not known, and they
may even vary over time depending on $\boldsymbol{Z}_i(t)$.

Hernan et al. (2000) described a *marginal structural* version of the Cox proportional hazards regression model, modified to account for each patient's time-varying
treatment history, say $H(t)$, up to time follow up time t. For example, $H_i(t)$ may
account for an adaptive treatment decision made by the physician for the ith patient
at time t, possibly based on $Z_i(t)$. The patient's treatment history up to t is denoted
by $\overline{H}(t) = \{H(s) : 0 \leq s < t\}$ and, similarly, the time-varying covariate history up
to t denoted by $\overline{Z}(t) = \{Z(s) : 0 \leq s < t\}$. The usual Cox model hazard is assumed
to take the more general time-varying form

$$\lambda(t \mid \overline{H}(t), X) = \lambda_0(t) \exp\{\beta_1 \overline{H}(t) + \beta_2 X\}.$$

Thus, β_1 is the effect of the possibly time-varying treatment, while β_2 is a vector
of the baseline covariate effects. Due to effects of the time-varying covariates, $\boldsymbol{Z}(t)$,
the conventional Cox model estimate $\hat{\beta}_1$ is asymptotically biased for the causal
treatment effect on survival, even after accounting for the effects of the baseline
covariates, X. To reduce bias, Hernan et al. (2000) proposed fitting the time-varying
Cox model given above. Temporarily assuming no administrative censoring, the risk
set contribution of patient i at time t is weighted by the *stabilized weight*

$$w_i(t) = \prod_{i=1}^{int(t)} \frac{\Pr\{H(k) = h_i(k) \mid \overline{H}(k-1) = \overline{h}_i(k-1), X_i\}}{\Pr\{H(k) = h_i(k) \mid \overline{H}(k-1) = \overline{h}_i(k-1), \overline{Z}_i(t)\}},$$

where $int(t)$ is the largest integer $\leq t$ and $\overline{H}(-1) = 0$. The key assumptions are that
1. $\boldsymbol{Z}(t)$ includes a risk factor for death that predicts subsequent treatment exposure.
2. The treatment exposure history $\overline{H}(t)$ predicts $\boldsymbol{Z}(t)$.
The idea underlying weighting by $w_i(t)$ is that this creates a pseudo-population in
which $\boldsymbol{Z}(t)$ no longer predicts treatment at time t; hence, it is not a confounder, and
the causal association between treatment and death is that of the population being
studied. The causal interpretation of $\exp(\beta_1)$ is that it is the hazard ratio at any time
of all patients continuously exposed to the treatment compared to all patients not
exposed. To account for right-censoring, Hernan et al. (2000) multiply $w_i(t)$ by a
second weight defined in terms of the usual independent censoring process.

In their analysis of the effects of AZT (zidovudine) on survival in HIV positive
men, Hernan et al. (2000) showed that the usual Cox model including only X_i gave
estimated relative risk of death as

$$\widehat{RR}^{unadjusted} = 2.32 \ (95\% \ ci \ 1.92 - 2.87),$$

which says that AZT was harmful. In sharp contrast, the time-varying Cox model
with stabilized weights to correct for bias gave the estimator

$$\widehat{RR}^{IPTW} = 0.74 \quad (95\% \ ci \ 0.57 - 0.96),$$

which says that AZT was beneficial, the opposite of the conclusion of the conventional analysis that AZT was harmful. This inferential reversal was due to the fact that the treating physicians were more likely to give AZT to patients with worse prognosis based on their time-varying covariates, such as CD4 counts, which were affected by previous treatment history. The model and estimator of Hernan et al. (2000) accounted for this, while the conventional model did not.

Sugihara (2009) describes IPTW survival estimation using a preliminary data analysis to estimate the propensity scores, $\pi_i, \dots, \pi_n$, based on the available patient covariates. To do this, an additional model $\pi_i = \pi(X_i, \beta) = \Pr(Z_i = 1 \mid X_i)$ where $Z_i = 1$ for A and $Z_i = 0$ for B for the treatment assignment probabilities must be assumed, such as a logistic regression model where, if $X_i = (X_{i,1}, \dots, X_{i,m})$,

$$\text{logit}\{\pi(X_i, \beta)\} = X_i \beta = \beta_0 + \beta_1 X_{i,1} + \cdots + \beta_m X_{i,m}.$$

As a preliminary analysis, this model is fit to the treatment assignment and covariate data $\{(Z_i, X_i), i = 1, \dots, n\}$ to obtain an estimate $\hat{\beta}$ and thus estimates $\hat{\pi}_i = \pi(X_i, \hat{\beta})$ for each $i = 1, \dots, n$. The data $\{(Y_i, Z_i, X_i), i = 1, \dots, n\}$ are then used to compute the IPTW estimator

$$\hat{\Delta}_{A,B}^{IPTW} = \frac{1}{N} \sum_{i=1}^{N} \left\{ \frac{Y_i Z_i}{\hat{\pi}_i} - \frac{Y_i (1 - Z_i)}{1 - \hat{\pi}_i} \right\}.$$

Since the model for π_i may be wrong, some sort of goodness-of-fit analysis may be required to obtain a good fit to the treatment assignment data. In any case, the variance of the estimator $\hat{\beta}$ must be accounted for when figuring out how to estimate $\text{var}(\hat{\Delta}_{A,B}^{IPTW})$. Indexing patients by $i = 1, \dots, n$, denote the observed time of death or administrative right-censoring by Y_i^o, with $\delta_i = 1$ if Y_i^o is the time of death and $\delta_i = 0$ if Y_i^o is the time of right-censoring, the treatment indicator variable $Z_i = 0$ or 1, and covariates by $X_i = (X_{i,1}, \dots, X_{i,p})$. Given the estimates, $\hat{\pi}_i = \hat{\beta} X_i$ for each i, the ith patient's weight is $w_i = 1/\hat{\pi}_i$ if $Z_i = 1$, and $w_i = 1/(1 - \hat{\pi}_i)$ if $Z_i = 0$. That is, w_i is the (estimated) inverse of the covariate-based estimated probability that the patient received their actual treatment.

To compute the IPTW-adjusted version of the usual Kaplan and Meier (1958) estimates, following Xie and Liu (2005), let $t_1 < t_2 < \cdots < t_D$ denote the D distinct times of death in the sample, with $d_{j,k}$ the number of deaths out of $R_{j,k}$ patients at risk of death just prior to t_j, for treatment $k = 0$ or 1. The inverse probability of treatment-weighted (IPTW) version of the observed number of deaths $d_{j,k}$ is

$$d_{j,k}^w = \sum_{i:Y_i^o = t_j} w_i \delta_i I(Z_i = k)$$

for treatment $k = 0$ or 1, and the IPTW version of the observed number at risk is

$$R_{j,k}^w = \sum_{i:Y_i^o \geq t_j} w_i I(Z_i = k).$$

The IPTW-adjusted Kaplan–Meier survival probability curve estimator is

$$\hat{S}^{IPTW}(t) = \prod_{t_j \leq t} \left(1 - \frac{d_{j,k}^w}{R_{j,k}^w}\right)$$

for $t \geq t_1$, and it equals 1 for $t < t_1$.

An IPTW-adjusted version of the log-rank statistic to compare survival in groups $Z = 1$ versus $Z = 0$ is

$$G^{IPTW} = \sum_{j=1}^{D} \left[d_{j,1}^w - R_{j,1}^w \left(\frac{d_{j,0}^w + d_{j,1}^w}{R_{j,0}^w + R_{j,1}^w}\right)\right].$$

This is the usual formula for the log-rank statistic, but replacing the usual counts for numbers of deaths and patients at risk of death by the inverse probability of treatment-weighted versions. An IPTW-adjusted version of the Cox model is obtained by modifying the usual score statistic $U(\beta)$ that is the basis for parameter estimation by solving $U(\beta) = 0$ under the Cox model by using the weights $w_1, \ldots, w_n$, as follows:

$$U^{IPTW}(\beta) = \sum_{i=1}^{N} \left[\delta_i w_i - \frac{\sum_u^N Y_u(t_i) w_u X_u \exp(\beta X_u)}{\sum_u^N Y_u(t_i) w_u \exp(\beta X_u)}\right].$$

Formulas for the variances for these IPTW bias adjusted statistics are given in Xie and Liu (2005), and Sugihara (2009). A comparison of different methods for estimating variances in IPTW-adjusted survival analysis is given by Austin (2016).

As an illustration of IPTW estimation in survival analysis, Lin et al. (2012) reported analyses of observational data from 676 esophageal patients who received treatment with one of two radiation therapy modalities, either intensity modulated radiotherapy (IMRT, n = 263) or the more conventional 3-dimensional conformal radiotherapy (3DCRT, n = 413). IMRT allows the radiation oncologist to plan the radiation dose based on tumor size, shape, and location, with the goal to deliver higher doses to the tumor while reducing exposure to healthy tissues. Patients were treated at a single institution during the period 1998–2008, but were not randomized between IMRT and 3DCRT. To analyze the data, IPTW was used to obtain bias-corrected comparisons of survival time, as well as the times to local and distant recurrence. The IPTW-adjusted estimates showed a larger advantage in survival time for IMRT versus 3DCRT than what conventional analyses would show. For example, the IPTW-adjusted Kaplan–Meier estimated probability (95% confidence interval) of 3-year survival was 0.53 (0.49–0.57) for IMRT compared to 0.43 (0.39–0.46) for

Table 6.2 Estimates of covariate effects and hazard ratios (HRs) in a fitted IPTW-adjusted Cox model for evaluating the effects of IMRT versus 3DCRT and prognostic covariates on survival time, based on observational data from 676 esophageal cancer patients

Variable	$\hat{\beta}(\hat{s}_\beta)$	$\widehat{HR}$
IMRT versus 3DCRT	−0.33(0.07)	0.72
Disease stage		
2 versus 1	0.54(0.34)	1.72
3 versus 1	0.93(0.33)	2.54
4 versus 1	1.10(0.36)	2.99
Lesion location		
Middle versus upper	0.54(0.20)	1.72
Low versus upper	0.51(0.13)	1.67
Surgery	−0.65(0.09)	0.52
Induction chemo	−0.27(0.08)	0.76
Age	0.007(0.004)	1.01
Performance status		
80 versus 90–100	0.12(0.08)	1.12
≤ 70 versus 90–100	0.44(0.11)	1.55
PET	−0.25(0.08)	0.78

3D-CRT. To account for effects of radiation modality and prognostic covariates, an IPTW-adjusted Cox model was fit, with the proportional hazards assumption first validated graphically and using a goodness-of-fit test given by Grambsch and Therneau (1994), Therneau and Grambsch (2000). The fitted IPTW-adjusted Cox model is summarized in Table 6.2. The fitted model showed that, accounting for the effects of known prognostic variables, and correcting for possible treatment selection bias by inverse probability of treatment weighting, IMRT appeared to give patients longer survival compared to 3D-CRT.

6.7 Bias Correction by Matching

Another practical way to reduce bias in order to make reasonably valid treatment comparisons based on observational data is to first preprocess the data by identifying pairs of patients, one from each treatment set, who have similar observed covariate vectors. Like IPTW, matching deals with two groups of patients, say one group who received an experimental treatment, E, indexed by $Z = 1$, and another who received a control treatment, C, indexed by $Z = 0$, where patients were not randomized between the treatments. Constructing matched pairs may be motivated by thinking about potential outcomes, $Y(1)$ and $Y(0)$, for a single patient. The inferential goal is to estimate the causal E versus C effect $\mu_1 - \mu_0 = E\{Y(1) - Y(0)\}$. As usual, for a patient treated with E, the outcome $Y = Y(1)$ is observed but $Y(0)$ is missing, and for a patient treated with C, $Y = Y(0)$ is observed but $Y(1)$ is missing. So the data suffer from the Fundamental Problem of Causal Inference, since with observational data this problem was not addressed by randomizing patients between E and C. The definitions of E and C can be quite general, depending on the setting and inferential goals. They need not be anti-disease treatments, but more generally may be two possibly nonexperimental conditions that one wishes to compare, such as whether one has suffered exposure to a potentially hazardous condition, such as a toxic agent or pollutant.

Seminal papers that established matching as a practical methodology for bias correction are Rubin (1973), Rosenbaum and Rubin (1983), and Rosenbaum and Rubin (1985). Discussions also are given in the books Rosenbaum (2001) and Rosenbaum (2010). To correct for bias when estimating $\mu_1 - \mu_0$ based on observational data, the idea of matching is to begin by identifying pairs of E and C patients who have similar distributions of their nontreatment covariates, $X = (X_1, \ldots, X_p)$, while ignoring their outcomes. A matched pair consists of one patient from the E group and one patient from the C group, chosen so that their X vectors are similar. Denoting the two samples of outcomes and covariates by $(Y_{1,1}, X_{1,1}), \ldots, (Y_{1,n_1}, X_{1,n_1})$ and $(Y_{0,1}, X_{0,1}), \ldots, (Y_{0,n_0}, X_{0,n_0})$, a matched pairs analysis is done in two steps. First, matched pairs are identified using only the covariates. The matching process may be carried out using various algorithms, but here is a simple, general approach. Index each sample in random order. Begin by first examining the covariates $X_{1,1}$ of the first E patient, say $i = 1$, and finding a C patient, with index j_1, whose covariates X_{0,j_1} are similar to $X_{1,1}$. There are several ways to define "similar" covariate vectors, which I will discuss below. The two patients treated with E and C, indexed by $(1, j_1)$ are the first matched pair. Ideally, one would like to have X_{0,j_1} identical to $X_{1,1}$, but in practice, this usually is not possible. One thus identifies the C patient j_1 so that these two covariate vectors are similar to each other, using some numerical metric. Subjects $i = 1$ and $j = j_1$ are removed from their respective samples, and this process is repeated for E patients $i = 2, \ldots, n_E$. The matched pairs thus are indexed by $(1, j_1), \ldots, (n_1, j_{n_1})$, determined so that each pair of covariate vectors $X_{1,i}$ and X_{0,j_i} are similar.

Finding a covariate match for each E patient is not trivial. In practice, the sample size n_0 of C patients should be substantially larger than the sample size n_1 of E patients. If the C sample is large enough, this process may be elaborated so that more than one C patient is matched to each E patient. For example, in 1:2 matching, each E patient may have two C patients with similar X vectors. The second step of the inferential process is to fit a parametric model to the resulting matched subsample $\{[(Y_i, X_i), (Y_{j_i}, X_{j_i})], i = 1, \ldots, n_1, \}$ with the goal to estimate the causal E versus C effect.

Rosenbaum and Rubin (1985) illustrated their methodology with a dataset in which 221 children had suffered prenatal exposure to barbiturates ($Z = 1$), matches were chosen from a set of 7,027 unexposed children ($Z = 0$), and Y measured psychological development. A naive analysis might simply compare the Y data between all 7,027 unexposed children and 221 exposed children. But such a comparison would be biased because the children were not randomized between being exposed or not to barbiturates prenatally, which obviously would be highly unethical. The reduced sample of 221 unexposed children matched to the 221 exposed children is much smaller than the 7,027 from which it was chosen. The trade-off inferentially is reliability is lost due to the smaller sample size, but the selected and matched subsample of unexposed children provides a less biased estimate of the effect of prenatal exposure to barbiturates on psychological development than a naive comparison based on all the data would give.

Rosenbaum and Rubin (1985) proposed that matching should be done based on the patients' propensity scores. Indexing treatments (or exposure groups) by $Z = 1$ for E and $Z = 0$ for C, recall that the propensity score is $\pi(Z, X) = \Pr(Z = 1 \mid X)$. A model for $\pi(Z, X)$ is assumed, such as

$$g\{\pi(Z, X \mid \beta)\} = \eta(Z, X \mid \beta) = \gamma Z + \sum_{j=0}^{p} \beta_j X_j = \gamma Z + \beta X,$$

where $X_0 = 1$ and g is a link function such as a logit, probit, or complementary log-log, possibly determined by a preliminary goodness-of-fit analysis of the fitted models for $\pi(X \mid Z)$ using each of the three links. The model is fit to the (Z, X) data to obtain estimates $\hat{\beta}$ and thus the propensity scores $\hat{\pi}(Z, X) = \pi(Z, X \mid \hat{\beta})$, and $\hat{\gamma}$.

Rosenbaum and Rubin (1985) described three alternative methods for obtaining matched pairs. The nearest-propensity score method finds the C subject $j = j_1$ for which $\hat{\eta}(Z_j = 0, X_j)$ is closest to $\hat{\eta}(Z_1 = 1, X_1)$, forms the first pair $(1, j_1)$, removes the subjects $i = 1$ and $j = j_1$, and repeats this for $i = 2, \ldots, n_E$. A second method is based on the Mahalanobis metric

$$d_M(\mathbf{u}, \mathbf{v}) = (\mathbf{u}, \mathbf{v})^t \hat{\Sigma}(\mathbf{u}, \mathbf{v}),$$

where $\mathbf{u}$ and $\mathbf{v}$ are vectors of $(X^t, \hat{\eta})^t$ values from the two samples and $\hat{\Sigma}$ is the sample covariance matrix based on the C sample. The matching algorithm proceeds as before, but using the d_M metric rather than closeness of the $\hat{\eta}(Z_j, X_j)$'s. A third method hybridizes the first two by (i) first requiring that the E and C patients match perfectly on one or two important categorical covariates, (ii) determining a set of such C patients ($Z = 0$) with

$$\left| \eta(Z = 0, X \mid \beta) - \eta(Z = 1, X \mid \beta) \right| < \kappa$$

for some fixed real-valued distance $\kappa > 0$, the "caliper," and (iii) selecting the matching C patient from this subset using the d_M metric.

Ho et al. (2007) discuss how first constructing matched pairs based on covariates while ignoring outcomes variables, and then using a parametric model fit to the matched pairs to estimate causal effects, reduces the dependency of one's inferences on parametric modeling choices and specifications, which may result from such things as fitting different parametric models for $(Y \mid Z, X)$ and performing goodness-of-fit analyses. The basic idea, as in Rosenbaum and Rubin (1985), is to decouple the matching process and subsequent model-based analyses to estimate causal effects, in order to make causal inferences less model-dependent. This is a version of the more general problems that one never knows the "correct" model for $(Y \mid Z, X)$, and that fitting several different parametric models but then only reporting the final fitted model gives estimators having distributions that depend on the entire model selection process. Ideally, the goal is to obtain matched pairs for which the empirical covariate distributions are identical, $\hat{p}(X \mid Z = 1) = \hat{p}(X \mid Z = 0)$. In practice, this equality usually is not exact, but approximate.

One may use counterfactuals to define two types of causal effects. The first is the *average treatment effect*

$$\text{ATE} = \frac{1}{n} \sum_{i=1}^{n} E[Y_i(1) - Y_i(0) \mid X_i] \tag{6.4}$$

$$= \frac{1}{n} \sum_{i=1}^{n} [\mu_i(1) - \mu_i(0)] \tag{6.5}$$

which is the sample average of the difference of the means of the two potential outcomes $Y_i(1)$ and $Y_i(0)$ defined as functions of the patients' covariate vectors. The second is the *average treatment effect on the treated*,

$$\text{ATT} = \frac{\sum_{i=1}^{n} Z_i \, E\{Y_i(1) - Y_i(0) \mid X_i\}}{\sum_{i=1}^{n} Z_i} \tag{6.6}$$

$$= \frac{\sum_{i=1}^{n} Z_i \, \{\mu_1(X_i) - \mu_0(X_i)\}}{\sum_{i=1}^{n} Z_i}. \tag{6.7}$$

This is the average causal effect of treatment $Z = 1$ versus $Z = 0$ among patients in the sample who received treatment ($Z = 1$), and may be regarded as the causal effect of the treatment $Z = 1$ versus $Z = 0$ for patients who actually could have received either treatment.

Under a parametric model with link function g for which

$$\mu_1(X_i) = g(\alpha + \gamma + \beta X_i) \quad \text{and} \quad \mu_0(X_i) = g(\alpha + \beta X_i)$$

if one has randomized then any connection between treatment Z_i and the covariates X_i is broken, and $\text{ATT} = g(\alpha + \gamma) - g(\alpha)$. A likelihood-based, possibly Bayesian analysis yields an unbiased estimator of γ. Similarly, matching on X_i before fitting a parametric model, while not perfect, reduces possible association between Z_i and X_i. A key point is that exact matching on X_i is not required, but rather that, after matching, the empirical distributions $\hat{p}(X \mid Z = 1)$ and $\hat{p}(X \mid Z = 0)$ should be as close as is reasonably possible. If matching on X_i reduces the extent of the relationship between Z_i and X_i substantially, but not completely, then dependence of final inferences on the parametric model is not eliminated entirely, but it is greatly reduced. A very useful computer program for obtaining matches with this property is *MatchIt*, Ho et al. (2011), which is freely available from CRAN-R.

An illustration of matching for comparative inference is the following analysis of observational data from 212 patients with AML or MDS who received either a reduced-intensity conditioning (RIC) regimen or a myeloablative conditioning (MAC) regimen for allogeneic stem cell transplantation (allosct), reported by Alatrash et al. (2019). Both regimens contained IV Busulfan (Bu), with Bu in the RIC regimen dosed to a systemic exposure represented by the area under the concentration versus time curve (AUC) = 4,000 μMol-min or total course AUC = 16,000 Mol/min. MAC regimens were dosed at AUC = 5,000–6,000 Mol-min or total course AUC = 20,000–24,000 Mol-min. Because RIC regimens may provide lower toxicity, but at the cost of reduced efficacy, compared with MAC regimens, the goal of the analysis was to compare survival in the MAC and RIC groups. Because patients were not randomized between MAC and RIC, propensity score matching was done to obtain bias-corrected comparisons. The propensities were estimated using a logistic regression model for treatment probability ($Z = 1$ for MAC versus $Z = 0$ for RIC), using the following covariates, chosen due to their established prognostic status in allosct: age, co-morbidity index (0, 1–3, ≥4), complete remission with no minimal residual disease versus not, donor type (10/10 matched related versus 10/10 matched unrelated), source of stem cells (marrow versus peripheral blood), and disease status (secondary treatment versus *de novo*). Nearest neighbor matching was done using a caliper of 0.10 standard deviation of the propensity score, with pairs falling outside this distance discarded. A total of 56 matched pairs (112 patients) were obtained. Balance was verified by assessing standardized differences between the RIC and MAC groups for all variables in the matched cohort, with a target difference <20%. To assess the MAC versus RIC effect on overall survival time, since the proportional hazards assumption was violated based on the test of Grambsch and Therneau (1994), the usual approach of fitting a Cox model stratified on the matched pairs was not followed. Instead, based on a preliminary assessment of how well each of several accelerated failure time models fit the data using the Bayesian information criterion, see Claeskens and Hjort (2008), and Kaplan–Meier plots of survival distributions, the lognormal distribution was chosen. The matched pairs dataset was fit using function aftreg in the R v.3.3.2 package eha (event history analysis), stratifying on the matched pairs, with the shape parameter fixed at the value

Table 6.3 Fitted lognormal model, stratified on the 56 matched pairs (112 patents, 59 deaths) comparing survival of AML/MDS patients undergoing allogeneic stem cell transplantation with either a myeloablative conditioning (MAC) regimen or a reduced-intensity conditioning (RIC) regimen

Covariate	Hazard ratio	95% ci	p-value
MAC versus RIC	0.62	0.32–1.20	0.16
Age	1.03	0.96–1.10	0.41
Female versus male	1.86	0.61–5.71	0.28

0.42 obtained from the fitted unstratified model, but allowing the scale parameter to differ between matched pairs. The results are summarized in Table 6.3.

6.8   A Bayesian Rationale for Randomization

The motivation for randomization and methods for bias correction given above may all appear to be frequentist. Given the fundamental differences between Bayesian and frequentist statistics, an important scientific and practical question is why a Bayesian would want to randomize to make comparative inferences. In a landmark paper, Rubin (1978) provided a clear Bayesian rationale for randomization, stating "$\cdots$ randomization plays a central role in Bayesian inference for causal effects." Most of the concepts given in this paper already have been described in this chapter. Rubin's explanation began with the key ideas of potential outcomes and counterfactuals, the definition of causal effects in terms of expected values of potential outcomes that could have been observed under possible treatment assignments, and the Fundamental Problem of Causal Inference. Rubin's Bayesian rationale for randomizing is based on the idea that, to do causal inference, one must assume three models, for (1) the prior distribution of parameters in the distributions of potential outcomes, (2) the way that treatments are assigned to sampling units, such as patients in a clinical trial, and (3) the procedure for choosing values for data analysis. To deal with actual applications, Rubin also introduced the missing data indicators for covariates, $M_{ki}^x = 1$ if the covariate X_{ki} is recorded and 0 if X_{ki} is missing, and a vector of similar missingship indicators for a multivariate d-dimensional version of the potential outcomes, $Y_{i,1}(Z_i), \ldots, Y_{i,d}(Z_i)$, with $M_{j,i}^z = 1$ if $Y_{i,j}(Z_i)$ is recorded and 0 if not for each $j = 1, \ldots, d$.

Suppressing the subject index i and considering the simplest case where $d = 1$, that is, there is only one outcome variable, a key idea is that Bayesian inference considers the observed values of (Y, X, M, Z) to be realizations of random variables, and missing values to be unobserved random variables. Denoting the overall model parameter vector by θ, the joint distribution of these random variables can be written as the product likelihood

$$f(Y, X, Z, M \mid \theta) = f_1(Y, X \mid \theta) \, f_2(Z \mid Y, X, \theta) \, f_3(M \mid Y, X, Z, \theta).$$

The Bayesian model is completed by specifying a prior, $p(\theta)$. The first likelihood component can be expressed in the familiar form

$$f_1(Y, X \mid \theta) = f_{1,Y}(Y \mid X, \theta) \, f_{1,X}(X \mid \theta),$$

where $f_{1,Y}$ is a usual model for regression of Y on X, and $f_{1,X}$ is the distribution of the covariates considered as random variables. The distribution f_2 describes the treatment assignment mechanism, and f_3 is the distribution of the missingship mechanism.

Including the treatment assignments Z and missingship indicators M as random variables in the model is essential for explaining Bayesian causal inference. If the treatment assignment mechanism distribution $f_2(Z \mid Y, X, \boldsymbol{\theta})$ is under the experimenter's control then, for example, this may describe how one randomizes patients to treatments. Otherwise, the data are observational, and one must make key assumptions about treatment assignments and missingship to make any progress. If, for any given observed (Z, Y, X) and prior $p(\boldsymbol{\theta})$, $f_2(Z \mid Y, X, \boldsymbol{\theta})$ takes the same known probability, then the treatment assignment mechanism is called *ignorable*. Ignorability is defined similarly for the missingship mechanism distribution.

This Bayesian structure can be exploited to address the Fundamental Problem of Causal Inference, that one can never observe both $Y_i(1)$ and $Y_i(0)$ for any subject i. Recall that a very useful property of any Bayesian model is that is can be used to compute the predictive distribution of any observable variable, given the data actually observed. Here is the key to Bayesian causal inference:

Bayesian causal inference addresses The Fundamental Problem of Causal Inference by computing the predictive distributions of all potential outcomes that were not observed.

For example, if $Z_i = 1$, so that, by consistency, $Y_i = Y_i(1)$ since treatment A was assigned to patient i, then one can exploit the Bayesian model to compute the predictive distribution of the counterfactual $Y_i(0)$, which corresponds to what would have been observed if treatment B had been given to patient i instead. Denoting the posterior of $\boldsymbol{\theta}$ given all of the observed (Y_i, X_i, Z_i, M_i) data by $p(\boldsymbol{\theta} \mid data)$, this predictive distribution is

$$f^{pred}(Y_i(0) \mid Y_i, X_i, M_i, Z_i = 1, data)$$

$$= \int f(Y_i(0) \mid Y_i, X_i, M_i, Z_i = 1, \boldsymbol{\theta}) \, p(\boldsymbol{\theta} \mid data) d\boldsymbol{\theta}.$$

If $Z_i = 0$ then the roles of $Y_i(0)$ and $Y_i(1)$ are reversed and the predictive distribution of $Y_i(1)$ is computed.

The practical implication is that, while one cannot observe the causal effect $Y_i(1) - Y_i(0)$ for any subject, one can compute the predictive distribution of $Y_i(t)$ for the treatment t not given to subject i and use it to obtain a distribution for $Y_i(1) - Y_i(0)$. Rubin proved that, if the treatment assignment mechanism is ignorable, *which is the case if one randomizes*, then Bayesian inference for causal effects is completely determined by the observed data and the particular model assumptions made for the distributions of $[Y \mid X, \boldsymbol{\theta}]$ and $[\boldsymbol{\theta} \mid X]$. Notice that the Bayesian rationale for randomization is very different from the rationale for the frequentist estimator given by Eq. (6.1). Rubin (1978) gave many specific examples, and also discussed what to do if data are missing or the treatment assignment mechanism is not ignorable. The book by Little and Rubin (2002) provides detailed discussions and illustrations of how one may address these issues in practice. There is an immense published literature, and there are many software packages, on imputation methods for various types of missing data, and for making inferences in the presence of non-ignorable treatment assignment mechanisms.

6.9   Outcome-Adaptive Randomization

There's a sucker born every minute.
P. T. Barnum

Given the above discussion of methods for obtaining unbiased estimators by randomization, and methods to correct for bias in non-randomized observational data, it is important to discuss a statistical methodology that actually introduces bias. This is especially important because, recently, this methodology has become quite fashionable, but its properties often are not well understood, either by statisticians or non-statisticians. It is called "outcome-adaptive randomization."

Many decades before the widespread use of randomized comparative clinical trials, a college professor at Yale, Thompson (1933), published a paper that discussed estimating the probability, $\pi_{A<B}$, that one treatment B is better than another, A, based on binomial data. He suggested that, in an experiment to compare A and B, one might use a monotone increasing function f of $\pi_{A<B}$ to determine randomization probabilities $f(\pi_{A<B})$, computed adaptively from the accumulating data and used for each new subject. He wrote, given in terms of the notation used here,

". . . to fix the fraction of such individuals to be treated in the first manner" (what I call treatment B), "until more evidence may be utilised, where $0 < f(\pi_{A<B}) < 1$, the remaining fraction of such individuals $(1 - f(\pi_{A<B}))$ to be treated in the second manner" (what I call treatment A), "or we may establish a probability of treatment by the two methods of $f(\pi_{A<B})$ and $1 - f(\pi_{A<B})$, respectively."

Since, by today's standards, numerical methods for computing probabilities like $\pi_{A<B}$ were primitive in 1933, most of the paper was devoted to deriving formulas for doing the computations by hand.

Today, assigning treatments this way is known as "Thompson Sampling." Although this paper was published decades before the widespread use of randomized clinical trials to compare treatments in humans, in recent years many people have advocated using this idea to do outcome-adaptive randomization (AR) in comparative clinical trials. The published literature on AR methods is immense. Sverdlov (2015) edited a book in which experts explain various AR methods. In general, AR is implemented by using an estimate of $\pi_{A<B}$, or of some similar function that compares the empirical success rates of A and B, to compute a probability $f(\pi_{A<B})$ of assigning treatment B to the next patient based on the current data at an interim point in a clinical trial, with $1 - f(\pi_{A<B})$ the probability of assigning A. For binomial data, as Thompson considered, this may be done using a Bayesian formulation by assuming, to start, that the response probabilities θ_A and θ_B follow non-informative beta priors, and defining $\pi_{A<B} = \Pr(\theta_A < \theta_B \mid data)$. For non-binomial data, such as survival times, with means μ_A and μ_B, a similar posterior probability such as $\pi_{A<B} = \Pr(\mu_A < \mu_B \mid data)$ may be used, computed under a Bayesian survival time model. For the binomial case, denoting the number of successes by R_j, sample size by n_j, and the empirical success rates by $\hat{\theta}_j = R_j/n_j$ for $j = A, B$ at any interim point in the trial, a non-Bayesian approach might be to define $\pi_{A<B} = \hat{\theta}_B/(\hat{\theta}_A + \hat{\theta}_B)$. More generally, one may assign B using the AR probability

$$f(\pi_{A<B}) = \frac{(\hat{\theta}_B)^c}{(\hat{\theta}_A)^c + (\hat{\theta}_B)^c},$$

and A with probability $1 - f(\pi_{A<B})$, where $c = 1/2$ or some other positive constant.

The main argument for using AR rather than randomizing fairly with constant probabilities 1/2 for each treatment arm is ethical. It is argued that, with AR, on average more patients enrolled in the trial will be given the treatment that truly is the better of the two. Also, as soon as one begins to treat patients and accumulate data, it is virtually certain that, at some interim point, the empirical response rates (or estimated mean survival times, etc.) of the two arms will not be equal so, for anyone who looks at the interim data, equipoise will no longer strictly be true. For example, if interimly during a trial, one observes 9/10 successes with B and 7/10 successes with A, then it may be argued that randomizing the next patient with equal probabilities 1/2 to A or B is unethical, since B has a higher observed success rate. If one takes the myopic viewpoint that the only thing that matters is what happens to the next patient, then it may seem that one should "play the winner" (PTW) and treat the next patient with B. Instead, following Thompson's suggestion, if one starts with beta(0.50, 0.50) priors on both θ_A and θ_B, then $\Pr(\theta_A < \theta_B \mid data) = 0.87$, and one might randomize the next patient to B with probability 0.87 and to A with probability 0.13. Thompson randomization may be considered a compromise between fair randomization and play the winner. It should be obvious that, even if in fact A and B are perfectly equivalent with identical response probabilities, an imbalance such as 7/10 versus 9/10 in the empirical success rates is very likely to occur quite often during the trial purely due to the play of chance.

Another motivation for AR may come from the disconnect between fair randomization and medical practice. The unbiased comparison provided by fair randomization serves the needs of future patients, but flipping a coin to choose each patient's treatment looks strange, and may be unsettling to many physicians. In their day-to-day practice, a physician chooses each patient's treatment based on their observed prognostic variables. But in terms of parameter estimation, this practice is inherently biased so, in this sense, fair randomization is the antithesis of routine medical practice.

If the goal of a clinical trial is to maximize the benefit to the patients in the trial, then the best action may appear to be to give each new patient the treatment having the larger current estimated response rate or mean survival time. As mentioned above, this strategy is known variously as a "greedy," "PTW," or "myopic" decision algorithm. It turns out that, perhaps counterintuitively, a PTW algorithm may be bad for the patients enrolled in the trial. To see why this is so, consider comparing response probabilities θ_A and θ_B when in fact the true but unknown values are $\theta_A = 0.30$ and $\theta_B = 0.50$. Say you start by treating three patients on each arm, and thereafter you use the PTW rule. Suppose that, for example, 0/3 responses are observed with B and at least one response is observed with A. Since $\hat{\theta}_B = 0$ and $\hat{\theta}_A > 0$, the PTW rule is certain to treat all future patients in the trial with A, despite the fact that A has a much lower true response probability than B. This is a simple example of the general problem that, simply due to random variation in the data, the PTW rule easily can get stuck giving the inferior treatment to all but a small number of patients. So a PTW rule actually has a substantial risk of being bad for the patients in the trial. This problem with greedy or myopic decision rules has been well known for many years in sequential analysis, and is discussed by Sutton and Barto (1998). It sometimes is called the "exploitation versus exploration" problem, or "stickiness," since one can get stuck using a suboptimal decision rule.

A key point about AR is that $\pi_{A<B}$ is a function of the current data, so it is a statistic. Early on, people noticed that, however $\pi_{A<B}$ and the randomization probabilities $f(\pi_{A<B})$ are defined and computed, the statistic $\pi_{A<B}$ tends to be highly variable. This makes the AR probabilities rather unstable, and they often may take on extreme values close to 0 or 1, especially for small to moderate sample sizes early in the trial. The additional variability added by AR, compared to fair randomization, can be quite large. To understand why this is true, first recall that fair randomization introduces variability, beyond the inherent randomness of the observed outcomes. But AR introduces much more variability because the randomization probabilities $f(\pi_{A<B})$ are statistics, rather than fixed values. A major inferential problem caused by AR is that, because it unbalances the sample sizes based on the data, it produces biased estimates of the parameters. That is, AR defeats the reason for randomizing in the first place, which is to obtain unbiased comparative estimates. I will give numerical examples of this below.

Various fix-ups for the problems that AR produces have been proposed. One common method is to enforce an initial "burn-in" period where fair randomization is used for a predetermined number of patients enrolled at the start of the trial, after which the AR procedure is started. Another type of fix-up is to use some function of $\pi_{A<B}$, such as

$$f(\pi_{A<B}) = \frac{\{\pi_{A<B}\}^c}{\{\pi_{A<B}\}^c + (1 - \{\pi_{A<B}\})^c} \tag{6.8}$$

for given $c > 0$, to shrink $\pi_{A<B}$ toward 1/2. This reduces the chance of extreme randomization probabilities close to 0 or 1. For example, with the popular value $c = 0.5$, the empirical probabilities $(0.87, 0.13)$ become $(0.72, 0.38)$, which are still in favor of B over A but less extreme. As more AR methods were proposed, and people began to use AR in actual clinical trials, e.g., Giles et al. (2003) and Kim et al. (2011), some less obvious, highly undesirable properties of AR became more apparent. The use of AR methods soon became very controversial. Many authors argued about both the ethics and practicality of AR, including Chappell and Karrison (2007), Korn and Freidlin (2011), Yuan and Yin (2011), Karrison et al. (2003) and Hey and Kimmellman (2015).

While there are numerous AR methods, to study the properties of AR in two-arm trials and compare them to fair randomization (FR) with fixed probabilities 1/2 for each arm, Thall et al. (2015) conducted a computer simulation study of four Bayesian

AR methods. The first two AR methods were AR(1) which uses $\pi_{A<B} = \Pr(\theta_A < \theta_B \mid data)$, and AR(0.5) which uses the randomization probabilities given by (6.8) with $c = 0.5$. For both AR methods, the trial is stopped with B declared superior to A if $\Pr(\theta_A < \theta_B \mid data) > 0.99$, and the symmetric rule is applied with the roles of A and B reversed. These two rules are applied continuously. They also studied modified versions that use stopping probability cutoffs 0.995 for AR(1)^{mod} and 0.9985 for AR(0.5)^{mod}, which ensure that they have overall false positive probabilities ≤ 0.05. As a comparator that uses FR, a Bayesian group sequential design (FR-GS) was simulated, using the treatment comparison rule to stop and declare B superior to A if

$$\Pr(\theta_A + 0.20 < \theta_B \mid data) > 0.95 - 0.80(n/200),$$

also using the symmetric rule with the roles of A and B reversed, applied at $n = 50, 100, 150,$ and 200 patients. The values 0.95 and 0.80 used in the stopping rule were derived to ensure overall false positive probability ≤ 0.05. Patients were randomized in blocks of size 8 to avoid interim sample size imbalance. Each case studied in the simulations using each of the five designs was replicated 10,000 times.

Table 6.4 summarizes the simulation results. In the null case where $\theta_B = \theta_A = 0.25$, by symmetry the Type I error probability for each method is twice the tabled value "Prob conclude $\theta_B > \theta_A$," so the false positive probabilities are 0.18 for AR(1), 0.24 for AR(1/2), and 0.048 for Equal GS. Thus, the simulations show that both AR(1) and AR(1/2) have extremely large false positive rates compared to the FR-GS design. The AR methods also have much smaller power when $\theta_B = 0.45$. The mean sample size imbalances $N_B - N_A$ in favor of B are very large for all AR methods when $\theta_B = 0.35$ or 0.45 and $\theta_A = 0.25$. These differences are the most common rationale for arguing that AR is ethically superior to FR. But only looking at these means is misleading. When $\theta_B = 0.35$ and $\theta_A = 0.25$, AR(1) has a 14% chance that $N_A \geq N_B + 20$, that is, there will be at least 20 more patients given the inferior treatment A rather than B. This is the opposite of the claimed ethical advantage of AR. This property, that AR may behave pathologically, is due to the fact that AR produces highly disperse sample size distributions, so when $\theta_A < \theta_B$ and the difference is 0.10, there is a nontrivial risk of creating a large sample size imbalance $N_B < N_A$ in favor of the inferior treatment.

All designs with interim adaptive rules introduce bias in final parameter estimators, but the bias is much larger with these AR methods. Both AR(1) and AR(0.5) are more disperse, and less precise, compared to FR-GS. When $\theta_B^{true} = 0.45$, so $\Delta^{true} = 0.45 - 0.25 = 0.20$, the mean of the estimate of $\Delta_{B,A}$ is 0.29 for AR(0.5), which gives Bias $= 0.29 - 0.20 = 0.09$ (45%), whereas the mean of the estimate of $\Delta_{B,A}$ is 0.24 for the FR-GS design, so its Bias $= 0.24 - 0.20 = 0.04$ (20%). Thus, AR(0.5) more than doubles the bias in the final estimator of $\Delta_{B,A}$, compared to the FR-GS design. The result is that, at the end of a trial conducted using AR(0.5), it is likely that the final estimates will overstate the actual benefit of treatment B over A by two fold or more.

The modified AR methods AR(1)^{mod} and AR(0.5)^{mod} were included in the simulation study to address the potential criticism that, because the AR(1) and AR(0.5)

Table 6.4 Comparison of four AR methods and fair randomization group sequential (FR-GS), for a clinical trial with a maximum of 200 patients, comparing treatments A and B, where the true response probabilities are $\theta_A^{true} = 0.25$ and $\theta_B^{true} = 0.25, 0.35$, or 0.45. The achieved sample sizes of the two treatment arms are denoted by N_A and N_B. $\gamma_{20} = \Pr(N_A \geq N_B + 20)$, the probability of a sample size imbalance of at least 20 patients in the wrong direction when $\theta_B^{true} > \theta_A^{true}$. Bias $= \hat{\Delta}_{B,A} - \Delta_{B,A}^{true}$. The methods AR(1)^{mod} and AR(0.5)^{mod} are versions of AR(1) and AR(0.5) modified to have false positive probability ≤ 0.05

θ_B^{true}	Method	Prob conclude $\theta_B > \theta_A$	Mean(25th, 95th %-iles) of $N_B - N_A$	γ_{20}	Bias
0.25	AR(1)	0.090	0 (−186, 186)	0.43	0.00
	AR(0.5)	0.120	0 (−100, 100)	0.33	0.00
	AR(1)^{mod}	0.025	0 (−186, 186)	0.45	0.00
	AR(0.5)^{mod}	0.025	−1 (−132, 132)	0.40	0.00
	FR-GS	0.024	0 (0, 0)	0.00	0.00
0.35	AR(1)	0.300	66 (−166, 188)	0.14	0.05
	AR(0.5)	0.410	37 (−50, 116)	0.07	0.06
	AR(1)^{mod}	0.140	83 (−182, 152)	0.14	0.05
	AR(0.5)^{mod}	0.130	64 (−68, 152)	0.08	0.03
	FR-GS	0.340	0 (−1, 2)	0.00	0.02
0.45	AR(1)	0.0590	80 (−62, 184)	0.05	0.09
	AR(0.5)	0.800	38 (−7, 116)	0.01	0.10
	AR(1)^{mod}	0.350	111 (−70, 192)	0.04	0.09
	AR(0.5)^{mod}	0.400	87 (2, 158)	0.01	0.07
	FR-GS	0.860	0 (−2, 2)	0.00	0.04

methods have the large type I error probabilities 0.18 and 0.16, it is unfair to compare them to FR-GS. But while the two modified AR methods do have Type I error probabilities ≤ 0.05 due to using larger, modified early stopping probability cutoffs, their power figures in the case $\theta_A^{true} = 0.25$ and $\theta_B^{true} = 0.45$ are the extremely low values 0.35 and 0.40, compared to 0.86 using FR-GS. So, if you use any of these four AR methods, you must settle for a design that either has (1) a greatly inflated Type I error probability, or (2) such low power that it is hardly worth running the trial at all. Considering the inflated Type I error, lower power, and much larger bias, along with the extremely worrying probability of a large sample size imbalance that is backward, in favor of the inferior treatment, in the intermediate case where $\theta_A^{true} = 0.25$ and $\theta_B^{true} = 0.35$, it appears that using AR introduces more problems than it supposedly solves. The large values of λ_{20} for all four AR methods in the null case where $\theta_A^{true} = \theta_B^{true} = 0.25$ illustrate the fact that AR produces much more disperse sample size distributions than FR. Thus, summarizing the behavior of an AR method in terms of mean sample sizes only while ignoring sample size variability, and ignoring statistical estimates of probabilities such as $\gamma_{20} = \Pr(N_A \geq N_B + 20)$, is misleading.

For the two-arm case, Korn and Freidlin (2011) considered properties of the Bayesian version of AR(c), where $\pi_{A<B} = \Pr(\theta_A < \theta_B \mid data)$, for the commonly used value $c = n/2N$, where n is the current sample size, as suggested by Thall and Wathen (2007). They noted that $AR(n/2N)$ stabilizes the randomization probabilities, stating that this may be used "to prevent the assignment probability from becoming too unbalanced, that is, being greater than 0.8 or 0.9 (extreme imbalances can create problems with the study interpretation if there are time trends ...)." They studied AR(n/2N) with randomization probabilities restricted to be ≤ 0.8, and compared this to randomization with fixed probabilities (1/2, 1/2) or (2/3, 1/3), which may be denoted by 1:1 and 2:1. They found that, for true response probabilities $\theta_A^{true} = 0.20$ and $\theta_B^{true} = 0.20$, 0.30, or 0.40 in a trial with maximum sample size 190, the above AR method gave, on average, 3 to 13 more nonresponders than fixed 1:1 randomization. The AR method also gave a smaller average number of nonresponders only in the extreme case $\theta_A^{true} = 0.20$ and $\theta_B^{true} = 0.50$, but there the numbers were 28.2 for 1:1 fixed randomization and 28.8 for AR, a trivial difference. So, the claimed ethical advantage of AR for the patients enrolled in the trial turned out to be false, since AR produced more nonresponders.

Korn and Freidlin (2011) also reported simulation results for trials with a time trend to evaluate the properties of AR. With a time trend, known as "drift," the values of both θ_A^{true} and θ_B^{true} change by the same amount over time, so $\theta_A^{true} - \theta_B^{true}$ remains the same. Drift might occur, for example, due to an improvement in the prognosis of enrolled patients over the course of the trial. To give the AR method a chance to perform well in this difficult but realistic case, they constructed a block stratified version of AR, with blocks of size 50. This blocking approach for dealing with drift when using AR was suggested by Karrison et al. (2003). In this case, for $\theta_A^{true} = 0.80$ and $\theta_B^{true} = 0.80$, 0.85, or 0.90, they found that the block stratified AR method had 45, 33, and 15 more nonresponders, respectively, compared to 1:1 randomization. Korn and Freidlin (2011) concluded

"Adaptive randomization is inferior to 1:1 randomization in terms of acquiring information for the general clinical community and offers modest-to-no benefits to the patients on the trial, even assuming the best-case scenario of an immediate binary outcome."

Discussing the paper of Korn and Freidlin (2011), Berry (2011) wrote

"Adaptive randomization has the greatest potential in multi-armed trials in which many questions are addressed, but limited usefulness in the context of two-armed trials, fixed sample size, and no biomarkers."

With all this discussion and controversy, there still appeared to be no published systematic evaluation of how AR methods actually behave in multi-arm clinical trials. To explore this mystery, Wathen and Thall (2017) conducted an extensive computer simulation study. But designing such a simulation study is a complex problem, mainly because there are numerous different ways to design and conduct a multi-arm comparative clinical trial. To actually construct a trial design, one must make many decisions. As noted by Wathen and Thall (2017), a multi-arm randomized trial may

1. include three, four, five, or some larger number of treatment arms,

2. have whatever sample size one can afford or rationalize,
3. include a control arm C, or not,
4. restrict the number of patients randomized to C if it is included, or not,
5. use as outcome a binary indicator of treatment success, ordinal variables representing the severities of one or more adverse outcomes, survival or progression-free survival time, or some combination of these,
6. include any of a wide variety of possible rules for making interim between-arm comparisons or stopping an arm early,
7. when some arms are terminated early, enrich the remaining arms with larger sample sizes, or not,
8. select either one best or several experimental treatments, and
9. include two stages, more than two stages, or monitor continuously.

Given all of these possibilities, to obtain simulation results for clearly specified trial designs, Wathen and Thall (2017) decided to evaluate properties of several different but specifically defined Bayesian AR methods. They studied five-arm randomized clinical trials, either with or without a control treatment arm, C and maximum overall sample size N either 250 or 500. They studied designs for a trial comparing experimental treatments E_1, E_2, E_3, E_4 and control treatment C, or five experimental treatments $E_1, \ldots, E_5$ without a control arm, based on a binary success/failure outcome. The designs included three types of interim and final decision rules. For the version of the design including a C arm, these rules were as follows:

Futility: For each $k = 1, 2, 3, 4$, arm E_k is terminated early due to futility if $\Pr(\theta_k > \theta_C + 0.20 \mid data_n) < 0.01$. If all four experimental arms are terminated, the trial is stopped.

Enrichment: If an arm E_k is terminated early due to futility, then the remaining patients, up to N, are randomized among the remaining open arms.

Selection: If an arm E_k is not terminated early due to futility, then at the end of the trial E_k is selected if $\Pr(\theta_k > \theta_C + 0.20 \mid data_n) > a_U$.

If the futility rule terminates all E_k's early, then the trial is stopped before N is reached with no E_k selected. The selection rule allows more than one E_k to be selected, and the numerical value of the decision cutoff a_U was calibrated to ensure overall false positive (Type I error) probability 0.05 for the trial, where a false positive was defined as selecting any E_k in the global null case where all true $\theta_k^{true} = \theta_C^{true} = 0.20$.

Defining AR probabilities in a multi-arm trial is a bit trickier than for two arms, and it can be done in many different ways. To do this, Wathen and Thall (2017) first defined the probabilities, in general for K treatment arms,

$$r_{k,n} = \Pr(\theta_k = \max\{\theta_1, \ldots, \theta_K\} \mid data_n)$$

for interim sample size n. Given this, they defined AR(c) to be the version of the design using the adaptive randomization probabilities

$$r_{k,n}^{(c)} = \frac{(r_{k,n})^c}{\sum_{j=1}^{K} (r_{j,n})^c}$$

for $c > 0$. The AR methods studied included $AR(1)$, $AR(0.5)$, $AR(n/2N)$, and also a modification of $AR(1)$ with the randomization probabilities restricted to the interval $[0.1, 0.9]$, denoted by $AR(1, 0.1)$. The comparator was a design using the same interim decision rules, but with equal randomization probabilities throughout, denoted by ER. All designs assumed beta(0.2, 0.8) priors for all θ_k's, and included an initial burn-in with the first 50 patients randomized equally among the arms, restricted so that exactly 10 patients were assigned to each of the five arms.

For a trial with all experimental arms, $E_1, \ldots, E_K$ and no control arm as a comparator, the futility stopping and selection rules were as follows:

Futility, No Control Arm: For each $k = 1, \ldots, K$, arm E_k is terminated early due to futility if $\Pr(\theta_k > \max\{\theta_r : r \neq k\} \mid data_n) < 0.01$.

Selection, No Control Arm: If arm E_k is not terminated early due to futility, then E_k is selected if $\Pr(\theta_k > \max\{\theta_r : r \neq k\} \mid data_n) > a_U$, with at most one E_k selected.

The simulations studied the designs in each of the following three cases for $K = 5$ arms:

Global Null Case: All $\theta_k^{true} = \theta_C^{true} = 0.20$.

LFC: One $\theta_k^{true} = 0.40$ and all other $\theta_j^{true} = \theta_C^{true} = 0.20$ for $j \neq k$, called the "least favorable configuration."

Staircase Scenario: The true response probabilities of the E_k's follow a "staircase" pattern, 0.25, 0.30, 0.35, 0.40 if the control arm has true $\theta_C = 0.20$, or the five experimental treatment probabilities 0.20, 0.25, 0.30, 0.35, 0.40 if there is no C arm.

The staircase scenario may be considered much closer to what is likely to be encountered in practice than the LFC, which actually is a much easier case to deal with if there is some improvement over θ_C.

Summaries of the extensive simulation results were reported by Wathen and Thall (2017) in seven tables, which I will not repeat here. Together, they yielded the following general conclusions about the use of AR in multi-arm clinical trials.

For multi-arm randomized trials that include a control arm as a comparator:

1. In the global null case, ER has larger correct futility early stopping probabilities than any AR method studied.
2. Under the LFC, AR(1, 0.1) and ER have nearly identical probabilities of correctly selecting the one truly superior E_k. All other AR methods, AR(1), AR(0.5), and AR(n/2N), have much lower correct selection probabilities.
3. AR(1), AR(.5), and AR(n/2N) should never be used.
4. In the staircase scenario, AR(1, 0.1) provides a sample size imbalance in favor of a superior E_k, but it has lower probabilities of correctly selecting either the best or second best E_k, compared to ER.

For multi-arm selection trials that do <u>not</u> include a control arm as a comparator:

Such a trial should never be conducted, regardless of whether ER or some form of AR is used. This is because, in the "staircase" scenario, such a trial has extremely low probabilities of selecting a superior E_k, even for sample size $N = 500$ with five treatments. Conducting a multi-arm selection trial without a control arm as a comparator is a complete waste of time and resources.

Some of these simulation results are not intuitively obvious. The large published literature on AR methods that often touts its desirability, and the paucity of systematic simulation studies of the actual properties of AR methods, appear to be the main reasons that many people still believe that using AR in comparative clinical trials is good statistical practice. Another important problem, that I have not discussed, is how AR behaves when patient heterogeneity is taken into account. Some chapters in Sverdlov (2015) discuss AR methods that incorporate patient covariate information. However, once covariate effects are considered, things become much more complex inferentially. This is because accounting for treatment–covariate or treatment–subgroup interactions greatly increases the number of treatment effect parameters; hence, it reduces reliability in general, regardless of randomization method. I will discuss clinical trial designs that account for patient heterogeneity and treatment–subgroup interactions in Chap. 11.

Recall that the two main goals of a clinical trial are to treat the patients in the trial, and to obtain data that provide a reliable basis for inferences that may benefit future patients. A great deal of thought has been devoted to deriving randomization methods to obtain unbiased estimators, as well as methods to correct for bias in observational data. From this perspective, based on the results of the simulation studies reviewed above, outcome-adaptive randomization appears to throw out the baby with the bathwater.

In the interest of full disclosure, I must admit that I once designed a clinical trial to compare two chemotherapies for sarcomas by doing outcome-adaptive randomization, even adjusting the AR probabilities to account for each patient's covariates. The design is described by Thall and Wathen (2005), and the trial results were published by Maki et al. (2007). Yes, it's true. I might attempt to mitigate this giant mistake by saying that I was led astray by someone who was in the adaptive randomization business, but that would only shirk the responsibility that I must bear. At the time, I believed that the design was terribly clever, and I was quite proud of myself for constructing it. Based on this experience, and my more recent investigations of AR, I have come to believe that, if God did not want me to conduct simulation studies, She would not have given me computers.

Advice

1. If you are planning a randomized comparative clinical trial and someone proposes that you use outcome-adaptive randomization, *Just Say No*.
2. If you are planning a randomized comparative clinical trial of two or more experimental treatments, and an active control treatment for the disease exists, always include a control arm as a comparator.

Chapter 7
All Mixed Up

Intuition is a poor guide when faced with probabilistic evidence.
Dennis V. Lindley

Contents

Abstract Possible relationships between the probability of early response and expected survival time with a given treatment are at the heart of the conventional paradigm for using phase II response data to plan phase III trials. These relationships often are misunderstood, however, which can lead to very bad decisions. To illustrate this, I will present a simple probability computation which gives numerical results that may seem surprising. I then will give an example of how a trial effect can be mistaken for a treatment effect if one compares data from different trials rather than randomizing. A method will be described for computing the predictive probability that a future phase III trial will be successful given observed phase II data, and this will be illustrated by a numerical example. An example will be given of a randomized trial in which between-arm comparisons of the 90-day response probabilities and the 12-month progression-free survival probabilities gave opposite conclusions regarding which treatment was superior. These examples illustrate the facts that probability often can be counterintuitive, and that basing treatment comparisons on early outcomes can be very misleading.

© Springer Nature Switzerland AG 2020
P. F. Thall, *Statistical Remedies for Medical Researchers*, Springer Series
in Pharmaceutical Statistics, https://doi.org/10.1007/978-3-030-43714-5_7

7.1   The Billion Dollar Computation

This chapter provides yet another Magic Formula that could save your friendly neighborhood pharmaceutical company somewhere in the neighborhood of $1,000, 000,000. That's a pretty snazzy neighborhood. What follows is an illustration of the fact that probability, if used sensibly, can be a very powerful tool for figuring out what is going on with a new drug or treatment. The Billion Dollar Computation actually is an application of elementary probability. It may help you think about how data from phase II trials may be used, or misused, to decide whether or not to conduct a large, expensive, randomized phase III trial that may take years and cost a vast amount of money to complete. It provides a good starting place for computing the probability that a phase III trial will yield positive results. This computation is so clever, you might use it to impress your friends, relatives, and learned colleagues. Like many of the examples in this book, I have made it simple in order to make some key points. It can be elaborated in many different ways, some of which I will present later.

For a given cancer, let S denote an existing standard treatment, and E an experimental agent or treatment regimen that one hopes will turn out to be better than S. Denote the indicator of tumor response by Y, and survival time by T. Let π_E = Pr(tumor response with E) and π_S = Pr(tumor response with S), and denote the 12-month survival probabilities

$$\mathcal{F}_1 = \Pr(T > 12 \mid Y = 1) \text{ for the responders}$$

and

$$\mathcal{F}_0 = \Pr(T > 12 \mid Y = 0) \text{ for the nonresponders.}$$

Suppose that $\mathcal{F}_1$ = 0.60 and $\mathcal{F}_0$ = 0.40. This says that responders have a 50% larger probability of surviving 12 months than nonresponders, so achieving a tumor response is very beneficial. Next, suppose that you know, from historical data, that S has response probability $\pi_S = 0.20$, and a recent phase II trial has shown that $\pi_E =$ 0.40. So E *doubles the tumor response rate compared to S*. This is a very exciting new result! If you tell a science reporter about how well E performs and the story gets on The Evening News, the stock of whatever lucky pharmaceutical company manufactures E should shoot up. The company executives would be crazy not to conduct a randomized phase III trial of E versus S to get E approved by the FDA for treating this cancer. They might even apply for special accelerated approval of E based on the exciting phase II results, in order to get started heroically saving the lives of desperate cancer patients sooner. Making money and saving lives is definitely a "Win-Win" scenario. The marketing people at the pharmaceutical company might get a jump on things by planning their media and sales campaigns. The inventor of E might go on a speaking tour, appear on some late-night talk shows, and buy a tuxedo in case those people in Sweden decide to award a Nobel Prize.

But before you rush off and spend many years and a vast amount of money conducting a big phase III trial, talk to a science reporter, or buy a tuxedo, you might do the following elementary probability computation. It can save you time, money, and embarrassment. This computation also is useful for people who work at a federal regulatory agency when they are being pressured to give a drug accelerated approval based on early phase clinical trial results.

Denote the unconditional probabilities of 12-month survival with S and E, if you don't know whether the response variable is $Y = 0$ or 1, by $\mathcal{F}_S$ and $\mathcal{F}_E$. Statistics and probability actually have quite a few laws that cannot be broken. One of them is *The Law of Total Probability*. In this case, it says that, for S, the 12-month survival probability is

$$\mathcal{F}_S = \Pr(T > 12 \mid S, Y = 1) \times \pi_S + \Pr(T > 12 \mid S, Y = 0) \times (1 - \pi_S)$$
$$= 0.60 \times 0.20 + 0.40 \times (1 - 0.20) = 0.44.$$

In words, this says that the probability of surviving 12 months with S is the weighted average, also called a "mixture", of the probabilities for patients who respond and patients who do not respond, with the weights being the probabilities of response and nonresponse. Since response and nonresponse are complementary events, that is, one or the other must happen for each patient, the probability of nonresponse is with S is $1 - \pi_S$. Repeating the above computation for E gives

$$\mathcal{F}_E = \Pr(T > 12 \mid E, Y = 1) \times \pi_E + \Pr(T > 12 \mid E, Y = 0) \times (1 - \pi_E)$$
$$= 0.60 \times 0.40 + 0.40 \times (1 - 0.40) = 0.48.$$

Oops. It turns out that E only increases the 12-month survival probability from 0.44 to 0.48. If we assume, for simplicity, that the death rates for the two treatments are constant over time, so that survival time follows an exponential distribution then E increases mean survival time from 15.18 to 17.25 months. This whopping 2.07 months of additional expected survival time is not exactly thrilling news for someone who has the particular cancer being targeted by E. So, *a new treatment that doubles the response for a disease where response increases the 12-month survival probability by 50% actually provides a trivial improvement in expected survival time.*

This tiny improvement in mean survival may seem surprising, and a bit counter-intuitive. What is going on is that thinking about only the 12-month survival probabilities and the response rates with E and S ignores the nonresponders. The overall 12-month survival probability, or mean survival time, is what matters, and it is a mixture over both the responders and nonresponders. But with either treatment, most of the patients will be nonresponders, specifically 80% with S and 60% with E. *Consider the extreme case where all patients treated with E respond,* that is, $\pi_E = 1$. Then the 12-month survival probability with E would be $0.60 \times 1 + 0.40 \times 0 = 0.60$. So, *even of you have a new treatment with a 100% tumor response rate,* and you give it to all of your patients, for each patient the probability of death within 12 months is still $1 - 0.60 = 0.40$. Put another way, tumor response does not guarantee that the patient is cured.

Next, suppose that the investigators still want to do a phase III trial, even if the expected improvement in mean survival is only 2.07 months. After all, another 63 days of expected life is better than nothing. In 63 days, someone can write their last will and testament, say goodbye to their friends, family, and dog, and re-watch some of their favorite movies. So, let's think about the statistical problem of designing a randomized trial of E versus S. Consider a standard two-arm, frequentist group sequential phase III clinical trial design with up to three successive two-sided log-rank tests, overall Type I error probability 0.05, and power 0.90 to detect the alternative hazard ratio $(1/17.25)/(1/15.18) = 0.88$. This would require a sample size of about 2900. With an accrual rate of 20 patients per month, the trial would take about 14 years to complete. If you have a child in the third grade when the trial is started, you might expect your child to graduate from college when the trial is completed. So maybe conducting a phase III trial is not such a great idea, unless you make your living working in a clinical trial coordinating center. Of course, anyone familiar with this sort of computation knows that it never makes sense to target such a small improvement in mean survival time, i.e., 15.18 versus 17.25. Of course, someone might simply ignore the estimated improvement in mean survival, that is, ignore the data, and optimistically pick a larger targeted alternative value, like 20 months, to get a smaller planned sample size and shorter phase III trial. This new 20-month alternative mean survival time value gives a targeted hazard ratio of about 0.75, a convenient Straw Man to avoid a 14-year trial. This gives a trial with 670 patients and expected duration of slightly over 33 months. The new alternative mean survival value of 20 months has nothing to do with the value 17.25 months obtained from the mixture computation, except that it is larger than 15.18. This sort of thing also may be called "optimism," a well-known suboptimal strategy. But running this shorter trial really is a false economy. Why spend 33 months and millions of dollars conducting a trial of a new treatment E that, based on its estimated mean survival, is expected to provide a tiny improvement over S?

Of course, before we did the mixture computation and saw its implications, E certainly did look quite promising. Without the perspective provided by The Billion Dollar Computation, someone might be likely to conduct the 33-month trial, but end up with the unhappy conclusion that E is not so wonderful compared to S, after all. And people wonder why so many phase III trials have negative results. Really?

In practice, evaluating a new treatment actually is more complex. The probability computations that I gave above all assumed that all of the effect of treatment on survival is mediated by Y, that is, the possible effects take the causal form

$$\text{Treatment} \; \rightarrow \; \text{Tumor Response} \; \rightarrow \; \text{Survival Time.}$$

But what if there is another, direct treatment effect,

$$\text{Treatment} \; \rightarrow \; \text{Survival Time,}$$

that is not mediated by tumor response? This would say that, *due to some direct biological effect of E on survival time that is not mediated by tumor response,*

$\mathcal{F}_E(12 \mid Y = 1)$ may be larger than 0.60 and, similarly, $\mathcal{F}_E(12 \mid Y = 0)$ may be larger than 0.40. For example, suppose that $\mathcal{F}_E(12 \mid Y = 1) = 0.75$ rather than 0.60, and $\mathcal{F}_E(12 \mid Y = 0) = 0.55$ rather than 0.40. In this new example, E gives larger 12-month survival probabilities than S among responders, but also among nonresponders. So, response does not tell the whole story of how E affects survival. In this case,

$$\begin{aligned} \mathcal{F}_E(12) &= \mathcal{F}_E(12 \mid Y = 1)\pi_E + \mathcal{F}_E(12 \mid Y = 0)(1 - \pi_E) \\ &= 0.75 \times 0.40 + 0.55 \times (1 - 0.40) = 0.63, \end{aligned}$$

which is a nice improvement over 0.44. So maybe there is hope for E. But it should be clear from this example that *it is very important to determine whether there is a direct biological effect of E on survival that is not mediated by tumor response.*

Here are some one elaborations of this data structure and mixture computation. To apply this to your own particular setting, you might consider some combination of the following cases. The early outcome need not be a simple binary variable. Disease response often is ordinal, say using $y = 0, 1, 2, 3$ to numerically identify, respectively, progressive disease, stable disease, partial response, complete response. If we denote $\pi_{\tau,y} = \Pr(Y = y \mid \tau)$ for treatment $\tau = E$ or S and outcome $y = 0, 1, 2, 3$, then this would give the survival mixture probabilities as

$$\mathcal{F}_\tau(12) = \sum_{y=0}^{3} \Pr(T > 12 \mid \tau, Y = y) \times \pi_{\tau,y}.$$

The slightly simpler mixture computations given earlier might be obtained by dichotomizing the ordinal response variable, and defining response to be the event $[Y \geq 3]$, but of course this reduces the available information unnecessarily. Now one must keep track of survival distributions, $\mathcal{F}_\tau(12 \mid \tau, y)$ that vary with the four levels of Y for each treatment. Since there are lots of possible values of the probability vectors $\pi_\tau = (\pi_{\tau,0}, \pi_{\tau,1}, \pi_{\tau,2}, \pi_{\tau,3})$ for $\tau = E$ and S, comparison of these two vectors becomes more complex. That is, in terms of the probability vectors, π_S and π_E, when is E considered a promising improvement over S based on phase II data? One might assign weights to the four outcomes to quantify their relative desirabilities, and compute-weighted averages of π_S and π_E. But the survival probabilities $\Pr(T > 12 \mid \tau, Y = y)$ actually are the weights that matter most. So, what this implies is that survival probabilities such as $\mathcal{F}_E(12)$ and $\mathcal{F}_S(12)$, or possibly mean survival times, μ_E and μ_S, are what should be considered when evaluating phase II data. This implies, in turn, that phase II trials should be designed to account for both early outcomes and survival time data. That is, the conventional paradigm of [phase II based on tumor response rate] $\rightarrow$ [phase III based on survival time] really does not work very well.

7.2   Accounting for Uncertainty and Bias

But wait, there's more. Now that you know how to do mixture computations, you still need the numbers to plug into the formulas. In all of the above computations, for the sake of illustration, I assumed that all of the probabilities were known. But in real life, you don't know any of the probabilities. We know that it is a big mistake to confuse a statistical estimator with the parameter that is estimating, because any statistic has uncertainty associated with it. In the mixture computation, the probabilities π_S, π_E, $\mathcal{F}_{E,1}$, $\mathcal{F}_{E,0}$, $\mathcal{F}_{S,1}$, $\mathcal{F}_{S,0}$, all must be estimated from data. The values used in the computations above actually should be thought of as point estimates, $\hat{\mathcal{F}}_E(12 \mid Y = 1) = 0.60$ or 0.70, and so on. Since you must estimate all of the probabilities in the mixture, they all have associated uncertainty. The relevant final answers are not simply point estimates like $\hat{\mathcal{F}}_E(12) = 0.63$ and $\hat{\mathcal{F}}_S(12) = 0.44$ alone. Since the reliabilities of such estimates depend on the sizes and variability in the samples from which they were computed, some indices of uncertainty about the estimates are needed.

For example, if one observes 24 responses in 60 patients, for an empirical response rate of 40%, then starting with a $beta(0.50, 0.50)$ prior on π_E, a posterior 95% credible interval is $[0.28, 0.56]$, so just using the value $\pi_E = 0.40$ ignores the uncertainty quantified by the credible interval. That is, the phase II sample is not infinitely large. Likewise, in the example where there was a direct biological effect on survival not mediated by Y, the 12-month survival probabilities $\mathcal{F}_E(12) = 0.63$ and $\mathcal{F}_S(12) = 0.44$ actually are point estimates obtained from survival time data. Accounting for uncertainty due to the fact that the estimates are obtained from data, suppose the 95% posterior credible intervals for these probabilities are, say, $[0.45, 0.80]$ for $\mathcal{F}_E(12)$ and $[0.35, 0.60]$ for $\mathcal{F}_S(12)$. Since these intervals have quite a lot of overlap, it is far from certain that E will improve survival compared to S.

A big problem that casts a shadow over all these computations is that, because patients were not randomized between E and S, any comparison of parameters between these two treatments is biased due to treatment-trial confounding. That is, the conventional paradigm of conducting a single-arm trial of E and then comparing its results to historical data on S is scientifically flawed. What $\hat{\mathcal{F}}_E(12)$ actually is estimating is the 12-month survival probability with E *in the phase II trial*, and $\hat{\mathcal{F}}_S(12)$ actually is estimating is the 12-month survival probability with S *in the historical data*. The same goes for estimates of mean survival times, μ_E and μ_S, as well as π_E and π_S. Remember, the whole motivation for randomizing is to obtain unbiased estimators of things like $\pi_E - \pi_S$ and $\mu_E - \mu_S$. Computing comparative estimates based on data from different trials of E and S and pretending that the data were from a randomized trial is a big mistake since, due to treatment-trial confounding, between-study effects may be misinterpreted as between-treatment effects. In practice, trial effects may as large as, or larger than, treatment effects. *What all of this implies is that the phase II trial should have randomized patients between E and S in the first place, and moreover, survival data should have been recorded, as well as the early response data.*

Here is a simple example of what can happen when you conduct single-arm trials and use their data to compare treatments. A single-arm phase II clinical trial of combination chemotherapy A (n $= 44$) for acute myelogenous leukemia (AML) was conducted at a large cancer center in 1995. Subsequently, combination chemotherapy B for AML was studied later at the same cancer center in 1996–1998. A Bayesian survival time regression model was fit to the combined data, accounting for the B versus A treatment effect and known prognostic covariates, including performance status, age, and type of cytogenetic abnormality. The fitted model gave posterior probability $\Pr(RR_{B-A} > 1 \mid data) = 0.87$, where RR_{B-A} = relative risk of death with B versus A. So, based on the fitted model, after accounting for covariates known to affect survival, the estimated odds were roughly seven to one in favor of A over B. But here is the catch. *A and B actually were the same chemo combination*: fludarabine + idarubicin + cytarabine + GCSF + ATRA. The apparent "treatment effect" was computed by comparing this chemo combination to itself. Since both trials were conducted in the same institution, *the estimated RR_{B-A} actually was a covariate-adjusted between-trial effect.*

If A and B had been different chemo combinations, but with identical actual effects on survival time, this difference might have been misinterpreted as a treatment effect. This sort of thing, comparing results of a single-arm trial to historical data as if a randomized study had been done, is published in the medical literature all the time. A common misconception is that a standard regression analysis can be used to correct for between-study bias. This is not true. Such incorrect analyses never should be trusted. Any comparative statistical analysis must use IPTW, stratification, matching, or some other reasonable method to correct for bias due to between-study effects.

If you think about the data needed to estimate the survival probabilities such as $\mathcal{F}_E(12 \mid Y = 1)$ and $\mathcal{F}_E(12 \mid Y = 0)$, one would need to follow patients for 12 months or longer and record their survival times, or how long they were followed without dying. Similarly, if you are interested in estimating 36-month survival probabilities, you would need to follow patients for at least 36 months. Of course, the choice of follow-up time would depend on the survival time distribution with S. Shorter times are meaningful for rapidly fatal diseases like septic shock, acute respiratory distress syndrome, relapsed/refractory acute leukemia, or metastatic pancreatic cancer. Longer follow-up times are needed for diseases where patients tend to live longer, such as with early stage breast cancer or prostate cancer. But regardless of what follow-up time is meaningful for a given disease, in the end what is needed is long enough follow-up to do a survival analysis reliably. This is precisely the sort of long-term follow-up that is needed to conduct a phase III trial.

7.3 Predicting Phase III Success

If you still want to conduct phase II and phase III as separate, consecutive trials, here is a simple, potentially very useful Bayesian computation that you might perform to help you decide whether to conduct phase III. It requires a model, such as one of the mixture models given above, that relates early phase II outcomes to survival time, or PFS time if that is what you intend to use as your primary outcome in phase III. The computation uses the idea of a future event, which in this case is F_{III} = [the phase III trial will conclude that E is superior to S]. For this event to be meaningful, you need to specify a phase III design so that, given the model parameter θ, you can simulate phase III data to feed into the design and see whether its decision rules conclude F_{III}. The probability $\Pr(F_{III} \mid \theta)$ that the phase III trial will conclude that E is superior to S depends on the model parameter θ. If $f(\theta \mid data_{II})$ is the posterior as the end of phase II, then the *predictive probability* of F_{III} given the phase II data is the average

$$\Pr(F_{III} \mid data_{II}) = \int_{\theta} \Pr(F_{III} \mid \theta) f(\theta \mid data_{II}) d\theta.$$

A bit of thought shows that $data_{II}$ should include both response and survival data, and θ should include all parameters needed to compute the mixture survival distribution, which is the basis for the phase III design. For this to be of much use, $data_{II}$ should include a reasonable amount of survival time data, i.e., the times of death or follow-up without death, from the phase II patients. One also might include historical data on response and survival time with S, $data_{hist,S}$, i.e., use the distribution $f(\theta \mid data_{II} \cup data_{hist,S})$, although phase II versus historical trial effects may muddy the results. If the predictive probability $\Pr(F_{III} \mid data_{II})$ is small, say 0.10, then one may not want to start a phase III trial. The numerical computation may be done by simulating a sample $\{\theta^{(1)}, \ldots, \theta^{(M)}\}$ of parameters from the posterior $f(\theta \mid data_{II})$ based on the phase II data and, for each simulated parameter $\theta^{(m)}$, repeatedly simulating a phase III trial and recording whether the future event F is concluded. Then, $\Pr(F_{III} \mid \theta^{(m)})$ is computed as an empirical proportion from the simulated phase III trials under $\theta^{(m)}$ for each $m = 1, \ldots, M$. It is computationally intensive, but straightforward to do.

This probability computation is a special case of *predictive power*, which can be computed at any interim point in the phase II → phase III process. At the start of phase II, when there are no data, the computation can be done using the prior $f(\theta)$ in place of $f(\theta \mid data_{II})$. This computation does not include any interim data, and is based on the prior and the design only. Later on, one can carry it out to obtain an informed idea of what is going on once phase III has begun, at some interim point during the phase III trial, using the current $data_{interim}$ with $f(\theta \mid data_{interim})$ in place of $f(\theta \mid data_{II})$. If, for example, $\Pr(F_{III} \mid data_{interim}) = 0.01$, then it might be a good idea to stop the phase III trial. This can be done quite generally for any future event or decision F, based on the current data.

A detailed numerical example of a predictive power (PP) computation is given by Schmidli et al. (2006), involving comparison of three experimental immunosuppressive regimens for use in kidney transplantation, compared to a standard regimen. In their example, a four-arm randomized phase II trial is considered with 80 patients per arm, for 320 total, based on the 6-month rate of failure (acute rejection, graft loss, or death). The aim is to randomize one of the three experimental agents against the standard control in a subsequent two-arm phase III trial, to show either non-inferiority

Table 7.1 Illustration of posterior phase II 6-month event rates and corresponding predictive phase III power of concluding non-inferiority or superiority, for each of three experimental treatments compared to a control. The numerical values are based on one simulated trial, and each is accompanied by the 5th and 95th quantiles of its distribution

Treatment	Posterior 6-month phase II event rate (5th–95th quantiles)	Predictive probability of phase III decision	
		Non-inferiority	Superiority
Control	0.31 (0.26–0.36)	–	–
Experimental 1	0.27 (0.21–0.34)	0.98 (0.89–1.00)	0.11 (00–0.40)
Experimental 2	0.33 (0.26–0.39)	0.52 (0.08–0.94)	0.02 (0.00–0.08)
Experimental 3	0.40 (0.33–0.47)	0.07 (0.00–0.29)	0.00 (0.00–0.01)

or superiority, with 230 patients per arm, for a total of $320 + 460 = 780$ patients in phase II and phase III combined. Table 7.1 is a reduced version of Table II of Schmidli et al. (2006), which summarizes two types of predictive phase III decision probabilities. The prior 6-month event probabilities (5th quantile, 95th quantile) are 0.30 (0.23, 0.38) for the control and 0.30 (0.15, 0.46) for each of the three experimental arms. The numerical results from one simulated illustrative trial in Table 7.1 provide a basis for making decisions after phase II. For example, if experimental treatment 1 is selected at the end of phase II for evaluation in phase III, then the predictive probability that the phase III trial will conclude it is not inferior to the control is 0.98, while the predictive probability phase III will conclude it is superior to the control is 0.11. The results indicate that, based on the phase II data, if concluding non-inferiority is the goal, then experimental treatment 1 is the best choice for randomized comparison to the control in phase III. If concluding superiority is the goal, then it is best to stop the process and not conduct phase III at all.

This sort of predictive analysis can be repeated for any hypothetical interim data, or using actual data, in the context of virtually any design, as a basis for decision-making about what to do in phase III. A more elaborate version of this is to incorporate interim predictive power computations as part of a phase II/III design, and calibrate design parameters by simulation. For example, Inoue et al. (2002) provide a seamless phase II/III design based on a binary or categorial early response variable Y and survival time T, assumed to follow a model depending on $p(T \mid Y, Z)$ and $p(Y \mid Z)$ for treatments $Z =$ experimental E or control C. The survival distribution is represented as the mixture

$$p(T \mid Z) = \sum_y p(T \mid Y = y, Z) \, p(Y = y \mid Z).$$

In the design, decisions are based on simulated predictive probabilities given (1) the data available at future time $t + 12$ if accrual is terminated at t and patients are followed for 12 more months, and (2) the data available at a final evaluation time D if the maximum number of patients are accrued and follow-up continues to D.

Possible decisions based on predictive probabilities include stopping early and concluding that E is superior to C, or that E is inferior to C, continuing phase II, or organizing phase III.

7.4   A Paradoxical Clinical Trial

The following example is a disguised version of a randomized comparative clinical trial that actually was conducted. Patients with multiple myeloma (MM) undergoing autologous stem cell transplantation (autosct) were randomized between two preparative regimens, the standard S and $S + X$, where X was an additional agent motivated by the possibility that it might improve patient outcomes. The trial was designed using Complete Response evaluated at day 90 post-transplant, CR, with PFS time as a secondary endpoint. As with many oncology trials, the idea was that X would increase the probability of CR, and that this in turn would increase PFS. The trial was nominally "phase II" since its primary outcome was the binary event CR. The design included a Bayesian rule to stop the trial early if the interim data showed that a large difference between the CR probabilities in the two treatment arms was likely.

The trial design was as follows. Denoting the probabilities of CR at day 90 in the two treatment arms by π_S and π_{S+X}, under a Bayesian model it was assumed that these both followed non-informative $beta(0.20, 0.80)$ priors, based on an assumed historical CR probability of 0.20 A maximum of 190 patients were to be randomized between the two arms. The following two symmetric interim decision rules were to be applied when totals of 65 and 130 patients had been evaluated for CR, and at the end for 190 patients if the trial was not stopped early:

(1) Stop the trial and conclude that $S + X$ is superior to S in terms of 90-day CR rate if
$$Pr(\pi_S + 0.10 < \pi_{S+X} \mid data) > 0.77.$$

(2) Stop the trial and conclude that X is superior to $S + X$ in terms of 90-day CR rate if
$$Pr(\pi_{S+X} + 0.10 < \pi_S \mid data) > 0.77.$$

The cutoff 0.77 was determined by preliminary computer simulations to ensure an overall false positive conclusion probability of 0.10 if $\pi_S^{true} = \pi_{S+X}^{true} = 0.20$, that is, if the two preparative regimens had the same historical mean 90-day CR rate.

Here is what happened. Accrual went past the first interim sample size of 65, that is, accrual was not suspended at 65 patients to apply the rule, and 76 patients were treated by the time the 90-day CR data could be collected, analyzed, and the rules applied. In the $S + X$ arm, the observed data were

$$\frac{\#\text{ patients with CR}}{\#\text{ patients treated and evaluated}} = \frac{6}{44} \ (13.6\%).$$

In the S arm, the observed data were

$$\frac{\text{\# patients with CR}}{\text{\# patients treated and evaluated}} = \frac{13}{32} \ (40.6\%).$$

So, part of the way through the trial, the empirical 90 day CR rate in the S arm was triple the corresponding rate in the $S + X$ arm. Based on these interim data, the posterior probability, given above, that the S only arm was superior to the $S + X$ arm by 0.10 in terms of CR probability was 0.95, well above the decision cutoff of 0.77. If the 0.10 improvement is replaced by 0 in the above posterior probability, then the posterior probability that S has a higher 90-day CR rate is $Pr(\pi_{S+X} < \pi_S \mid data)$ = 0.996. From a frequentist viewpoint, the two-sided Fisher's exact test p-value is 0.0144. This was reported to the appropriate institutional clinical trial monitoring committee and, following the protocol's design, the trial was terminated. At that point, the conclusion was the unexpected outcome that the S arm actually had a higher CR probability than the $S + X$'arm, so it appeared that adding X to S had made things worse.

But something very surprising was discovered when analyzing the PFS time data. Since no patients had progressive disease or died prior to day 90, in order to evaluate both the 90-day CR probabilities and PFS time together, PFS time was measured starting at day 90 post-allosct, in a landmark analysis. After waiting to obtain longer follow-up in order observe some failure (disease progression or death) events, so that there would be enough data to evaluate and compare PFS between the two arms, surprisingly, the $S + X$ arm patients had longer PFS time, on average, than the S arm patients. So, the 90-day CR rates implied that S was better than $S + X$, but the PFS time data implied that $S + X$ was better than S.

To see how this apparent paradox can be true, denote PFS time, starting from day 90 when CR was evaluated, by T, and let NCR denote the event that a patient did not achieve CR at day 90. For the two treatment arms, denote the probabilities that a patient survived at least 12 months past day 90 without disease progression by

$$\mathcal{F}_{S+X} = \Pr(T > 12 \mid S + X) \quad \text{and} \quad \mathcal{F}_S = \Pr(T > 12 \mid S).$$

Since this was a landmark analysis starting at day 90 post-transplant, to be included a patient had to survive 90 days without relapse. The data gave the estimated values $\widehat{\mathcal{F}}_{S+X} = 0.90$ and $\widehat{\mathcal{F}}_S = 0.77$. The 12-month interval was chosen as a meaningful reference time, but the same advantage for $S + X$ over S was seen for other reference times. A similar difference was seen for PFS time recorded from date of transplant. That is, analysis of the PFS time data gave the conclusion that $S + X$ was superior to S, which was the opposite of the conclusion based on the 90-day CR data analysis. This statistical result was even more perplexing because, within each treatment arm, on average patients who achieved CR at day 90 had a longer PFS time, so achieving CR was advantageous in terms of longer PFS. That is, the estimates satisfied the inequality

$$\widehat{\Pr}(T > t \mid CR, \tau) \ > \ \widehat{\Pr}(T > t \mid NCR, \tau)$$

for any time $t > 0$ and both treatments $\tau = S$ or $S + X$. These perplexing, apparently contradictory statistical results led to the following obvious question.

Obvious Question: If the S arm had a significantly higher 90-day CR probability than the $S + X$ arm, and achieving CR was associated with longer PFS time in both arms, how could the $S + X$ arm have longer average PFS time than the S arm?

These results left the clinical investigators somewhat confused. How could this be? The answer to this conundrum is provided by applying the *Billion Dollar Computation*. It didn't actually save \$1,000,000,000, but it did help to avoid making a silly conclusion about the efficacy of $S + X$ compared to S. First, remember that the distribution of T in each arm, starting at day 90, is a mixture of two distributions, one for patients who achieved 90-day CR and another for patients who did not achieve 90-day CR. To compute the mixture distribution for each arm of the trial requires a bit more notation. For the four combinations of treatment arm $\tau = S$ or $S + X$, and whether 90-day CR was achieved (CR) or not (NCR), denote the probabilities that a patient survived at least 12 months past day 90 without disease progression by

$$\mathcal{F}_{\tau, CR} = \Pr(T > 12 \mid \tau, CR) \quad \text{and} \quad \mathcal{F}_{\tau, NCR} = \Pr(T > 12 \mid \tau, NCR).$$

The Billion Dollar Computation says that the distribution of T in each arm is a probability-weighted average of the distributions of T for patients who did and did not achieve 90-day CR,

$$\mathcal{F}_S = \mathcal{F}_{S, CR} \times \pi_S + \mathcal{F}_{S, NCR} \times (1 - \pi_S),$$

and similarly

$$\mathcal{F}_{S+X} = \mathcal{F}_{S+X, CR} \times \pi_{S+X} + \mathcal{F}_{S+X, NCR} \times (1 - \pi_{S+X}).$$

Plugging the estimators of all quantities obtained from the data into the above formulas gives

$$\widehat{\mathcal{F}}_S(12) = \widehat{\mathcal{F}}_{S, CR}(12) \times \pi_S + \widehat{\mathcal{F}}_{S, NCR}(12) \times (1 - \pi_S)$$
$$= 0.91 \times 0.406 + 0.68 \times (1 - 0.406) = 0.77,$$

and

$$\widehat{\mathcal{F}}_{S+X}(12) = \widehat{\mathcal{F}}_{S+X, CR}(12) \times \hat{\pi}_{S+X} + \widehat{\mathcal{F}}_{S+X, NCR}(12) \times (1 - \pi_{S+X})$$
$$= 0.98 \times 0.136 + 0.89 \times (1 - 0.136) = 0.90.$$

So, contrary to what might be expected based on intuition, while the $S + X$ arm had a lower estimated 90-day CR probability, and despite the fact that achieving CR at day 90 increased the 12-month survival probability in both arms, the $S + X$ arm did indeed have a larger estimated 12-month PFS probability.

Of course, a proper analysis should provide additional information about the uncertainties associated with these 12-month PFS estimators, so each estimator should be accompanied by a posterior credible interval. But that is beside the point. The results were not definitive, but they suggested that adding X to S might be beneficial in terms of PFS. The statistical analyses, including the mixture computation, were reported to the Institutional Review Board (IRB), along with a request to reopen the trial using a new design with PFS time as the primary outcome, a larger sample, and longer follow-up, hence a more reliable, more definitive comparison between the two arms in terms of PFS distributions. This request was granted by the IRB, and the trial was reopened and completed using a new PFS time-based design.

Things Learned From This Clinical Trial

1. The estimated 90-day CR probability was higher in the S arm compared to the $S + X$ arm. This difference actually was quite large, and could be considered significant from either a Bayesian or a frequentist point of view.

2. On average, patients who achieved 90-day CR has a longer PFS time than those who did not, within each treatment arm, so in general achieving CR was clinically beneficial in terms of PFS.

3. Despite the above two results, on average, the PFS time, either starting either at day 90 or starting at day 0, was longer in the $S + X$ arm compared to the S arm.

A general point is that what happened with this trial provides a concrete illustration of Lindley's warning. Probability computations often lead to answers that may be counterintuitive. Mixture computations of the sort given above are a simple but very common example. To provide some insight into what actually is going on, what people often miss is that the mixture component coming from the NCR patients actually has a much larger probability weight than the CR component for both treatments. About 60% of the S patients and 86% of the $S + X$ patients did not achieve CR, so the NCR components $\widehat{\mathcal{F}}_{S,NCR}(12) = 0.68$ and $\widehat{\mathcal{F}}_{S+X,NCR}(12) = 0.89$ get most of the weight in the mixture computations. Of course, this trial echoes the toy example given in Sect. 7.1

It is useful to consider what these examples say about the conventional phase II $\rightarrow$ phase III clinical trial paradigm. With the conventional approach, data on some early outcome like $Y =$ I[tumor response] obtained from a small to moderate-sized phase II trial is used to decide whether to conduct a phase III trial based on survival time or PFS time. The usual practical rationale for first conducting a phase II trial based on an early "response" outcome, is that, since Y can be observed relatively quickly, it is a feasible way to obtain preliminary data to decide whether E is sufficiently promising to conduct phase III. Moreover, a higher response rate is associated with longer mean survival time or mean PFS time. This gets at the idea of "surrogacy" of the early outcome in place of the later outcome. The idea is that, since phase III is based on survival or progression-free survival time, it is sensible to first evaluate the effect of E on an early surrogate outcome in a smaller trial that takes less time to complete. The computations given above show that, if you really care how long patients live, or how long they live without suffering a disease progression, conducting a phase II trial based on an early binary response outcome alone may be worse than useless. It may lead to a conclusion that is just plain wrong. Despite opinions to the contrary, there is no free lunch.

While I have characterized phase II trials as being based on an early binary response indicator, it is important to note that a fair number of investigators base their phase II trials on PFS time, or more generally time to disease worsening. The time to progression, or last follow-up time without progression, are used as the patient's outcome data. But a hidden issue now is that, in the revised design, PFS time is used as a surrogate for survival time. So the issue of surrogacy still arises.

Takeaway Messages:

1. The relationship between early response and survival time may be counterintuitive. A much higher response rate $\hat{p}_E > \hat{p}_S$ with E does not necessarily imply a large improvement in survival time or PFS time.
2. The $1,000,000,000 computation is just a start. In practice, things usually are a lot more complex.
3. The conventional approach of using an early outcome, such as $\geq 50\%$ tumor shrinkage, from a single-arm trial as the only basis for deciding whether to initiate a phase III trial is deeply flawed. It confounds treatment effects with trial effects, and it also ignores possible relationships between early outcomes and survival time.
4. In addition to early outcome data, survival time data or PFS time data also are needed in phase II. This, in turn, will require longer follow-up.
5. Typical phase II sample sizes are too small to make truly reliable inferences. Larger sample sizes are needed.
6. In early phase trials that use historical tumor response as an outcome, it is important to look for direct biological effects of E or S on survival time that are not mediated by tumor response.
7. To avoid confounding and biased E versus S comparisons, conduct a randomized phase II trial of E versus S, rather a singe-arm trial of E with comparison of its data to historical data on S, which instead might be used to help design this randomized trial.
8. When comparing E to S, do not treat estimators as if they are known parameter values. Account for uncertainty properly, by computing confidence intervals or posterior credible intervals, or doing a Bayesian analysis with comparisons and any other inferences based on posterior quantities.
9. Sometimes, doing what you believe is the right thing can lead you to make incorrect conclusions and take dysfunctional actions.

Chapter 8
Sex, Biomarkers, and Paradoxes

Who ya gonna believe, me or your own eyes?
Groucho Marx, as Rufus T. Firefly in the movie "Duck Soup"

Contents

Abstract This chapter will begin with an example of Simpson's Paradox, which may arise in tables of cross-classified categorical data. An explanation of how the paradox may arise will be given, and a method for using the tabulated data to compute correct statistical estimators that resolve the paradox will be illustrated. A second example will be given in the context of comparing the batting averages of two baseball players, where the paradox cannot be resolved. An example of cross-tabulated data on treatment, biomarker status, and response rates will be given in which there appears to be an interactive treatment–biomarker effect on response rate. This example will be elaborated by also including sex in the cross-classification, which leads to different conclusions about biomarker effects. A discussion of latent variables and causality will be given. Latent effects for numerical valued variables in the context of fitting regression models will be illustrated graphically. The importance of plotting scattergrams of raw data and examining possible covariate–subgroup interactions before fitting regression models will be illustrated. An example will be given of data where a fitted regression model shows that a patient covariate interacts with a between-treatment effect, with the consequence that which treatment is optimal depends on the patient's covariate value.

© Springer Nature Switzerland AG 2020   163
P. F. Thall, *Statistical Remedies for Medical Researchers*, Springer Series
in Pharmaceutical Statistics, https://doi.org/10.1007/978-3-030-43714-5_8

8.1 A Paradox

If you haven't seen an example of this paradox before, you might find it entertaining. Table 8.1 shows data on success rates for two treatments, A and B, cross-classified by the sex of the patient. Each cell of the table gives [number of responses]/[number of patients treated], followed by the resulting empirical response probability in parentheses. Comparing the response probabilities for each of the two sexes shows that B was better than A in men, and B also was better than A in women. But when sex is ignored and the data from men and women are combined, the last column shows that A had a higher overall response probability than B. So, if you just ignore sex, A was better than B in people. This is an example of the famous "Simpson's Paradox." It can occur quite easily when tabulating data to do comparisons and there are subgroups involved.

To find out what actually is going on with the data in this table will require us to engage in sex discrimination. First, notice that women had a much higher overall response rate of 72% ($100 \times 355/490$), compared to 48% ($100 \times 244/510$) for men. The next thing to notice is that 80% of the people who got treatment A were women, but only 13% of people who got treatment B were women. So the sample is heavily biased in favor of A over B, since A was given disproportionately to women, who had a higher overall response rate. That is, the comparison to A to B is unfair.

If what you actually want is to do is estimate the overall response rates for each of the two treatments, so that you can compare them fairly, here is a way to resolve the paradox. To do this, you must correct for the treatment assignment bias by properly weighting for the effects of sex. To do this, first stratify the sample by sex. Then compute the proportions of men and women in the entire sample, and use these as weights to compute the *sex-weighted estimator of the overall B versus A effect*. In what follows, everything is computed from data, so it is a statistical estimator, not a parameter:

Overall B versus A effect $=$

(B versus A effect in men) $\times$ Pr(man) + (B versus A effect in women) $\times$ Pr(woman)

$$= \left(\tfrac{200}{400} - \tfrac{44}{110}\right) \times \tfrac{510}{1000} + \left(\tfrac{54}{60} - \tfrac{301}{430}\right) \times \tfrac{490}{1000}$$

$$= (0.50 - 0.40) \times 0.51 + (0.90 - 0.70) \times 0.49 = 0.15.$$

Table 8.1 Comparing treatment success rates in men and women

	Man	Woman	Overall
Treatment			
A	44/110 (0.40)	301/430 (0.70)	345/540 (0.64)
B	200/400 (0.50)	54/60 (0.90)	254/460 (0.55)

So, once the estimation is done correctly, B actually was better than A in men, in women, and in people. Recalling the discussion in Chap. 6, this is just the causal estimate, obtained by using sex to define subgroups. I just left out all the detailed notation.

The key thing about the correct sex-weighted estimator given above is that Pr(male) and Pr(female) = $1 -$ Pr(male) were estimated using the overall sample proportions, $510/1000 = 0.51$ and $490/1000 = 0.49$. The innocent looking values $345/540 = 0.64$ and $254/460 = 0.55$ that appear in the last column of Table 8.1 actually are the culprits that cause the paradox. They are incorrect because they are computed using the within-treatment male/female weights, rather than the overall sample weights. In arm A, these weights are $110/(110 + 430) = 110/540 = 0.20$ for men and $430/(110 + 430) = 430/540 = 0.80$ for women. Similarly, in arm B, the weights are 0.87 for men and 0.13 for women. Looking at the arithmetic, the overall ratios are $(44 + 301)/(110 + 430) = 0.65$ for A and $(200 + 54)/(400 + 60) = 0.55$ for B. These are what you get by completely ignoring sex. But weighting in this incorrect way introduces sex-related treatment selection bias, so it gives the wrong answer. The point is that you must not only weight for sex, but to avoid sex-related bias you must weight correctly using the overall sample proportions.

An important thing to notice about this dataset is that it could not have come from a study in which the patients were randomized fairly between A and B. If this had been done, a simple computation using the binomial distribution with parameters $\pi = 1/2$ and $n = 490$ shows that the probability is nearly 0 that 490 women would be distributed as unevenly as 430 and 60 between the two treatments. Likewise, the subsample sizes of 110 and 400 for the 510 men also would be virtually impossible if they had been randomized fairly. So, the data must have been *observational*. When confronted by data like this, in addition to using stratum-weighted estimators to correct for this sort of biased treatment assignment, if possible you should find out *why* each of the two treatments was given to men and women with such different probabilities. Something suspicious is going on that caused sex-biased treatment assignment, but you can't figure out what it was by just looking at the table.

But wait, there's more. In the above example, the inferential goals were to compare the response rates of treatments A and B for men, for women, and for people. This required us to obtain estimators of various parameters, which I did not bother to name. These are the four conditional probabilities

$$\theta_{B,M} = Pr(Response \mid the\ subject\ was\ a\ Man\ treated\ with\ B)$$
$$\theta_{A,M} = Pr(Response \mid the\ subject\ was\ a\ Man\ treated\ with\ A)$$
$$\theta_{B,W} = Pr(Response \mid the\ subject\ was\ a\ Woman\ treated\ with\ B)$$
$$\theta_{A,W} = Pr(Response \mid the\ subject\ was\ a\ Woman\ treated\ with\ A).$$

Denote the sex probabilities by $\lambda_M = $ Pr(Man) and $\lambda_W = 1 - \lambda_M = $ Pr(Woman). Using this notation, the B versus A effect in men is $\theta_{B,M} - \theta_{A,M}$, the B versus A effect in women is $\theta_{B,W} - \theta_{A,W}$ and the gender-weighted average effect is

$$\theta_B - \theta_A = (\theta_{B,M} - \theta_{A,M})\lambda_M + (\theta_{B,W} - \theta_{A,W})\lambda_W.$$

So, $\theta_B - \theta_A$ is a parameter, which characterizes the B versus A effect in a population of people that the data in the table might be considered to represent. This is the causal parameter, and $\theta_{B,M} - \theta_{A,M}$ and $\theta_{B,W} - \theta_{A,W}$ are the respective gender-specific causal parameters. To obtain an estimate of $\theta_B - \theta_A$, I substituted the statistical estimators $\hat{\theta}_{B,M} = 200/400$, $\hat{\theta}_{A,M} = 44/110$, etc., computed from the data in the table. So, these are examples of causal estimators.

Takeaway Messages:

1. In a case where you want to estimate a comparative parameter, Simpson's Paradox can be solved by computing the subgroup-weighted (stratified) causal estimate.
2. Make sure that you compute the subgroup weights using the entire sample, and not the within-treatment subsamples.
3. It is important to keep in mind the idea that the estimator really is used to make inferences about a larger population of men and women who might be treated with A or B.

8.2 Batting Averages

No book on medical statistics can be complete without an example from baseball. Here is another way that Simpson's Paradox can arise, but in a case where the data in the table are all that matters, and there is no larger population to make inferences about. To see how this may work, just change the labels in the example so that A and B represent two baseball players, Alonzo and Bubba, and change "Man" to "First Half of the Baseball Season" and "Woman" to "Second Half of the Baseball Season." Now suppose that the numbers represent hits and at bats. For example, Alonzo had 16 hits in 100 at bats during the first half of the season, and so on. I have changed the numbers so that they look more like batting averages, but we have yet another case of Simpson's Paradox. Table 8.2 says that Bubba had a higher batting average than Alonzo in the first half of the season, 180 versus 160, and Bubba also had a higher batting average than Alonzo in the second half of the season, 330 versus 310. But, looking at the entire season without breaking it into halves, Alonzo had a higher batting average than Bubba, 260 versus 230. So, based on batting averages over the entire season, Alonzo's agent could argue that he should get a higher salary than Bubba. But using the separate batting averages in the two halves of the season, the team owner could argue that Alonzo should get a lower salary than Bubba.

Now that you know about Simpson's Paradox, you probably took a closer look at the data, and may have noticed that Alonzo and Bubba had very different numbers of at bats in the two halves of the season. There are lots of possible explanations. Maybe Bubba suffered an injury in midseason and he could not play as much in the second half, so Alonzo had to take up the slack and play a lot more games. But

Table 8.2 Comparing batting averages of Alonzo and Bubba. Each entry is [number of hits]/[number of at bats]

	Baseball season		
	First half	Second half	Overall
Alonzo	16/100 (0.160)	62/200 (0.310)	78/300 (0.260)
Bubba	36/200 (0.180)	33/100 (0.330)	69/300 (0.230)

maybe midseason was also when the team hired a new batting coach, and everybody's batting average went way up. Or maybe the difference in number of at bats was due to some coaching decision. Maybe batting averages went up because the players felt more motivated later in the season. A few at bats can make a big difference in a player's batting average, and these data don't reflect hits made in clutch situations, runs batted in, or whether a hit was a home run or a single. But unlike the treatment comparison example, the batting averages for Alonzo and Bubba are not estimators of some larger population. The table gives all of the hits and at bats data on Alonzo and Bubba for the season. So, it does not make sense to stratify on the two halves of the season and compute a weighted batting average. For the season, Alonzo actually batted 260 and Bubba actually batted 230. Simpson's paradox still applies. Even the sacred game of baseball is not immune to this counterintuitive numerical paradox.

8.3 A Magic Biomarker

Since this is not a book about baseball, I will return to the knotty problem of treatment evaluation. In the new age of targeted therapy, all sorts of biomarkers have become important variables. The idea is that a patient's biomarkers may be used to tailor their treatment, which now is called *Precision Medicine*. Laboratory researchers, pharmaceutical marketing reps, and science reporters love to use this phrase. A "biomarker" may identify a particular genetic abnormality, the expression level of a protein produced by a particular gene, a signature involving several genes that has been identified using microarrays, or the presence of a cell surface marker believed to be associated with the disease being treated. For example, in oncology, an experimental treatment may be a molecule designed to kill a particular type of cancer cell by bonding to a receptor on the cell's surface to disrupt its activity and thus kill it, and the biomarker identifies whether a cell has the receptor. In some settings, there may be thousands of biomarkers. For the sake of illustration, consider the simplest case of one biomarker that is either positive (present) or negative (absent) in each patient. An illustrative dataset is given in Table 8.3.

The table appears to provide very exciting results. Treatment A has a higher response rate in biomarker positive patients (65% versus 50%), while B has a higher response rate in biomarker negative patients (71% versus 55%). This tells physicians how they might personalize treatment for patients with the disease, to help stack

Table 8.3 Comparing treatment success rates in biomarker subgroups

	Biomarker		
	Positive	Negative	Overall
Treatment			
A	312/480 (0.65)	33/60 (0.55)	345/540 (0.64)
B	179/355 (0.50)	75/105 (0.71)	254/460 (0.55)

the deck in favor of the patient based on their biomarker. For each patient, first evaluate their biomarker, and then use that information to choose the treatment that is more likely to achieve a response. If this exciting discovery leads to development of a molecule to target the biomarker and kill the cancer cell, the biomarker might seem like magic to a person with the disease or a practicing physician. For whoever discovered the magic biomarker, maybe it is time to shop for a tuxedo.

But let's take a closer look at the numbers. The fact that you know about Simpson's Paradox doesn't help here, since correctly estimating the overall response rate is beside the point, which is that there is a treatment–biomarker interaction. To help see what is going on, Table 8.4 gives additional information, in the form of a more refined classification of these data that accounts for three factors: treatment, biomarker, and a third variable, sex. A bit of detective work shows that the biomarker actually is sexy, since 329/490 (67%) of women are biomarker +, compared to only 162/510 (32%) of men. This new table actually is a refinement of both Table 8.3 and the original Table 8.1, which can be obtained from Table 8.4 by ignoring either the biomarker or sex. So, this new table contains information about sex, biomarker, and treatment, in a three-way cross-tabulation. To summarize, here is we know at this point:

1. The treatment × biomarker cross-tabulation in Table 8.3 ignores sex.
2. Women had a much higher response rate than men, regardless of treatment, 72% versus 48%.
3. Women were much more likely to be biomarker + than men, 93% versus 75%. This might be due to some sort of sex-linked biology having to do with X and Y chromosomes, but exploring that would go far beyond what is in the more elaborate Table 8.4.
4. The data obviously did not come from a randomized study. Some sort of sex discrimination was used to assign treatments.

The apparent treatment × biomarker interaction in Table 8.3 may not be magic after all, since the three-way classification in Table 8.4 shows a strong sex–biomarker interaction.

If you have been keeping track of general principles, you may have noticed that I did something very misleading in *both* Table 8.3 and Table 8.4. In each table, the proportions given in last column labeled "Overall" are computed by weighting the subgroup columns incorrectly. In Table 8.3, the data from the biomarker Positive and Negative columns are weighted incorrectly when computing the ratios for *A* and *B* in the "Overall" column. Similarly, in Table 8.4, the data from the Men and Women columns are weighted incorrectly when computing the ratios in the "Overall" column for each of the (Treatment, Biomarker) subgroups. So, these tables lead to incorrect

Table 8.4 Comparing treatment success rates in all four biomarker × sex subgroups

Treatment	Biomarker	Men	Women	Overall
A	+	32/80 (0.40)	280/400 (0.70)	312/480 (0.65)
A	−	12/30 (0.40)	21/30 (0.70)	33/60 (0.66)
B	+	130/300 (0.43)	49/55 (0.89)	179/355 (0.50)
B	−	70/100 (0.70)	5/5 (1.00)	75/105 (0.71)

"Overall" estimates, since they incorporate sampling bias. This mistake may not be as obvious as it is when you get a ridiculous answer as in Simpson's Paradox in Table 8.1, but the same weighting mistake is being made. If you caught this, then good for you. If you didn't catch it, don't feel bad. The point is that you need to understand mistakes in order to learn how to avoid them. Now we can think about how to analyze and interpret the data correctly.

If all you have is Table 8.3, and you do not know the sexes of the patients, then sex is a *lurking variable* or *latent variable*. It explains what appears to be a biomarker effect, but you don't know that because the variable is not in your data. A bit of thought shows that, by the same reasoning, if you only have Table 8.1, which ignores the biomarker, then the biomarker might be the lurking variable that explains the apparent sex effect. This illustrates a very important general point about how people often summarize data. If someone ignores one of the variables in Table 8.4 by collapsing it into a simpler table, as if they never had the variable in the first place, then it is very easy to be misled. But people do this sort of thing all the time when analyzing data.

This also illustrates a very deep problem that may occur in almost any scientific study where you want to make inferences about relationships between variables. If Table 8.1 is all that you have, and you don't know that the biomarker exists, then it looks like there really is an interaction between sex and treatment in terms of response rate. Figure 8.1 illustrates all of the possible causal pathways among treatment, sex, biomarker, and clinical outcome. If you know about all of these variables and none of them are unknown factors lurking in the background, you still need a fundamental reason to explain why being female causes a higher response rate regardless of treatment. If not, then inferentially you at least should estimate the overall response rate by weighting males and females correctly. In general, the larger problem is that there always may be one or more lurking variables, also known as "external confounders," that you simply don't know about. Unfortunately, what you don't know can kill you. This is why a fundamental assumption in causal inference is that there are no external confounders, since otherwise one cannot make any progress inferentially. This was one of the important assumptions given in Chap. 6. Of course, it is impossible to verify the assumption of no external confounders. If you want to establish causality, you must make this and other key assumptions as articles of faith. Yasuo Ohashi, who served as Chairman of the Department of Biostatistics at the University of Tokyo, once mentioned to me that, to him, the field of casual inference seemed to be like a religion.

Fig. 8.1 Possible causal
pathways between treatment,
gender, biomarker, and
clinical outcome

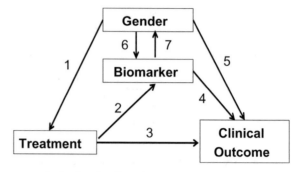

1 = Gender-biased treatment assignment

2,3 = Actual treatment effects

4,5 = Gender and biomarker affect outcome

6,7 = The Biomarker is sexy

Table 8.5 Rearranging the data by (Biomarker, Sex) subgroups to facilitate a stratified analysis

Stratum		Treatment		
Biomarker	Sex	A	B	Stratum prob.
+	Men	32/80 (0.40)	130/300 (0.43)	0.380
−	Men	12/30 (0.40)	70/100 (0.70)	0.130
+	Women	280/400 (0.70)	49/55 (0.89)	0.455
−	Women	21/30 (0.70)	5/5 (1.00)	0.035

A useful way to rearrange Table 8.4 is in terms of the four (Biomarker, Sex) combinations, as the rows of the table, and treat these as strata. Summarizing the data this way facilitates a stratified analysis to estimate the overall B versus A effect, so it is useful to compute and include the correct stratum probability estimates. This is given in Table 8.5. To obtain an overall estimator of the B versus A effect, you can use the same sort of weighting method applied before where there were two subgroups, men and women. In Table 8.5, there are four (Sex, Biomarker) subgroups (strata), but the method is the same. From the table, the causal (Sex, Biomarker) weighted estimate of the overall A versus B effect is

$$(B \text{ versus } A \text{ effect in } + \text{Men}) \times \Pr(\text{Biomarker} +, \text{Man}) +$$

$$(B \text{ versus } A \text{ effect in } - \text{Men}) \times \Pr(\text{Biomarker} -, \text{Man}) +$$

$$(B \text{ versus } A \text{ effect in } + \text{Women}) \times \Pr(\text{Biomarker} +, \text{Woman}) +$$

$$(B \text{ versus } A \text{ effect in } - \text{Women}) \times \Pr(\text{Biomarker} -, \text{Woman})$$

$$= \left(\frac{130}{300} - \frac{32}{80}\right) \times .380 + \left(\frac{70}{310} - \frac{12}{30}\right) \times 0.130 + \left(\frac{49}{55} - \frac{280}{400}\right) \times 0.455+$$
$$\left(\frac{5}{5} - \frac{21}{30}\right) \times .035 = 0.087.$$

Here are some general takeaway messages from this example:

Takeaway Messages:

1. The paradox can arise from non-randomized data where there is a strong association between some variable, like sex, that is unbalanced between treatments, and the outcome.
2. If you are designing a comparative experiment, whenever possible randomize subjects fairly between treatments since, on average, it balances both known and lurking variables between treatments.
3. To compare treatments fairly based on observational data, if possible, compute stratum-weighted estimates of treatment effect.
4. The paradox can arise from the effects of one or more latent variables that you don't know about.
5. If you have observational data, in your statistical analyses do not assume or pretend that it came from a randomized study.
6. If you ignore a variable, for example, by collapsing a three-way classification to a two-way table, effects of the variable that you have ignored may appear to be effects of the variables that you have retained.

8.4 Plotting Regression Data

There is a saying that most of statistics either is regression, or would like to be. The field of graphical methods and residual analyses to determine functional forms and assess goodness-of-fit in regression modeling is immense. Some books on regression graphics are Cook (1998) and Atkinson and Riani (2000), and other books are Ibrahim et al. (2001) on Bayesian survival regression, Royston and Sauerbrei (2008) on fractional polynomial regression models, and general treatments of a variety of different regression models by Harrell (2001) and Vittinghoff et al. (2005).

In this section, I will give some simple examples of plots of regression data that can provide useful visual guidance for formulating a model. The ideas are very similar to those described above for categorical variables, but illustrated here graphically for quantitative covariates. The main point is that easily constructed plots can reveal patterns that suggest functional relationships between an outcome variable Y and a covariate X when formulating a regression model, and that using plots to explore possible treatment–covariate interactions can help you avoid basing inferences on a regression model that may be completely wrong.

It is well known that, after a given dataset has been used to determine a model that fits the data well, say by selecting a subset of predictive covariates or transforming some covariates in a multiple regression model, the selected model invariably does a worse job of prediction when applied to a new dataset having the same form. That is, a "best" fitted model is only best for the dataset that was used to determine it. A more proper way to proceed is to use some form of cross-validation. In its simplest form, one might randomly separate a dataset into two halves, apply graphical analyses and other model determination methods in one half and then fit the resulting regression model in the other half. Then reverse the roles of the two halves, repeat the process, and average the results. There is an immense literature on assessing model accuracy, cross-validation, and model selection methods, including K-fold cross-validation, "leave one observation out" (jackknife) methods, and the bootstrap, Efron (1979). A very thorough account is given in the book by James et al. (2013). Here, I will only give a few simple examples of ways to take a preliminary graphical look at a regression dataset, with the goal to avoid a model that is grossly wrong.

Figure 8.2 gives an example of why you should plot your data before fitting a regression model that includes one or more numerical covariates. In this example, the data have the simple form $(X_1, Y_1), \ldots, (X_n, Y_n)$. The scattergram shows that Y obviously is a decreasing, curvilinear function of X. The pattern is obvious once you draw the scattergram. But if someone does not bother to do this and just assumes the usual linear model $Y_i = \beta_0 + \beta_1 X_i + \epsilon_i$, $i = 1, \ldots, n$ with independent $N(0, \sigma^2)$ error terms $\{\epsilon_i\}$, then the plot given on the left shows that the fitted straight line $\hat{Y} = \hat{\beta}_0 + \hat{\beta}_1 X$ does a very poor job of describing the data. It ignores that fact that, for small X_i, the outcome variable Y_i takes on much larger values than the linear model would predict. The plot on the right shows a fitted quadratic line, $\hat{Y} = \hat{\beta}_0 + \hat{\beta}_1 X + \hat{\beta}_1 X^2$, which fits the data much more accurately. This is not rocket science, but the point is that if you don't plot your regression data, you may miss some obvious pattern and fit a model that clearly is wrong. Other nonlinear functions also should give a good fit, but a quadratic function works well here. A useful refinement of the model is to first center X on its sample mean $\overline{X}$ and divide by the sample standard deviation s_X. This helps reduce collinearity between the X and X^2 terms, which can produce an unstable fitted model. So, one would fit the model $Y_i = \beta_0 + \beta_1 (X_i - \overline{X})/s_X + \beta_2 \{(X_i - \overline{X})/s_X\}^2 + \epsilon_i$ for independent $N(0, \sigma^2)$ residuals $\{\epsilon_i\}$. Centering and scaling this way works generally for any order polynomial to reduce collinearity. A further refinement is motivated, for this example, by the observation that the scattergram suggests the variance of Y_i may increase with X_i. So it might be useful to examine the fit of a model in which you assume a variance function such as $\text{var}(Y_i \mid X_i) = \text{var}(\epsilon_i \mid X_i) = |X_i|\sigma^2$ or maybe $\sqrt{|X_i|}\sigma^2$, and use weighted least squares to obtain a fitted model.

Figure 8.3 shows a scattergram with the fitted least squares straight line, which shows that Y may increase very slightly with X, but if there is any actual relationship it is very weak. The small positive slope easily could be due to random variation in the data. In this example, there is a third, lurking binary variable, $Z = 0$ or 1, that identifies two subgroups, but it is not identified in Fig. 8.3 since the linear regression model assumes $E(Y \mid X, Z) = \beta_0 + \beta_1 X$ and thus ignores Z. A regression model that accounts for both Z and X, and includes interaction between X and Z, is

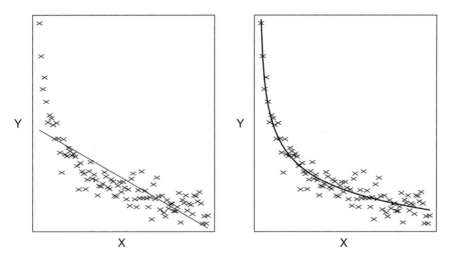

Fig. 8.2 Scattergram of a dataset with the least squares fitted straight line (left) and the fitted quadratic line (right)

$$E(Y \mid X, Z) = \beta_0 + \beta_1 X + \beta_2 Z + \beta_3 X Z.$$

Under this more general model, the line for $Z = 1$ is $E(Y \mid X, Z = 1) = (\beta_0 + \beta_2) + (\beta_1 + \beta_3)X$, and the line for $Z = 0$ is $E(Y \mid X, Z = 0) = \beta_0 + \beta_1 X$. This implies that the comparative effect of $Z = 1$ versus $Z = 0$ on Y is $E(Y \mid X, Z = 1) - E(Y \mid X, Z = 0) = \beta_2 + \beta_3 X$, which is a function of X. Figure 8.4 shows the same scattergram, but using different plotting symbols for the (X_i, Y_i) pairs for $Z_i = 1$ or 0. The two fitted lines obtained from this second model are very different from the single line in Fig. 8.3 from the first model, showing that Y decreases with X within each subgroup, and moreover, the two subgroup-specific lines are very different from each other. For example, if $X = $ Age and Z identifies two different treatments, with a higher value of Y corresponding to a more desirable outcome, then Fig. 8.4 shows that (1) there is a clear superiority of one treatment over the other, and (2) treatment benefit decreases with older Age for both treatments.

If one were to analyze this dataset by fitting *both* the line assuming $E(Y \mid X, Z) = \beta_0 + \beta_1 X$ and the line assuming $E(Y \mid X, Z) = \beta_0 + \beta_1 X + \beta_2 Z + \beta_3 X Z$, then these two fitted models, taken together, might lead to the apparently paradoxical conclusions that (1) Y decreases with X for $Z = 1$, (2) Y decreases with X for $Z = 0$, and these are two different lines, but (3) if Z is ignored then Y increases linearly with X. This is a more elaborate version of Simpson's Paradox, but for numerical valued variables rather than categorical variables, in the presence of a binary indicator Z for treatment or subgroup. Of course, the resolution of this apparent paradox is simply that it is impossible for both models to be correct. Figure 8.4 strongly suggests that the model including the interaction term $\beta_3 X Z$ fits the data much more closely than the model that excludes Z. The fact that ignoring Z leads to an erroneous conclusion should not be surprising.

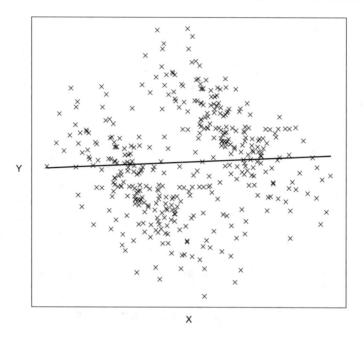

Fig. 8.3 Scattergram of dataset 2 with one fitted least squares straight line

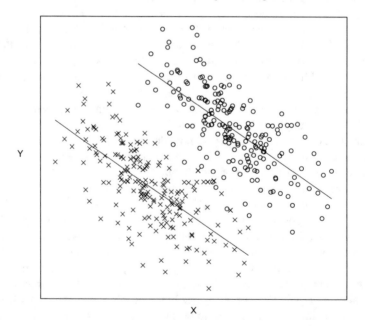

Fig. 8.4 Scattergram of the second dataset, identifying the two subgroups by using different plotting symbols, with the two subgroup-specific fitted least squares straight lines

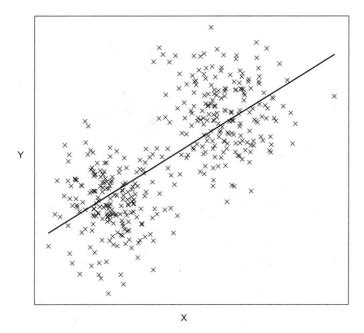

Fig. 8.5 Scattergram of the third dataset, with one fitted least squares straight line

Figure 8.5 shows a scattergram for a third dataset, with a fitted least squares straight line showing that Y increases steeply with X. But, as in the previous example, there is a lurking binary variable $Z = 0$ or 1. If a model with an X-by-Z interaction taking the same form as that given above is fit, then Fig. 8.6 shows that the subgroup-specific fitted lines indicate X and Y actually have no relationship within either subgroup. The apparent relationship shown in Fig. 8.5 is due entirely to the lurking variable Z. This sort of thing can occur, for example, if a laboratory experiment is repeated on two different days, and the values of X and Y both increase on the second day, possibly due to the way a machine that measured X and Y was calibrated, or there was some lurking variable like a higher room temperature on the second day that affected both X and Y. A famous example of this type of lurking variable effect led to the belief in the 1940s that drinking soda pop caused polio. If X = soda pop consumption and Y = incidence of polio, then Fig. 8.5 may seem to indicate that higher soda pop consumption caused an increase in the incidence of polio. But if Z identifies the seasons summer (upper right) and winter (lower left), then once one accounts for the season effect, Fig. 8.6 shows that soda pop consumption and polio incidence both were higher in warmer weather, but there was no relationship between them during either season. So, the apparent relationship in Fig. 8.5 actually was a seasonal effect acting on both soda pop consumption and incidence of polio. Soda pop consumption was an Innocent Bystander. It also illustrates the important general scientific fact that association, apparent or real, does not imply causation.

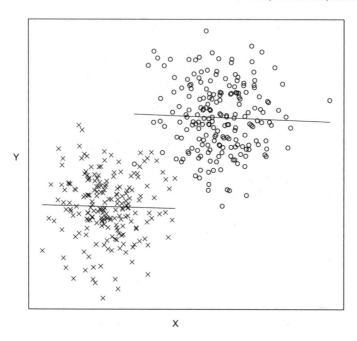

Fig. 8.6 Scattergram of the third dataset, identifying the two subgroups by using different plotting symbols, with two subgroup-specific fitted least squares straight lines

Figure 8.7 shows plots of $Y =$ percent cells killed on $X =$ dose, for each of four replications of the same experiment, in a case where X and Y are completely unrelated. That is, the agent being evaluated has no effect on percent cells killed, so varying its dose does not matter. In this case, Y is just the baseline cell death rate for the particular experimental setup, and it varies randomly around an average rate of about 50%. In other words, Y is completely random noise. But if you repeat an experiment enough times, eventually you will see a pattern, purely due to the play of chance. This is another example of the "Ducks and Bunnies" effect. The lower right scattergram and fitted least squares line in Fig. 8.8 shows a replication of the experiment where the fitted least squares line happened to have a positive slope. If whoever repeated this experiment cherry-picks this outcome and publishes the plot, without mentioning the other preceding repetitions of the experiment, and claims that percent cells killed increases with dose of the agent, then simple random variation is being misinterpreted as an actual dose effect. This is an example of a reported experimental effect that other researchers will not be able to replicate in the future, unless they proceed similarly by repeating the experiment until they obtain a fitted (dose, percent cells killed) line that has a positive slope. This behavior may be motivated by the honest belief that someone who repeats an experiment must be getting better at it with each repetition, so each successive result must be more reliable than those that preceded it. This is like someone who repeatedly flips a coin

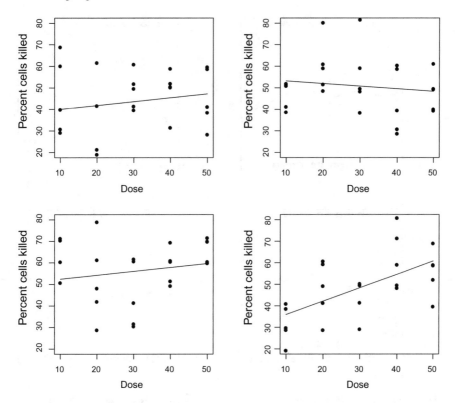

Fig. 8.7 Four scattergrams of data obtained from repeating an experiment in which percent cells killed is completely unrelated to dose

until they get 10 heads in a row, and then says "This coin must be unbalanced, because I flipped it 10 times in a row and got a head every time."

Figure 8.8 gives (X, Y) scattergrams of four different datasets, where X is numerical valued but the outcome is the binary variable $Y = 0$ or 1, that is, the indicator of whether an event, like treatment response, occurred or not for each subject. This is the common data structure in which a logistic or probit regression model usually is fit to estimate $\Pr(Y = 1 \mid X)$. For each dataset in Fig. 8.8, the local smoother LOWESS (locally weighted scatterplot smoothing) of Cleveland (1979) was fit, and the smoothed line was drawn along with the scattergram to provide a visual representation of how $\pi(X) = \Pr(Y = 1 \mid X)$ varies with X. The point of this example is very simple. Without using some sort of local smoother, like LOWESS, just looking at a plot of 0's and 1's on X gives almost no idea of what the probability function $\pi(X)$ looks like. It is important to bear in mind that using smoothing plots of this sort to figure out what the form of the linear term $\eta(X)$, such as logit$\{\pi(X)\}$, should be in a particular model may be a bit tricky, since $\pi(X)$ is a probability but $\eta(X)$ is a real-valued function of X that determines $\pi(X)$. In the upper left plot of Fig. 8.8, there does not appear to be much pattern, so it seems that Y probably does not vary

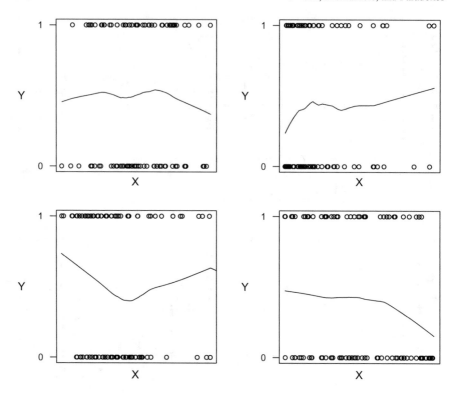

Fig. 8.8 Smoothed scattergrams of four different simulated datasets with binary outcome $Y = 0$ or 1 and continuous covariate X

with X. The upper right plot shows that $\pi(X)$ increases with X, but the function may not be linear in X. The lower left plot suggests that $\pi(X)$ may be non-monotone in X, and the lower right plots suggest that $\pi(X)$ is decreasing in X. These plots illustrate how a non-model-based smoother may be used to provide a preliminary diagnostic plot to suggest what sort of parametric function should be fit. Some statisticians argue that the locally smoothed plot is all that is required, rather than using it as a tool to help determine a parametric model. Drawing a picture of regression data with binary outcomes to compute and plot an estimate of $\pi(X)$ is very easy to do using any of a variety of widely available local smoothers. Another method for dealing with binary outcome plots is to "jitter" each Y_i by replacing it with $Y_i + u_i$ where the u_i's are small, simulated $N(0, \sigma^2)$ variables for some small fixed variance σ^2. Another graphical approach is to start by fitting a simple logit-linear parametric model with $\eta(X) = \beta_0 + \beta_1 X$, plot the residuals $r_i = (Y_i - \hat{\pi}_i)/\{\hat{\pi}_i(1 - \hat{\pi}_i)\}^{1/2}$ on X_i for $i = 1, \ldots, n$, create a locally smoothed line based on the residual plot, use this plot to help identify how X might be transformed, and possibly modify the parametric form of $\eta(X)$ on that basis. These are part of an extremely large array of models and methods for fitting binary outcome and other generalized linear models (GLMs). Some useful papers on GLMs are Nelder and Wedderburn (1972), Pregibon

(1980), Pregibon (1981), Fowlkes (1987), and Breslow and Clayton (1993), and the books by McCullagh and Nelder (1989), Dobson and Barnett (2008), and Dey et al. (2000).

In the next example, the regression data consist of 300 simulated values of a binary response indicator Y, $X_1 = $ age, $X_2 = $ (creatinine level $- 140$), and a binary indicator $Z = 1$ if the treatment was B and $Z = 0$ if A. The regression model is for the probability of response, $\pi(X, Z)$, for a patient with prognostic variables $X = (X_1, X_2)$ treated with Z. The data were simulated for ages ranging from 19 to 65, and creatinine values ranging from 110 to 150. The model used to simulate the data was $\mathrm{logit}\{\pi(X, Z, \epsilon)\} = \eta + \epsilon$, where

$$\eta = \beta_0 + \beta_1 X_1 + \beta_2 X_2 + \beta_3 Z + \beta_4 X_2 Z \tag{8.1}$$

using the numerical values $\beta_0 = -19.5$, $\beta_1 = 0.5$, $\beta_2 = 0.05$, $\beta_3 = 0.15$, $\beta_4 = 0.15$, with the error term $\epsilon \sim N(0,5)$ added to increase the variability in the data. To generate each binary outcome, values of age, creatinine, and ϵ_i first were simulated, the resulting $\pi(X_i, Z_i, \beta, \epsilon_i) = \exp(\eta_i + \epsilon_i)/[1 + \exp(\eta_i + \epsilon_i)]$ was computed using the fixed parameter values, and then the outcome was determined as $Y_i = 1$ if $\pi(X_i, Z, \epsilon) > 0.5$ and $Y_i = 0$ otherwise.

Under the "true" model with all $\epsilon_i \equiv 0$, due to the treatment–creatinine interaction, the B versus A treatment effect is

$$\Delta(X_2) = \mathrm{logit}\{\pi(X, 1)\} - \mathrm{logit}\{\pi(X, 0)\} = \beta_3 + \beta_4 X_2.$$

Thus, the treatment effect $\Delta(X_2)$ depends on the patient's creatinine level. For this simulated dataset, using the assumed true parameter values, since $\beta_3 = \beta_4 = 0.15$, $\Delta(X_2) = 0$ if creatine $= 139$. Thus, for patients with creatinine level below 139 the B versus A treatment effect $\Delta(X_2) < 0$, so A is better than B. If creatinine level is above 139 then $\Delta(X_2) > 0$ so B is better than A.

Of course, in practice you do not know the true model and parameter values, so you must proceed statistically by making inferences based on the available data. To do this with the simulated dataset in this illustration, two Bayesian logistic models of the form $\mathrm{logit}\{\pi(X, Z)\} = \eta$ were fit, assuming a $N(0, 5)$ prior on each β_j. The first model assumes the correct linear term given by Eq. (8.1) that includes a treatment–creatinine interaction term, and the second model incorrectly assumes the simpler additive model $\eta = \beta_0 + \beta_1 X_1 + \beta_2 X_2 + \beta_3 Z$. Under the interaction model, the posterior probability that B is superior to A for a patient with creatinine variable X_2 is

$$\Pr(B \text{ is better than } A \mid X_2, \ data) = \Pr(\beta_3 + \beta_4 X_2 > 0 \mid data).$$

A larger value of this posterior probability corresponds to greater superiority of B over A for a patient with creatine variable X_2. To illustrate this, Table 8.6 gives this probability, and the corresponding 95% posterior credible interval for the B versus A effect $\beta_3 + \beta_4 X_2$, for each of a several creatinine levels, which also are plotted in Fig. 8.9.

Table 8.6 Posterior probabilities that treatment B is superior to A, and 95% credible intervals for the B versus A effect, $\beta_3 + \beta_4 X_2$, as a function of creatinine level

Creatinine Level	Pr(B is superior to A)	95% posterior ci for $\beta_3 + \beta_4 X_2$
110	0.004	−2.754, −0.401
120	0.032	−1.940, 0.063
130	0.301	−1.423, 0.824
140	0.669	−1.164, 1.854
150	0.832	−0.997, 3.022

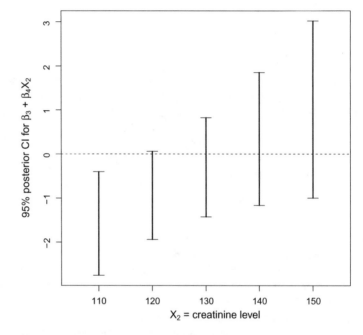

Fig. 8.9 Posterior 95% credible intervals for the comparative B versus A treatment effect as a function of creatinine level

If one assumes the simpler additive model, or simply ignores X entirely, the fact that treatment interacts with creatinine level in its effect on the probability of response will be missed. Under the simpler additive model, the B versus A treatment effect is the coefficient β_3 of Z, and the posterior probability that B is superior to A is $\Pr(\beta_3 > 0 \mid data) = 0.027$. Since this posterior probability is very small, corresponding to 37 to 1 odds in favor of A over B, the inference under the incorrect model would be that A is very likely to be a much better treatment than B for all patients. This conclusion, if published, could lead physicians to give a treatment that is suboptimal to patients with high creatinine levels, making it more likely that those patients will fail to respond and consequently die. This example, while hypothetical, illustrates how an incorrectly simplified statistical model, and the resulting incorrect inferences, might harm patients and possibly cost them their lives. Most practicing physicians are aware of settings where the effects of a particular treatment may change and possibly be harmful for certain subgroups of patients. The above example is a statistical formalization of how this sort of knowledge may be obtained inferentially from regression data.

It is important to bear in mind that, if data of the above form are not from a randomized study, but rather are observational, then any of the analyses given above may be misleading due to treatment selection bias. Ideally, some method for bias correction, as discussed in Chap. 6, should be applied.

Takeaway Messages

1. Determining a regression model that fits a given dataset reasonably well is very important. Preliminary plots and goodness-of-fit analyses for fitted models should be part of the process.
2. It may be a serious mistake to assume that patients are homogeneous if they are not, since this may lead to incorrect inferences about treatment effects. What is good for one patient may be bad for another.
3. For any regression model involving treatment indicators and numerical or categorical covariates, it is worthwhile to examine fitted models that include possible treatment–covariate interactions.
4. Regression analysis is not just a *pro forma* process. A flawed data analysis and resulting incorrect treatment evaluation can kill patients.

Chapter 9
Crippling New Treatments

> *The conventional view serves to protect us from the painful job of thinking.*
> John Kenneth Galbraith

Contents

Abstract Conventional phase I clinical trials, in which a dose is chosen using adaptive decision rules based on toxicity but ignoring efficacy, are fundamentally flawed. This chapter will provide several illustrations of this important fact. The worst class of phase I "toxicity only" designs are so-called $3 + 3$ algorithms, which are widely used but have terrible properties. Regardless of methodology, the conventionally small sample sizes of phase I trials provide very unreliable inferences about the relationship between dose and the risk of toxicity. More generally, the paradigm of first doing dose-finding in a phase I trial based on toxicity, and then doing efficacy evaluation in a phase II trial, is fundamentally flawed. This chapter will provide numerical illustrations of all of these problems. It will be explained, and illustrated by example, why the class of phase I–II trials, which are based on both efficacy and toxicity, provide a greatly superior general alternative to the conventional phase I $\rightarrow$ phase II paradigm. The EffTox design of Thall and Cook (2004) and Thall et al. (2014a) will be reviewed and compared to a $3 + 3$ algorithm and the continual reassessment method of O'Quigley et al. (1990) by computer simulation. It will be argued that, because conventional methods do a very poor job of identifying a safe and effective dose, it is very likely that, at the start of clinical evaluation, they cripple many new treatments by choosing a suboptimal dose that is unsafe, ineffective, or both.

© Springer Nature Switzerland AG 2020
P. F. Thall, *Statistical Remedies for Medical Researchers*, Springer Series
in Pharmaceutical Statistics, https://doi.org/10.1007/978-3-030-43714-5_9

9.1   Phase I Trials

Phase I trials first arose years ago in oncology when it was recognized that most chemotherapies have severe adverse side effects, that is, they are cytotoxic. Essentially, chemotherapy is poison that kills cells, which means that it kills both cancer cells and non-cancer cells. The basic idea motivating the use of chemotherapy in oncology is that, hopefully, it will kill enough cancer cells to bring the patient's cancer into remission, or achieve a "response." Since both Pr(response | d) and Pr(toxicity | d) increase with d = dose of a cytotoxic agent, it is good to have a higher dose in order to achieve higher Pr(response | d), but the dose cannot be so high that Pr(toxicity | d) is unacceptably high. In oncology and other areas of medical research, the primary goal of a conventional phase I trial is to find a so-called *maximum tolerable dose* (MTD), which is a dose that may be considered to have an acceptable level of toxicity. Unfortunately, it became conventional practice to take the extremely myopic strategy of first determining an MTD based on toxicity only in phase I, and later evaluating Pr(response | d = MTD) in phase II. As the examples given in the following text will show, this conventional phase I → phase II strategy is terrible.

To make matters even worse, the most widely used phase I methods are $3 + 3$ algorithms (Storer 1989), which have many different variants, but appear to make sense intuitively and seem to be simple to apply. However, numerous computer simulation studies have shown, again and again, that all $3 + 3$ algorithms have very bad properties when evaluated under a range of possible dose–toxicity scenarios. But you don't really need to do computer simulations to identify the flaws in $3 + 3$ algorithms. Discussions of the properties of various $3 + 3$ or other algorithm-based phase I designs, and comparisons to model-based phase I designs, are given, for example, by Korn et al. (1994), Lin and Shih (2001), Jaki et al. (2013), and Yan et al. (2018). Love et al. (2017) discuss how dose-finding designs are chosen and applied in practice, and argue in favor of model-based designs over algorithm-based designs like the $3 + 3$. Given their widespread use in phase I trials, however, it appears that the extremely poor properties of $3 + 3$ designs are not well understood in the medical oncology community, or maybe they simply are ignored. Since there are many different $3 + 3$ algorithms, to make the discussion precise, I will use the commonly used version of this family of algorithms given in Table 9.1. If you attempt to evaluate the properties of the algorithm in Table 9.1, or of any similar $3 + 3$ algorithm, by computer simulation, you quickly discover that the table does not provide enough detail to allow you to write a computer program that implements it. Many decisions that must be made in practice during trial conduct are left unspecified. Thus, it is impossible to carry out computer simulations to evaluate the algorithm's properties. To deal with this problem, I have given some additional General Rules needed to complete the specification of this $3 + 3$ algorithm.

Table 9.1 A 3 + 3 algorithm for phase I clinical trial conduct, subject to the General Rules

# Toxicities/ # Patients	Decision
0/3	Escalate one dose level, if allowed by General Rule 1, otherwise treat three more patients at the current dose level
0/3 + [0/3 or 1/3]	Escalate one dose level, if allowed by General Rule 1, otherwise treat three more patients at the current dose level
0/3 + [2/3 or 3/3]	Stop, choose one dose level lower than the current dose as the MTD, if allowed by General Rule 4
1/3	Treat three more at the current dose level
1/3 + 0/3	Escalate one dose level if allowed by General Rule 1, otherwise choose the current dose level as the MTD, if allowed by General Rule 4
1/3 + [1/3 or 2/3 or 3/3]	Stop, choose one dose level lower than the current dose as the MTD, if allowed by General Rule 4
2/3 or 3/3	De-escalate one dose level

Additional General Rules to Accompany 3 + 3 Algorithms

1. Never reescalate to a dose level after de-escalating from that dose level.
2. If the decision is to de-escalate or choose one dose level lower than the current dose level, but the current dose the level is the lowest being considered, stop and do not choose any dose level.
3. If the decision is to escalate above the highest dose level, stop and choose no dose level.
4. If the decision is to stop and choose one level below the current dose level, but only three patients have been treated at the next lower dose level, then treat three more patients at the next lower level.

These general rules can be changed, and in practice they often are. For example, if someone doesn't like General Rule 3, they might declare the highest dose level to be the MTD. If they don't like General Rule 4, then maybe they will treat 6 more patients at next lower level. There are numerous different versions of this algorithm. Most phase I protocols do not give nearly the amount of detail that I have provided above, so an actual algorithm often is not fully specified, and many decisions must be made subjectively during the trial. But if one wishes to evaluate the operating characteristics of a 3 + 3 algorithm, in order to write a computer program that can accommodate all possible cases that many arise, a decision must be specified and codified for each possible case.

The first major problem with 3 + 3 algorithms is that they are decision rules without any objective criterion for what constitutes a "good" or "safe" dose. If you look at the table closely, it does not say what to do in several possible cases. That is why I added the "General Rules," which are not included in any phase I protocol that I have ever seen, but seem to be what people do, or are likely to do. But things still are unclear. General Rule 2 was added to deal with the case where the lowest dose appears

Table 9.2 Illustration of the estimates for toxicity probability $\pi_T(d)$ at each dose d, that are provided by a typical $3+3$ design, in terms of 95% posterior credible intervals starting with beta $(0.50, 0.50)$ priors

d	[# Toxicities]/[# Treated]	95% posterior credible interval for $\pi_T(d)$
10	0/3	0.00–0.54
20	0/3	0.00–0.54
30	1/6	0.02–0.56
40	2/6	0.08–0.71

to be too toxic, since the algorithm itself does not say what to do in this case. With General Rule 4, suppose that you previously observed $0/3 + 1/3 = 1/6$ toxicities at the next lower dose level, and you de-escalate and treat three more patients at the next lower dose level, but then observe 2/3 toxicities. What should you do? It can't be the MTD since 3/9 toxicities have been observed. Or maybe it can be the MTD. Is 3/9 too many, or is it acceptable? It also is not uncommon to see "≥ 2" toxicities as a criterion for de-escalating, rather than "2/3 or 3/3." This leaves one wondering whether this means "$\geq 2/3$" or "$\geq 2/6$" or "$\geq 2/9$." Ambiguities are a by-product of the major problem that $3+3$ algorithms are ad hoc rules without any optimality criterion. Implicitly, this form of the algorithm seems to have the aim to choose a dose d with toxicity probability $\pi_T(d) \equiv 1/6$. But it is very likely to fail to achieve even this, for the following simple reason.

A major problem is that $3+3$ algorithms make a final decision to choose an MTD for later use of a new agent based on a tiny sample size, spread unevenly over several doses. That is, they ignore the basic statistical principle that any estimate computed from a very small sample has very low reliability. Table 9.2 illustrates the simple fact that, at the end of a phase I trial conducted using a $3+3$ algorithm, almost nothing is known about $\pi_T(d)$ for any dose d studied in the trial, including the MTD. In this trial, the chosen MTD was $d = 30$, since it had 1/6 toxicities, while $d = 40$ had 2/6 toxicities. This takes small sample statistical inference to the point of the absurd. The posterior 95% credible interval is consistent with $\pi_T(MTD) = \pi_T(30)$ anywhere between 0.02 and 0.56. This example is typical. That is, when someone gets done with a phase I trial using a $3+3$ algorithm, in most cases they know almost nothing about $\pi_T(MTD)$, other than that there is at least a 95% chance that it is somewhere below about 0.50. This is hardly conclusive evidence that the selected MTD is safe. Given these tiny sample sizes, essentially, it is nothing more than a convenient fiction to declare *any* dose "the MTD". Table 9.3 compares the usual fantasies about $3+3$ designs to the painful realities.

Many dose-finding designs have been proposed to overcome the many limitations and problems with $3+3$ algorithms. A famous design is the continuous reassessment method (CRM), O'Quigley et al. (1990), O'Quigley and Chevret (1992), which assumes a very simple one-parameter model for the dose–toxicity curve $\pi_T(d)$, and relies on a target toxicity probability, π_T^*, ideally specified by the clinical investigator.

Table 9.3 The awful truth about 3 + 3 designs

Fantasy	Reality
Intuitively appealing seems like the right thing to do.	Intuition is misleading. 3 + 3 is the wrong thing to do
Seems simple to apply	The 3 + 3 is not so simple. Many decisions are left unspecified
Does not require a statistician or a computer program	Ignoring statistical science is a recipe for disaster
The trial is finished quickly	Finishing quickly. $\implies$ n is very small at the MTD $\implies$ Any inferences about the MTD are very unreliable $\implies$ The MTD is likely to be either unsafe or ineffective
Everybody else does it, so that makes it the right thing to do	Behaving like sheep is a not a good idea when conducting medical research

The CRM defines the dose with estimated toxicity probability closest to π_T^* to be "optimal." Usually, some target value between 0.20 and 0.33 is assumed for π_T^*, but other values may be used. For each new cohort, the parameter that determines the dose–toxicity curve is reestimated using all current accumulated data, and this updated curve is used to determine the current optimal dose. At the end of the trial, the optimal dose based on the dose–toxicity curve estimated from the final data is selected. Various modifications and extensions of the CRM have been proposed, by Goodman et al. (1995), Faries (1972), O'Quigley and Shen (1996), de Moor et al. (1995), and Yin and Yuan (2009), among many others. Compared to any 3 + 3 algorithm, the CRM is more likely to correctly identify the dose that is optimal in the above sense. Essentially, this is because the CRM has an optimality criterion while any version of the 3 + 3 algorithm is just a set of rules without any criterion at all.

The main limitation of both the CRM and 3 + 3 algorithms is that they are based on toxicity alone, and ignore efficacy. An example of some of the strange consequences of this common practice in phase I trials is as follows. Suppose that the CRM is used with target $\pi_T^* = 0.30$. This implies that it considers a dose d_1 with toxicity probability $\pi_T(d_1) = 0.05$ to be inferior to a dose d_2 with toxicity probability $\pi_T(d_2) = 0.35$, since $|0.05 - 0.30| = 0.25 > 0.05 = |0.35 - 0.30|$. That is, the CRM says that *a dose with 35% toxicity is superior to a dose with 5% toxicity*. Of course, this is completely insane, unless you assume that there is some other, desirable treatment effect, such as $> 50\%$ tumor shrinkage or expected survival time, that becomes increasingly likely with higher dose. The same reasoning applies to 3 + 3 algorithms. Toxicity-only phase I designs only make sense if one assumes, *implicitly*, that $\pi_E(d)$ increases with d for some efficacy event E. Put another way, although efficacy is not used explicitly by 3 + 3 algorithms, the CRM, or other toxicity-only phase I methods, if it were not assumed implicitly that $\pi_E(d)$ increases with d for some desirable outcome E, then the best possible action would be to not treat the patient, since this would ensure that toxicity cannot occur.

9.2   Choosing the Wrong Dose in Phase I

Physicians are taught "First, do not harm" in medical school. This seems to make perfect sense, except that in oncology and other areas of medicine, some treatments have the potential to be very harmful, and possibly kill the patient. Most chemotherapies for cancers are essentially poisons, given as therapy with the hope that the cancer cell killing benefit will outweigh toxicity, which occurs when the chemotherapy kills normal, non-cancer cells. The same reasoning underlies radiation therapy, where the hope is that the benefit of burning away cancer cells will outweigh the harm from damaging normal cells. Surgery to cut out a solid tumor carries all sorts of risks, which the surgeon must believe are warranted by removing the cancerous tissue.

Statistical science enters this process in the following way, which requires one first to think about some elementary probability. For simplicity, assume that a desirable clinical outcome, E = Efficacy, may or may not occur, and likewise the undesirable outcome, T = Toxicity, may or may not occur. In actual trials, things usually are a lot more complex, but I have made things simple here to make some simple points. There are four possible elementary outcomes: [E and T], [E and no T], [no E and T], [no E and no T]. For convenience, I will index these, respectively, by (1, 1), (1, 0), (0, 1), and (0, 0). Since the marginal events are E = [E and T] $\cup$ [E and no T], and T = [E and T] $\cup$ [no E and T], where "$\cup$" is the set union symbol that means "or," the probability of toxicity at dose d is

$$\pi_T(d) = \pi_{1,1}(d) + \pi_{0,1}(d),$$

and the probability of efficacy at d is

$$\pi_E(d) = \pi_{1,1}(d) + \pi_{1,0}(d).$$

This is illustrated in Table 9.4. A central issue is that the event that E and T both occur at d, which has probability denoted by $\pi_{1,1}(d)$, is both a good outcome and a bad outcome. It is good because a dose d that increases $\pi_{1,1}(d)$ must increase $\pi_E(d)$. It is bad because a dose d that increases $\pi_{1,1}(d)$ also must increase $\pi_T(d)$. This simple but inescapable point, which is a consequence of elementary probability, is a central issue in treatment evaluation. It is a nice fantasy to hope that the outcome will be [E and no T], and that some dose d of a magic treatment will make $\pi_{1,0}(d) =$

Table 9.4 Efficacy and Toxicity probabilities

	Toxicity	No Toxicity	
Efficacy	$\pi_{1,1}$	$\pi_{1,0}$	π_E
No Efficacy	$\pi_{0,1}$	$\pi_{0,0}$	$1 - \pi_E$
	π_T	$1 - \pi_T$	

1, or very large, but optimism is a poor strategy. A higher dose of a cytotoxic agent often is likely to increase $\pi_{1,1}(d)$, and thus increase both $\pi_E(d)$ and $\pi_T(d)$.

Determining an MTD based on toxicity alone, while ignoring efficacy, can easily have very undesirable consequences. As an example, consider five doses (1, 2, 3, 4, 5) of a new targeted agent, which have the true toxicity probabilities (0.02, 0.06, 0.30, 0.40, 0.50). Both the $3 + 3$ algorithm and the CRM with target $\pi_T^* = 0.30$ are most likely to select $d = 3$ as the MTD, and use it for a subsequent cohort expansion or phase II trial. Here are three possible cases in terms of the efficacy probabilities.

Plateau Effects

The true efficacy probabilities are (0.20, 0.50, 0.51, 0.52, 0.52). In this case, $\pi_E(d)$ increases to a plateau of about 0.50 at $d = 2$ and thereafter increases very slightly for $d = 3$, 4, or 5. Although $d = 3$ is the most likely chosen MTD, $\pi_E(2) = 0.50$ and $\pi_E(3) = 0.51$ are virtually identical, but $d = 2$ is much safer than $d = 3$ since $\pi_T(2) = 0.06$ while $\pi_T(3) = 0.30$. Despite the fact that $d = 2$ is greatly preferable to $d = 3$, any toxicity-only-based phase I method cannot determine this because it ignores efficacy.

Steep Increase Effects

The true efficacy probabilities are (0.20, 0.25, 0.30, 0.60, 0.65). Escalating from $d = 3$ to $d = 4$ increases the toxicity probability from $\pi_T(3) = 0.30$ to $\pi_T(4) = 0.40$, but it doubles the efficacy probability, from $\pi_E(3) = 0.30$ to $\pi_E(4) = 0.60$. While this small increase of 0.10 in toxicity probability is a very reasonable trade-off for doubling efficacy by choosing $d = 4$ rather than $d = 3$, again, toxicity-only methods cannot determine this.

Harm Without Benefit

The true efficacy probabilities are (0.00, 0.01, 0.01, 0.02, 0.02). That is, the agent has almost no anti-disease effect. Obviously, the best decision is to not choose any dose, and treat as few patients as possible. But phase I methods still are most likely to choose $d = 3$, since they make decisions based on toxicity alone. Efficacy is ignored, based on the convention "First choose a safe dose in phase I, and then evaluate efficacy later on in phase II."

These three examples illustrate a simple, important fact. *Using toxicity but ignoring efficacy when choosing a "best" dose for future study or use in the clinic is a very bad idea.* This is particularly true for novel molecularly targeted agents and immunotherapy agents, for which the MTD that is selected based on toxicity often is not the optimal biological dose that yields the highest efficacy. This may be due to a plateau, or possibly even an inverted U-shape, in the dose–efficacy curve $\pi_E(d)$. In the discussion of phase I–II designs, in Sect. 9.3, I will revisit these three cases in a simulation study.

Many individuals who routinely conduct phase I trials using vaguely defined $3 + 3$ algorithms insist that these trials do not ignore efficacy. They often point out that they record efficacy while conducting their phase I trials. But they do not use efficacy in their decision rules to choose doses for successive patient cohorts. This is like a police officer who carefully observes crimes, but never takes any action to stop the perpetrators or make arrests.

Because phase I trials typically are small, the final number of patients treated at the dose selected as the MTD usually is very small, especially with $3 + 3$ algorithms. To obtain a more reliable estimate of toxicity, and possibly collect some data on efficacy at the selected MTD, it is common practice to treat an "expansion cohort," which is some specified number of additional patients to be treated at the MTD after it is selected. This may seem to make sense, since a larger sample size provides a more reliable estimator. The increasingly common practice of adding an expansion cohort to be treated at the selected MTD from a conventional phase I trial is based on the fallacious assumption that the MTD is known reliably, or with certainty, to be the best dose. Unfortunately, the common use of expansion cohorts has several logical, scientific, and ethical flaws. A detailed evaluation is given by Boonstra et al. (2015). The use of expansion cohorts raises many possible questions, depending on the various numbers of toxicities or responses that may be observed in the expansion cohort. Recall that the subsample size at the MTD from phase I is far too small to make reliable inferences about $\pi_T(MTD)$. So, as successive patients are treated at the MTD in an expansion cohort, it is not unlikely that the additional toxicity data will contradict the earlier conclusion that the selected MTD is safe. For example, starting with 2/9 toxicities at the MTD in phase I, suppose that 6 of the first 10 patients in an expansion cohort of size 100 have toxicity. This would give a total of $2/9 + 6/10 = 8/19$ (42%) toxicities at the MTD. These early toxicities in the expansion cohort suggest that the selected MTD may not be safe. So, should one treat 90 more patients at the MTD, as mandated by the protocol, or violate the protocol by abandoning the MTD and de-escalating? If one decides to de-escalate, the next question is what sort of rules should be applied to choose a dose, or doses, for the remaining 90 patients. If, instead, one continues to treat patients at the MTD, and observes 5 more toxicities in the next 10 patients, for a total of $2/9 + 11/20 = 13/29$ (45%), what should one do next? What if no clinical responses are observed in phase I, and only 1 of the first 20 patients of the expansion cohort responds? Denoting the probability of response at the MTD by $\pi_E(MTD)$, assuming a priori that $\pi_E(MTD) \sim beta(0.50, 0.50)$, it follows that $\Pr(\pi_E(MTD) < 0.10 \mid data) = 0.86$. In terms of efficacy, is it really worthwhile to treat 80 more patients at the MTD?

The general point is that the idea of treating a fixed expansion cohort at a chosen MTD may seem sensible, since more data are being obtained, but in practice, it can become very problematic quite easily. Ignoring the information from the expansion cohort as patients are treated and their outcomes are observed makes no sense whatsoever. In recent years, the sizes of expansion cohorts following phase I trials have exploded, to 100 or more in some protocols. What nominally is a large "phase I expansion cohort" actually is a phase II trial, but conducted without any design, other than a specified sample size. This practice fails to use the new data in the expansion cohort adaptively to change the MTD if appropriate, and thus fails to protect patient safety adequately or discontinue trials of experimental treatments that are ineffective.

Another approach is to conduct a phase II trial after phase I, typically using a Simon design, using the MTD. But the Simon design has no rules that deal with toxicity, so no way to deal with an MTD that turns out to be too toxic. Since 3 + 3 designs have a high risk of choosing an ineffectively low dose, if the agent may have been effective at a sufficiently high dose, the phase I → phase II design has a nontrivial chance of crippling the treatment by choosing an ineffective MTD before it gets to phase II.

As a typical example, a phase I trial of an agent may be conducted at dose levels labeled −1, 0, 1 using a vaguely defined 3 + 3 algorithm based on a maximum of 18 patients, the MTD "identified" in this way may then be used in a phase II trial based on "response" using a Simon design with Type I error probability 0.05, power 0.80, $p_0 =$ 0.05, and $p_a = 0.25$, which requires $n_1 = 9$ patients for stage 1 and $n_2 = 8$ patients for stage II. Thus, this phase I → phase II design requires at most 35 patients. If toxicity is excessive in phase II, however, a common approach is to continue the phase II trial at a lower dose and ignore the fact that the data actually correspond to a mixed sample treated at two different doses. Thus, the Simon design is no longer valid because there actually are two response probabilities, one for each dose. If, instead, a phase II trial is stopped due to excessive toxicity at the MTD, then either the agent must be abandoned or, alternatively, a second phase I dose-finding trial may be conducted to find a lower dose that is acceptably safe. This raises the question of how the data from the previous phase I and phase II trials may, or may not, be used in the design and conduct of such a third trial. If a patient discontinues treatment in phase II due to toxicity, then replacing this patient seems to imply that this simply can be ignored rather than counting that patient's treatment as a failure. For this phase II design, the null hypothesis that Pr(response) $= p = 0.05$ is rejected if 3 or more responses are observed in 17 patients. From a Bayesian viewpoint, assuming a beta(0.25, 0.75) prior for p, since Pr($p > 0.25 \mid 3/17$ responses) $= 0.20$, this criterion for rejecting the null hypothesis in favor of $p > 0.05$ would give unconvincing evidence of a $\geq 25\%$ response rate. Again, the putative target 0.25 used to compute the sample size to obtain 80% power is a Straw Man. A posterior 95% credible interval for p would be [0.045–0.382], which would convey very little about p, other than that it is unlikely to be smaller than 0.05. Many investigators wrongly believe that there is something advantageous about having small sample sizes in early phase clinical trials. The elementary fact that inferential reliability decreases with smaller sample sizes says that the opposite is true. In this example, the samples are so small that it is hardly worth conducting the study at all, since so little will be learned. At the end of this study, the investigators could know very little about safety of the MTD, what a good dose may be, what a best dose may be, or the response rate at the selected MTD.

Here is an example of a phase I trial that led a dead end. The goal was to optimize the dose of a new agent, which I will call X, added to an established preparative regimen for allogeneic stem cell transplantation (allosct) to treat patients with severe hematologic diseases. Six doses of X, rescaled as 1, 2, 3, 4, 5, 6, given for several days pretransplant, were studied. Dose-limiting toxicity (DLT) was defined, as commonly done for phase I trials in allosct, as graft failure or grade 4 (life-threatening or disabling) or grade 5 (fatal) non-hematologic noninfectious toxicity, mucositis, or diarrhea, occurring within 30 days. The time-to-event continual reassessment method (TITE-CRM) of Cheung and Chappell (2000) was used to determine an optimal dose, with target Pr(DLT) = 0.30, starting at the lowest dose. A total of 68 patients were treated, so the trial was much larger than most phase I trials. The TITE-CRM escalated rapidly to dose 6, which it chose as optimal, and the respective sample sizes at the six doses were (3, 3, 3, 4, 4, 51). This was due to the fact that the observed rates of DLT were extremely low at all doses, compared to the target 0.30. This is not surprising, given the severity of the defined DLT events and the fact that, in general, such severe events are unlikely in modern allosct. Since grade 3 (severe and undesirable) toxicity of any type was not counted as a DLT, if the definition had included grade 3 toxicities, the TITE-CRM might have chosen a lower dose of X.

Patients in the trial were followed longer than in most phase I trials, in order to obtain preliminary survival data. While the sample sizes are too small to make any sort of confirmatory conclusions, many survival times were right-censored, and patients were assigned to doses using the TITE-CRM's sequentially adaptive algorithm rather than being randomized, the Kaplan–Meier plots suggested that the 51 patients treated at dose 6 had shorter survival time than the 17 patients treated at the five lower doses. Perhaps longer follow-up may have provided more reliable, possibly different Kaplan–Meier estimates, but this was the survival data available. Thus, while dose 6 was nominally "optimal" based on the TITE-CRM criterion with target Pr(DLT) = 0.30, for the particular definition of DLT, it appeared that, in terms of survival time in this particular setting, a lower dose of X may be optimal. But this was far from clear.

To assess whether the apparent survival difference may have been due to the effects of patient prognostic covariates, a Bayesian piecewise exponential model was fit to the data, summarized in Table 9.5. This fitted model should be interpreted with caution since, again, patients were not randomized between dose 6 and the other five doses of X, the sample size was not large, and the survival times of many of the patients who received dose 6 were censored since they were alive at last follow-up. Again, any inferences are far from confirmatory. The fitted model indicates that, after accounting for effects of well-known prognostic variables, dose 6 had a posterior probability of a harmful effect PHE = 1 − PBE = 0.92. Since survival time was negatively associated with having a matched unrelated donor (MUD) rather than a matched sibling donor, PHE = 1 − PBE = 0.92, and positively associated with receiving post-allosct maintenance treatment, PBE = 0.98, the association between each of these well-known prognostic variables and [Dose = 6] also was evaluated, to investigate whether the apparently harmful effect of [Dose = 6] actually may have been due to an imbalance in either of these variables in the [Dose = 6] subgroup. That is, an important question was whether [Dose = 6] was merely an Innocent Bystander. It turned out that there was no imbalance, with 29/38 (76%) patients receiving [Dose = 6] in the MUD subgroup and 22/30 (73%) in the non-MUD (matched sibling donor) subgroup. Similarly, 14/19 (74%) patients received [Dose = 6] in the MT subgroup and 37/49 (76%) in the non-MT subgroup. The apparently worse survival with [Dose = 6] compared to [Dose < 6] also was seen in these subgroups. These

Table 9.5 Fitted Bayesian piecewise exponential regression model for survival time, for 68 AML/MDS patients treated in a phase I trial to determine an optimal dose of agent X added to the standard pretransplant conditioning regimen in allogeneic stem cell transplantation. MRD = presence of minimal residual disease, MUD = matched unrelated donor, MT = maintenance treatment given post-allosct, HR = hazard ratio, and PBE = posterior probability of a beneficial effect for larger values of the covariate. 22 (32%) deaths were observed

Variable	Posterior quantities		
	Median (HR)	95% CI for HR	PBE
Dose = 6	2.73	0.82 – 7.66	0.08
Age	1.03	0.10 – 1.04	0.11
MRD	3.63	1.17 – 11.63	0.02
Not in CR	0.77	0.21 – 3.46	0.64
High risk	1.40	0.39 – 5.52	0.26
Comorbidity score	1.05	0.78 – 1.40	0.33
MT	0.16	0.03 – 0.85	0.98
MUD	2.12	0.62 – 7.21	0.08
Graft = marrow cells	0.38	0.64 – 1.77	0.87

post hoc analyses may be considered to be data dredging, but early phase trials are inherently exploratory, and MUD and MT are well-established prognostic covariates.

While these inferences about the effect of dose 6 of X on survival are far from conclusive for the reasons noted above, they still are worrying. In particular, they make it ethically problematic to proceed with further study of agent X at dose 6 combined with the standard preparative regimen. The TITE-CRM did a good job of achieving its goal, which was to select the dose of X having Pr(DLT) closest to 0.30, for the given definition of DLT, but in terms of survival time, a lower dose may be more desirable. The problem is that these phase I trial results make it unclear how to move forward with clinical evaluation of X, say in a phase II trial. A major cause of this quandary is that this was a phase I trial with dose-finding based only on DLT, as defined. No efficacy outcome related to dose and survival time, other than the unlikely event of death within 30 days (grade 5 toxicity), was used along with 30-day DLT to assign patients sequentially to doses, or to define and select an optimal dose. This example illustrates the general problem that starting drug development with a conventional "toxicity-only" phase I trial to determine a nominally optimal dose, or MTD, of a new agent for use in a later study may be a serious mistake. In Sect. 9.3, the alternative approach of first conducting a phase I–II trial based on both efficacy and toxicity to determine an optimal dose, rather than doing a phase I trial, will be described.

Takeaway Messages About the Phase I → Phase II Paradigm

1. If the $\pi_E(d)$ curve has a plateau, a typical phase I design often chooses a dose that is too high, and too toxic.
2. If the $\pi_E(d)$ curve increases sharply for doses near the MTD, a typical phase I design often chooses an MTD that is too low, and thus inefficacious, and misses the truly best dose.
3. If all doses are ineffective, with very small $\pi_E(d)$, a typical phase I design will not stop the trial early for futility because it ignores efficacy.
4. A dose that is chosen as optimal in terms of toxicity alone in a phase I trial may be suboptimal in terms of early efficacy or survival time. That is, the usual phase I paradigm often does not work.

Disease recurrence is the major precursor to death in allosct, and most new treatments are aimed at extending recurrence-free survival (RFS) time. For dose-finding, no early efficacy outcome is available, except possibly defined as the absence of one or more adverse events. For example, efficacy might be defined as no grade ≥ 2 graft versus host disease (GVHD) within 100 days. If so, using time-to-toxicity and time-to-efficacy, the late-onset EffTox (LO-ET) phase I–II design of Jin et al. (2014) might have been used for dose-finding in the trial of X, instead of the TITE-CRM. This application of the LO-ET design would rely on the implicit assumption that a dose of X chosen to optimize the trade-off between 100-day toxicity and 100-day GVHD is likely to extend RFS time.

Depending on prior assumptions about whether the disease recurrence rate or probability of toxicity are monotone increasing or not in the dose of X, a third, very unconventional approach would be as follows:

An Alternative Experimental Approach

1. Randomize patients fairly among the six doses of X and a control arm with the established preparative regimen.

2. Use recurrence-free survival time and time-to-toxicity, both monitored over 100 days, as co-primary outcomes, with joint utilities defined using the approach of Thall et al. (2013).

3. Include a safety rule to terminate accrual to any dose found to have an unacceptably high probability of 100-day toxicity.

4. Enrich the subsample sizes of patients randomized to the remaining safe doses.

This certainly is not what typically is done in phase I, but it would (1) find a dose that is optimal in terms of the risk–benefit trade-off between toxicity and RFS, (2) protect patient safety, and (3) yield much more useful data for planning a later, survival time-based trial than the data described above. To obtain reliable safety monitoring and inferences, such a design would require a sample size much larger than typically is used in phase I, but that would be a good thing, not a bad thing. After all, protecting patient safety should never be a secondary consideration in a clinical trial, and if the trial's final results will not provide reasonably reliable inferences, then why bother to conduct it?

Takeaway Messages About the Conventional Phase I → Phase II Paradigm.

1. An expansion cohort after phase I is essentially a single-arm phase II trial without any safety or futility stopping rules.
2. Because little is known about the probability of toxicity of a new agent at an MTD chosen in typical phase I trial, treating an expansion cohort without adaptive rules for toxicity may be unsafe.
3. Because little is known about the probability of efficacy of a new agent at an MTD chosen in typical phase I trial, if the agent is ineffective at the MTD then treating an expansion cohort without adaptive futility stopping rules may be unethical.

9.3 Phase I–II Designs

All of the above problems with conventional dose-finding methods can be avoided by using phase I–II designs, which combine the basic elements of phase I and phase II by hybridizing them into a single trial. The goal of a phase I–II design is to choose an optimal dose based on both efficacy and toxicity. There is a very large and growing literature on phase I–II designs. This includes the early paper by Gooley et al. (1994), designs dealing with a wide variety of cases by Thall and Russell (1998), Braun (2002), Dragalin and Fedorov (2006), Yin et al. (2006), Thall et al. (2008, 2013, 2014b), Thall and Nguyen (2012), Mandrekar et al. (2010), Jin et al. (2014), Yuan and Yin (2009), Lee et al. (2015), the books by Chevret (2006) and Yuan et al. (2016), and medical papers on trials run using phase I–II designs by de Lima et al. (2008), Konopleva et al. (2015), and Whelan et al. (2008). A review is given by Yan et al. (2018).

In phase I–II trials, there is no abrupt switch from a toxicity-based phase I trial to a response-based expansion cohort or phase II trial. Instead, these two trials are hybridized by the phase I–II approach and run as a single trial. The main motivation is that response matters in phase I, and toxicity matters in phase II, so it makes sense to consider these two outcomes together. When a phase I–II trial is completed, there is no need for a subsequent phase II trial, since efficacy has been evaluated. The use of both the toxicity and efficacy data gives phase I–II designs several important advantages. Compared with conventional phase I designs, phase I–II designs are more efficient, and can reliably identify a dose that is optimal in terms of both safety and efficacy that very likely would be missed by a phase I design. Depending on the trial objectives, different strategies can be employed to use both toxicity and efficacy for choosing doses. Because a phase I–II trial replaces the two conventional phases, it should have a larger sample size than used conventionally in phase I. Moreover, using the (Dose, Efficacy, Toxicity) data is a much more informative basis for identifying an optimal dose, compared to only using (Dose, Toxicity).

To illustrate how a phase I–II design works, I will use the EffTox trade-off-based design of Thall and Cook (2004), including two important refinements given by Thall et al. (2014a). A freely available computer program with a graphical user interface for implementing the EffTox design is available at the website at https:// biostatistics.mdanderson.org/softwaredownload/. The main features of the EffTox phase I–II design are as follows:

1. It has an explicit criterion that quantifies the desirability of each dose d based on the trade-off between the probabilities $\pi_E(d)$ and $\pi_T(d)$.
2. It treats successive cohorts of patients by adaptively using the (Dose, Efficacy, Toxicity) $= (d, Y_E, Y_T)$ data from all previous patients to select the dose having the best efficacy–toxicity trade-off.
3. It includes two explicit criteria that specify whether a dose is acceptably safe, and whether it has sufficiently high efficacy. These criteria are updated based on the most recent data when a dose must be chosen for a next cohort.

To implement the EffTox design, the investigator first must explicitly define efficacy and toxicity. The investigator also must specify a fixed upper bound $\overline{\pi}_T$ on $\pi_T(d)$, and a fixed lower bound $\underline{\pi}_E$ on $\pi_E(d)$. A dose d is defined to be *unacceptable* if either of the following two conditions is met:

$$\Pr\{\pi_E(d) < \underline{\pi}_E \mid data\} > p_{U,E}$$

or

$$\Pr\{\pi_T(d) > \overline{\pi}_T \mid data\} > p_{U,T},$$

where the numerical probabilities 0.90 or 0.80 usually are used for the decision cutoffs $p_{U,E}$ and $p_{U,T}$. The first expression says that d is inefficacious. The second expression says that d is too toxic. Unacceptable doses are not used to treat patients. These two rules serve as gatekeepers to avoid treating patients with a dose that the interim data have shown is too toxic (the safety rule) or too inefficacious (the futility rule). If the interim data at some point show that all doses are unacceptable, then the trial is stopped with no dose selected. Thus, effectively, the trial has both safety and futility monitoring rules.

An efficacy–toxicity trade-off contour is used as a criterion to choose a best acceptable dose for each cohort. Figure 9.1 gives the family of efficacy–toxicity trade-off contours for a particular EffTox design. The trade-off contours are constructed so that all (π_T, π_E) pairs on each contour are equally desirable. In Fig. 9.1, the numerical desirabilities of the contours increase moving from upper left to lower right, as π_T becomes smaller and π_E becomes larger. The upper left corner is the least desirable case $(\pi_E, \pi_T) = (0, 1)$, where efficacy is impossible and toxicity is certain, and the right bottom corner is the most desirable case $(\pi_E, \pi_T) = (1, 0)$ where efficacy is certain and toxicity is impossible. To construct efficacy–toxicity trade-off contours for a trial, one may specify three probability pairs (π_E, π_T) considered equally desirable. A function is fit to these equally desirable pairs to construct the target trade-off contour and generate the family of contours. During the trial, to choose

Fig. 9.1 Illustration of the (π_E, π_T) trade-off contours for a particular EffTox phase I–II design

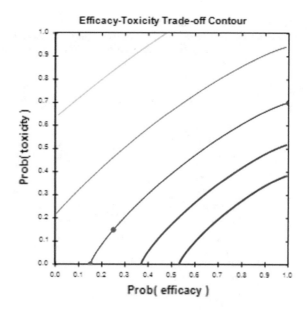

Efficacy-Toxicity Trade-off Contour

Prob(toxicity)

Prob(efficacy)

a dose for the next cohort, the posterior means of $(\pi_E(d), \pi_T(d))$ are computed for each acceptable d. The contour of this pair is identified, and the desirability of d is defined to be the desirability of that contour. The acceptable dose d having largest desirability is used to treat the next cohort. At the end of the trial, the dose with largest desirability is selected as optimal.

To illustrate the EffTox design and compare it to two commonly used phase I designs, I will consider a trial of an experimental agent administered at one of the five doses 25, 50, 75, 100 or 125 mg/m², given as part of a preparative regimen for cancer patients undergoing an autologous stem cell transplant (autosct). Toxicity is defined as any of the events regimen-related death, graft failure, or severe (grade 3, 4) atrial fibrillation, deep venous thrombosis, or pulmonary embolism, occurring within 30 days from transplant. Efficacy is defined as the outcome that patient is alive and in complete remission at day 30 after the autosct. The fixed upper limit on $\pi_T(d)$ is $\overline{\pi}_T = 0.30$, the fixed lower limit on $\pi_E(d)$ is $\underline{\pi}_E = 0.20$, and the equally desirable (π_E, π_T) pairs (0.15, 0), (0.25, 0.15) and (1, 0.70) are used to generate the target efficacy–toxicity trade-off contour. The maximum sample size is 60, with cohort size 3, for a maximum of $60/3 = 20$ cohorts. A "do-not-skip" rule is imposed that does not allow an untried dose to be skipped when escalating.

A computer simulation study to compare this particular phase I–II EffTox design with the $3 + 3$ design and the CRM with target toxicity probability $\pi_T^* = 0.30$ is summarized in Table 9.6. The simulations consider a trial with five dose levels, maximum sample size $N = 60$, and patients treated in successive cohorts of size 3. The CRM was run with dose-finding for 30 patients followed by a cohort expansion of 30 more patients at the selected dose to provide a fair comparison to the EffTox design with $N = 60$. For the $3 + 3$ design, since the sample size required to determine an MTD

Table 9.6 Simulation results for a 60-patient trial. The $3+3$ algorithm was run with an expansion cohort sized to achieve a total of 60 patients. The CRM was run with target $\pi_T^* = 0.30$ in the first 30 patients followed by an expansion cohort of 30 at the selected dose. The EffTox design was run with 60 patients

		Dose					$Pr(Stop)$
		1	2	3	4	5	
Scenario 1							
	$\pi_T(d)^{true}$	0.02	0.06	.30	0.40	0.50	
	$\pi_E(d)^{true}$	0.20	0.50	0.51	0.52	0.52	
	Trade-off	0.58	0.72	0.61	0.56	0.51	
$3+3$	Selection %	3.6	54.8	28.2	10.0	0.0	2.8
	# Patients	5.0	29.8	16.9	6.5	0.6	
CRM	Selection %	0.0	1.7	73.4	24.5	0.4	0.0
	# Patients	3.3	6.5	32.5	15.0	2.7	
EffTox	Selection %	2.0	66.0	30.0	1.0	0.0	1.0
	# Patients	4.8	30.9	20.7	3.0	0.7	
Scenario 2							
	$\pi_T(d)^{true}$	0.04	0.08	0.12	0.30	0.40	
	$\pi_E(d)^{true}$	0.10	0.20	0.60	0.35	0.20	
	Trade-off	0.52	0.56	0.75	0.54	0.41	
$3+3$	Selection %	6.9	12.6	43.5	24.3	0.0	11.7
	# Patients	6.9	9.6	22.8	14.0	1.7	
CRM	Selection %	0.0	0.2	8.4	72.6	18.8	0.0
	# Patients	3.5	4.8	10.7	28.5	12.5	
EffTox	Selection %	2.0	13.0	78.0	5.0	1.0	1.0
	# Patients	5.1	9.0	33.9	8.8	3.0	
Scenario 3							
	$\pi_T(d)^{true}$	0.08	0.12	0.16	0.20	0.30	
	$\pi_E(d)^{true}$	0.05	0.10	0.30	0.50	0.70	
	EffTox trade-off	0.48	0.50	0.58	0.66	0.70	
$3+3$	Selection %	11.3	16.5	19.4	23.0	0.0	29.8
	# Patients	9.4	11.3	11.8	12.1	1.9	
CRM	Selection %	0.0	1.6	16.0	34.3	48.1	0.0
	# Patients	4.6	7.6	12.8	15.5	19.5	
EffTox	Selection %	0.0	2.0	14.0	26.0	55.0	3.0
	# Patients	3.8	5.9	10.5	13.6	24.9	

varies after the MTD is selected, the expansion cohort sample size was determined to obtain a total sample size of 60. Four scenarios defined in terms of true $\pi_E(d)$ and $\pi_T(d)$ probabilities were simulated using each design. The simulation scenarios are constructed to be concrete examples of the effects described in Sect. 9.2.

In Scenario 1, $\pi_T(d)$ increases with dose, while $\pi_E(d)$ increases from $d = 1$ to $d = 2$ and then approximately levels off, so $d = 2$ is optimal because $\pi_T(d)$ increases for $d = 3, 4, 5$, with very little efficacy gain. In Scenario 2, the $\pi_E(d)$ curve has an inverted U-shape with the highest efficacy at $d = 3$, while $\pi_T(d)$ increases with dose, so $d = 3$ is optimal with the highest efficacy–toxicity trade-off. Scenario 3 is a case usually observed with conventional cytotoxic agents, and with some targeted agents, where both $\pi_E(d)$ and $\pi_T(d)$ increase with dose, and dose $d = 5$ is optimal with the highest efficacy–toxicity trade-off. Each design was simulated 10,000 times under each scenario.

Overall, the EffTox design has the highest probability of correct selection (PCS). In Scenarios 1 and 2, the CRM performs worse than the $3 + 3$ design, because the CRM is likely to achieve the goal of selecting the dose with $\pi_T(d)$ closest to 0.30, but in these two scenarios, this dose is suboptimal. In Scenario 1, the optimal dose is $d = 2$, but $\pi_T(3)$ is closest to 0.30. In Scenario 2, the optimal dose is $d = 3$, but $\pi_T(4)$ is closest to 0.30. Because the CRM ignores efficacy, it misses the optimal dose most of the time in these scenarios. When the dose with $\pi_T(d)$ closest to 0.30 happens to be optimal, as in Scenario 3, the CRM performs well essentially due to luck. The simulations also illustrate the fact that the $3 + 3$ algorithm has a dose selection distribution that is very disperse, with the percentages spread out among the doses. This is due to the facts that the $3 + 3$ has no criterion for what is a good dose, it cannot reescalate, and it stops with a relatively small sample size.

Scenario 4, which is not given in the table, is one where no dose is acceptable because the agent is ineffective, with efficacy probabilities (0.03, 0.04, 0.05, 0.06, 0.07) at the five doses, and toxicity probabilities (0.05, 0.10, 0.20, 0.35, 0.50). In Scenario 4, the PCS is defined as the percentage of times that no dose is selected. Because the CRM and $3 + 3$ both ignore efficacy, in Scenario 4 they have respective PCS values 5.5% and 0.9%, while EffTox has PCS 87%. That is, when no dose is effective, EffTox is very likely to recognize this and correctly stop the trial without selecting any dose. In contrast, in Scenario 4, the $3 + 3$ algorithm has dose selection percentages (9.1, 28.9, 37.5, 18.6, 0) and the CRM has dose selection percentages (2.7, 28.8, 53.3, 13.5, 0.6). That is, both the $3 + 3$ and CRM are very likely to select an ineffective MTD, simply because they ignore efficacy. This illustrates the fact that the EffTox dose admissibility criterion for $\pi_E(d)$ provides a more general version of what typically is called a futility stopping rule in phase II trials. This admissibility criterion uses the efficacy data, which is very likely to show little or no responses at any dose, to shut down the trial early due to futility. This is an important aspect of the fact that EffTox is a phase I–II design.

Takeaway Messages About Phase I–II Trials

1. Conducting a reasonably designed phase I–II trial is greatly superior to the conventional paradigm of first conducting phase I based on toxicity to choose an MTD and then conducting phase II at the chosen MTD to evaluate efficacy.
2. The design is based on a trade-off criterion that characterizes and quantifies the risk–benefit between the probabilities of efficacy and toxicity for each dose.

Chapter 10
Just Plain Wrong

Two things are infinite: the universe and human stupidity; and
I'm not sure about the universe.
Albert Einstein

Contents

Abstract This chapter will give examples of particular clinical trial designs that are fundamentally flawed. Each example will illustrate a fairly common practice. The first example is a futility rule that aims to stop accrual to a single-arm trial early if the interim data show that it is unlikely the experimental treatment provides at least a specified level of anti-disease activity. The rule is given in terms of progression-free survival time. An alternative, much more sound, and reliable futility monitoring rule that accounts for each patient's complete time-to-event follow-up data will be presented. The second example will show how the routine practice of defining patient evaluability can lead one astray when estimating treatment effects, by misrepresenting the actual patient outcomes. The next two examples pertain to the problems of incompletely or vaguely specified safety monitoring rules. The final example shows what can go wrong when one ignores fundamental experimental design issues, including bias and confounding, when evaluating and comparing multiple treatments. As an alternative approach, the family of randomized select-and-test designs will be presented.

© Springer Nature Switzerland AG 2020
P. F. Thall, *Statistical Remedies for Medical Researchers*, Springer Series
in Pharmaceutical Statistics, https://doi.org/10.1007/978-3-030-43714-5_10

10.1   Clinical Trial Design, Belief, and Ethics

If someone who is not a physician were to put on a white lab coat with a fake id badge, walk into a hospital, pretend to be a physician, and start to treat patients, things probably would go very wrong very quickly. If not stopped, this person likely would end up making all sorts of wrong decisions and killing people through sheer ignorance. Of course, no sane person would ever do such an outrageous thing. But I wonder. How long it would take for this faker to be discovered and stopped? After all, a white lab coat is a powerful symbol that says "This Person is a Doctor."

As outrageous as it would be for someone to actually do this, here is something similar that is done quite commonly. Many physicians with no formal training in statistics, and little or no understanding of basic statistical principles or scientific method, construct things that they call statistical designs for clinical trials. Some examples are given in the following text. I would not call them "clinical trial designs," but someone else might. All of the examples are hypothetical. You may judge whether there is any resemblance to real designs, past or present, appearing in actual clinical trial protocols that you have seen. It is important to keep in mind that the two main purposes of a clinical trial are to treat the patients enrolled in the trial and to obtain data that may benefit future patients. The statistical design of a clinical trial is critically important, both scientifically and ethically, because it has real consequences that affect the health and welfare of both the patients enrolled in the trial and future patients.

The question of whether or not a particular behavior is ethical may not have an obvious answer, since it hinges on what the individual engaging in that behavior knows or believes. This is especially true in clinical trials, where knowledge and belief often involve complex medical and statistical issues. It may be that, although two individuals engage in the same behavior, one is behaving ethically while the other is not. This could be due to the fact that the two individuals have different knowledge. For example, suppose that two physicians both give prognostically identical patients treatment A, but one of the physicians is aware of a clearly superior treatment B while the other is not. Then the physician with the knowledge of B may have behaved unethically, while the physician who is ignorant of B is behaving ethically. Of course, the argument that ignorance is a defense for apparently unethical behavior has its limits. Any competent physician should have at least a reasonable knowledge of current practice and available treatments for the diseases that they treat.

On a deeper level, knowledge ultimately depends on belief. Here is one semantic definition of "knowledge." Someone "knows" that something is true if they believe that it is true, they have good reason to believe that it is true and, furthermore, it's true. In the abstract, many particular actions may be categorized unambiguously as "ethical" or "unethical." But in real life, aside from cases where the distinction is obvious, the situation often is much more complex than this simple dichotomy. This is especially true in many clinical trials. The physicians involved in a trial should understand the key elements of the statistical design, and the statistician who constructed the design should understand the key elements of the medical practice, treatment effects, and the items in the list given in Chap. 3. But the world is not perfect. In most settings, difficult circumstances and practical imperatives force compromises to be made in the design and conduct of clinical trials. These compromises often are all that can be achieved in the real world. So, in clinical trials, ethics is a matter of belief, knowledge, circumstance, and practical necessity.

Depending on the disease and therapeutic regime, there may be many different clinical outcomes that possibly can occur once a patient's treatment regimen is started. If you plan to keep track of everything and make adaptive decisions on that basis, things can get very complicated very quickly. Decisions may include adjusting doses between or within patients, safety monitoring, futility monitoring, or making a final conclusion about treatment superiority. Most new treatments do not work exactly as intended. They may have some anti-disease effect that is too small to matter, or no anti-disease effect at all. Even if a new treatment does have a clinically meaningful anti-disease effect, it may have adverse effects so severe that, in terms of its risk–benefit profile, the treatment is undesirable. In many settings, the phrase "anti-disease effect" may oversimplify things, since this may refer to an early outcome that does not guarantee long-term benefit. Remember the Billion Dollar Computation? Things often are even more complicated, since there may be several different anti-disease effects, several different adverse effects, and these may occur at different times following the start of treatment.

Given all of these possibilities, and the need to evaluate a new treatment in one or more clinical trials, any design must include practical decision rules to protect patients from new treatments that are either unsafe or ineffective. Inevitably, if you wish to do this properly, this leads to specific rules that stop accrual if the interim data provide convincing evidence. I have discussed such rules in earlier chapters, but there still are plenty of examples of how things can go wrong. These are important to think about because, as usual, it may not be obvious why a particular rule or set of rules is flawed. Constructing early stopping rules can be tricky.

In earlier chapters, I have given examples of some common flaws in conventional clinical trial designs. I selected the following additional examples of flawed clinical trial designs from a much larger set. The aim is to reveal key aspects of what often is wrong with certain types of designs or decision rules that may not be obvious. Each is an example of something that, at least in my experience, is done quite commonly. My goal is to help you guard against making similar mistakes in your own trials.

10.2 A Futile Futility Rule

Here is an example of a futility stopping rule that does not work. Consider a phase II trial designed to evaluate a new cancer treatment that is aimed at reducing the rate of disease progression, hence improving survival. I won't bother with the details of the disease or design, other than the specification that 50 patients are to be treated, and T = progression-free survival (PFS) time is the primary outcome. The following futility early stopping rule is included in the protocol. First, "Response" is defined

Table 10.1 Example of numbers of patients accrued, and not responding, over the course of the trial

Months from start	6	12	18	24	30	36	42	48
#Accrued	9	18	27	36	45	50	50	50
#Nonresponders	2	3	5	9	15	18	22	27

as the event $[T > 12$ months], that is, that the patient will survive at least a year without disease progression. This is a laudable goal, but something very foolish has been done by defining "response" in this way. It reduces the continuous event time variable T to a binary variable. The problem with this is that, while a patient can be scored as a nonresponder at the time of progression or death (treatment failure), which could be any time between 0 and 12 months, a patient can only be scored as a responder after 12 months, if (s)he has been followed for 12 months without dying or suffering a disease progression. In terms of any rule to monitor the response rate, 12 months is a very long time to wait before scoring a patient as a "responder."

The protocol contains a futility rule saying that accrual will be stopped early if $\leq 23/50$ responses were observed. No statistical rationale or operating characteristics for this rule are given. An equivalent way to say this is that, as soon as it is known that $50 - 23 = 27$ or more patients progress or die before 12 months of follow-up, that is, are nonresponders as defined, then accrual to the trial should be stopped. Given this, a key fact is that an overall accrual rate of 18 patients per year, equivalently 1.5 patients per month, is anticipated. Table 10.1 gives an example of how this trial might play out over time. A bit of arithmetic shows that, on average, you should expect to accrue $1.5 \times 6 = 9$ patients in the first 6 months, $1.5 \times 12 = 18$ patients in the first 12 months, and so on. The numbers of nonresponders (patients who progress or die during their 12 months follow-up) illustrate how patients may be accrued to the trial and their outcomes may be observed as they are evaluated over time.

In this example, by the time you find out that there are 27 nonresponders in the 50 patients, at month 48, it has been 12 months since all 50 patients were accrued at month 36. So, logistically, the futility rule is complete nonsense, since it is likely that all patients will have been accrued before the rule, as given, can be applied to stop accrual. *The source of the problem is that a time-to-failure variable has been reduced to a binary indicator of an event that may take as long as 12 months to evaluate.* A patient cannot be scored as a "nonresponder" until they have been followed for 12 months and are alive without disease progression. For example, the binary "response" outcome of a patient who is alive without progression at 10 months of follow-up cannot be scored, despite the fact that there is a substantial amount of information about the patient's outcome. Given that you know $T > 10$, it is likely that you will observe $T > 12$ after two more months of follow-up. This sort of mistake can occur if someone writes down an arbitrary decision rule, and does not think through the logistics and possible consequences in terms of how the rule is likely to play out during the trial.

A straightforward, effective way to deal with this problem is to use all of the event time data. That is, do not reduce T to a binary indicator that may take 12 months to score. Intuitively, if one keeps track of the length of time that each patient has been followed and whether or not death or progression has occurred, this will provide a much richer amount of information than dichotomizing each patent's outcome as either "response" or "nonresponse." The following Bayesian design, which is an application of the method given in Thall et al. (2005), exploits the simple idea that a patient who is alive without progression, for example, at 10 months of follow-up is better off than a patient who is alive without progression at 2 months of follow-up. That is, the fact that a patient has been followed for any given time <12 months without treatment failure (disease progression or death) is useful information. Formally, it is very easy to prove that, for example,

$$\Pr(T > 12 \mid T > 10) > \Pr(T > 12 \mid T > 2).$$

This says that the probability of failure occurring after 12 months is larger for a patient who has lasted 10 months without failure than for a patient who has lasted 2 months. More generally, denote each patient's most recent follow-up time by T^o. The conditional probability $\Pr(T > 12 \mid T > T^o)$ must increase as the patient's follow-up time T^o without progression or death goes from 0 to 12. In words, this says that the probability a patient will go 12 months without progression or death, i.e., "respond," increases with the amount of time that the patient has been followed without either event occurring.

To formalize this, define the indicator $\Delta = 1$ if $T^o = T$ and $T^o = 0$ if $T^o < T$. Notice that, in this trial, T^o is at most 12 months. So, for a patient who has been followed for time T^o, $\Delta = 1$ if T^o is the time of progression or death, and $\Delta = 0$ if the patient is alive and has not suffered a disease progression. This is exactly the data structure that is used to construct a Kaplan and Meier (1958) curve to estimate survival probabilities. The pair (T^o, Δ) can be evaluated for each patient at any time, and it tells you a lot more than the binary response variable, which may not be known for many patients since it may take 12 months to evaluate. If, for example, $(T^o, \Delta) = (6, 1)$ then you know that the patient is a nonresponder, that is, had disease progression or died, at 6 months. If $(T^o, \Delta) = (6, 0)$ then you do not yet know whether the patient will turn out to be a responder or not, but you have the partial information that the patient is alive 6 months without disease progression. This elementary representation is the first thing that one learns in any statistics course on survival analysis, and it does not require any deep mathematical knowledge to understand. Still, if you give someone without statistical training a dataset in which 50 patients died, including their times of death, and 50 patients were alive at their last follow-up time, and ask them to use the data to estimate mean survival time, it is not at all obvious how to do this. Some people might just take the 50 known death times and compute their mean. This is completely wrong. For example, if a patient has survived, say, 18 months, then you know that $T > 18$, and this is important information. This patient's survival time T has been right-censored at time $T^o = 18$ months, with $\Delta = 0$. Kaplan and Meier (1958) were the first to solve the statistical problem of estimating a survival distribution based on right-censored survival data. Once the Kaplan–Meier estimate has been computed, it can use used to compute estimates of mean or median (50th percentile) of the survival time, or any other percentile of the survival time distribution.

The same basic ideas can be used to construct an adaptive Bayesian rule for monitoring this trial. Getting back to the futile futility rule described above, a design with a valid futility stopping rule that uses the actual event time follow-up data on all patients, constructed using the method of Thall et al. (2005), is as follows. Intuitively, what the following design does is give partial credit to a patient who has gone part way through the 12-month follow-up window without progressive disease or death, and this partial credit increases the farther they go. This design uses a rule for stopping a single-arm trial early based on the right-censored time-to-event data (T^o, Δ), as in the above example. It assumes that the event time T follows an exponential distribution that has median $\tilde{\mu}_E$ for patients given treatment E and historical median $\tilde{\mu}_S$ for patients given standard treatment S. Inverse gamma (IG) priors are assumed for these two means, with the prior on μ_S informative to reflect experience with S and the prior on μ_E non-informative, so that the (T^o, δ) data from the patients treated with E in the phase II trial dominate the early stopping decision. If δ is a specified targeted improvement in median PFS time (or survival time, depending on the application) over S, the rule is to stop the trial early if

$$Pr(\tilde{\mu}_S + \delta < \tilde{\mu}_E \mid data) < p_L.$$

In words, this probability inequality says that the trial is stopped early if, given the current data, it is unlikely that E provides at least a targeted δ improvement in median PFS time over S. For practical application, the stopping rule may be applied either continuously each time a new patient is eligible for enrollment, or periodically, say each month. Given a monitoring schedule, in practice the numerical value of the cutoff parameter p_L is calibrated by computer simulation so that, in the case where the target is achieved with $\tilde{\mu}_E^{true} = \tilde{\mu}_S^{true} + \delta$, the probability of stopping the trial early is small, say 0.10 or 0.05.

To apply this design to the above example, suppose that $\tilde{\mu}_S$ follows an IG prior with mean 12 months and standard deviation 2, which says that $\tilde{\mu}_S \sim$ IG(38, 444). We assume that $\tilde{\mu}_E$ follows an IG prior with the same mean of mean 12 months but it is much more disperse, with standard deviation 18 to reflect very little prior knowledge about how E may behave. Say we wish a $\delta = 8$ month improvement over S, so accrual to the trial will be stopped early if it is very unlikely that this improvement will be achieved. The early stopping rule is applied on a monthly basis.

Table 10.2 gives the operating characteristics of this design, for maximum accrual 50 patients, assuming an accrual rate of 1.5 patients per month. This design has the advantage that the stopping rule is applied each month, so it takes advantage of all the event time information (T^o, δ) on each patient in the accumulating data. The cutoff $p_L = 0.01$ was calibrated to ensure a small early stopping probability, pstop $= 0.09$, in the case where the targeted median PFS $\tilde{\mu}_E^{true} = 12 + 8 = 20$ months is achieved with E. The early stopping probability increases as the true median

Table 10.2   Operating characteristics of the Bayesian phase II design based on the time-to-failure futility stopping rule, with maximum $N = 50$ patients. The fixed true median PFS assumed in each simulated scenario is denoted by $\tilde{\mu}_E^{true}$. Time is in months

Scenario	$\tilde{\mu}_E^{true}$	Pr(Stop early)	Mean number of patients	Mean trial duration
1	8	0.99	17.4	24.0
2	12	0.72	31.6	33.7
3	16	0.25	43.4	41.3
4	20	0.09	47.1	43.8

PFS of E decreases, to pstop $= 0.72$ for $\tilde{\mu}_E^{true} = 12$, which is the prior mean of the median PFS with S. The rule is nearly certain to stop the trial early if E is worse than S with $\tilde{\mu}_E^{true} = 8$. These values illustrate the general advantages that, if E is no better than S, the design is likely to stop early and treat far fewer patients. An experienced statistician might criticize this methodology on the grounds that assuming an exponential distribution says that the event rate is constant over time. In theory, this is a valid criticism, but it turns out that the stopping rule actually works quite well if the actual distribution of T is not exponential, that is, the method is robust, as shown in Thall et al. (2005).

This design becomes more reliable with larger sample size, but to be feasible with larger N_{max} a faster accrual rate probably would be required. For example, if the accrual rate is doubled, from 1.5 to 3 patients per month, then $N_{max} = 80$ is feasible since it would take a maximum of roughly 27 months to accrue 80 patients. In this case, the design gives pstop $= 0.10$, mean number of patients $= 74.4$, and mean duration 37 months for the fixed targeted median PFS value $\tilde{\mu}_E^{true} = 20$ months. The corresponding values are pstop $= 0.85$, mean number of patients $= 42.8$, and mean trial duration $= 26.6$ months for $\tilde{\mu}_E^{true} = 12$, that is, if E is equivalent to S. These computations were done using the program *One Arm TTE simulator version 3.0.6*, freely available from the software download website https://biostatistics.mdanderson.org/softwaredownload/.

10.3   The Evaluability Game

A common practice is to define patients as being "evaluable" for the effect of treatment on their disease if they have received all, or some specified minimum amount, of the planned therapy. This may seem sensible, in terms of inference about the actual treatment effect. However, it can have some unintended consequences that may lead to flawed inferences.

Consider a trial to compare two cancer chemotherapies, with a protocol that contains the following statement: "Treatment will consist of 4 courses of chemotherapy. A patient will be considered evaluable if (s)he completes at least 2 courses

Table 10.3 Comparison of treatments A and B based on toxicity rate and response rate in "evaluable" patients, where evaluable is defined as the patient receiving at least two courses of chemo

Arm	Enrolled	Dropouts after one cycle	Toxicities in evaluable pats (%)	Responses in evaluable pats (%)
A	100	20	8/80 (10)	40/80 (50)
B	100	90	1/10 (10)	9/10 (90)

of chemotherapy." This seems to make sense. After all, if a patient receives only 1 course of chemotherapy rather than the planned four courses, that may not be enough to fairly assess the treatment's anti-disease activity. So, the motivation for defining evaluability in this way is the desire to assess the effect of the treatment fairly. The protocol also gives the safety rule "Stop a treatment arm if Pr(Toxicity) > 0.20 in the evaluable patients." In the protocol, "toxicity" is defined as severe, i.e., grade 3 or worse. But it is unclear, specifically, what the toxicity stopping rule actually may be. Does it refer to the empirical toxicity probability being larger than 0.20, or to a more sophisticated decision criterion, like a Bayesian posterior probability-based rule such as $\Pr(\pi_{TOX} > 0.20 \mid data) > 0.90$, or to a frequentist test of hypotheses? In any case, the criterion for evaluability obviously anticipates the possibility that some patients may not complete the planned four cycles of therapy, and the stopping rule for toxicity anticipates the possibility of severe adverse effects.

Table 10.3 gives an example of data from a two-arm randomized trial to compare treatments A and B, based on the rates of toxicity and response. In this trial, to be scored as a responder, a patient has to be evaluable, respond, and not suffer toxicity. This definition is conservative, in that if toxicity occurs it precludes response, that is, toxicity and response are disjoint events. Table 10.3 gives the initial sample sizes of 100 per arm, the numbers of early dropouts, who by definition were inevaluable, and the rates of toxicity and response among the evaluable patients. The table says that the two treatments have the same 10% rate of toxicity, but the response rate is 90% for B compared to only 50% for A. Since the sample sizes both are 100, this appears to give strong evidence that B is a much better treatment than A.

A closer look at the data shows that these conclusions are complete nonsense. First of all, dropout is an outcome, not an inconvenience. The key question is, why would a patient whose life is at risk from their cancer not complete their therapy? A patient discontinues their treatment for a reason. In cancer therapy, the ultimate purpose of treatment is to keep the patient alive by killing the cancer. The reason that a cancer patient chooses to discontinue chemotherapy, that is, "dropout," almost invariably is that the treatment is so toxic the patient can no longer endure it. In some cases, if the toxicity is causing permanent damage to the patient's organs, such as the liver, kidneys, or brain, the attending physician will discontinue the treatment. For example, what good is it to tell a patient, "The treatment achieved a response, which increases your chance of long term survival. Unfortunately, the treatment also destroyed your kidneys, so you must spend the rest of your life on a dialysis machine."

Here is a somewhat more revealing way to summarize the data from this trial. The first step is simply to admit that a dropout after 1 course of chemo actually is discontinuation of therapy due to toxicity. So, there actually were $20 + 8 = 28$ toxicities in Arm A and $90 + 1 = 91$ toxicities in Arm B. Table 10.4 summarizes the data in a way that avoids the fiction that dropouts after one cycle of therapy did not count, and classifies the toxicities as early (after cycle 1) or late. Table 10.4 does not bother with percentages, since each count is out of 100 patients in each arm. If you don't like scoring patients who discontinued therapy after one cycle as having toxicity, you might change the wording "early toxicity" to "discontinued therapy after one cycle." Of course, this begs the question to which one already knows the answer. The more detailed, much more revealing summary of the data given in Table 10.4 shows that A had a much higher response rate than B, 40% versus 9%, and also had a much lower toxicity ("treatment failure") rate, 28% versus 91%. So, it is obvious that A is a much better treatment than B in terms of both toxicity and response. *This is the opposite of the conclusion that Table* 10.3 *implies.* Table 10.4 also shows that 90 of the 91 toxicities with B were early, rather than calling them "dropouts."

To do a Bayesian analysis, I will assume that the probabilities $\pi_t = (\pi_{t,1}, \pi_{t,2}, \pi_{t,3}, \pi_{t,4})$ of the four possible elementary outcomes in each treatment arm $t = A$ or B follow a non-informative Dirichlet (0.25, 0.25, 0.25, 0.25) prior. Recall that a Dirichlet distribution is an extension of a beta distribution that accommodates more than two elementary outcomes. This prior assumes that the four elementary outcomes are equally likely, and it has an effective sample size $0.25 + 0.25 + 0.25 + 0.25 = 1$, so is non-informative. A nice property of the Dirichlet is that it is conjugate, and the posterior parameters are obtained very simply, by just adding observed count of each elementary outcome to the corresponding prior parameter. In this example, the posterior of π_A is Dirichlet (40.25, 20.25, 8.25, 32.25) and the posterior of π_B is Dirichlet (9.25, 90.25, 1.25, 0.25). Another nice property of the Dirichlet is that the distribution of any sub-event is beta with parameters obtained as sums of the corresponding Dirichlet parameters. For example, in Arm A, the probability $\pi_{A,Best}$ of the best possible outcome, Best=[response without toxicity], follows a beta(40.25, 60.75) posterior. Based on this, the 95% posterior credible intervals for the probabilities of Best are $0.31 - 0.50$ with A and $0.04 - 0.15$ with B. Not only do these two credible intervals not overlap, but the lower bound 0.31 of the ci for $\pi_{A,Best}$ is far above the upper bound 0.15 of the ci for $\pi_{B,Best}$. Two key comparative posterior probabilities are

$$\Pr(\pi_{A,Best} > \pi_{B,Best} \mid data) > 0.99999999,$$

and

$$\Pr(\pi_{B,Toxicity} > \pi_{A,Toxicity} \mid data) > 0.99999999.$$

You could bet your life that A is a better treatment than B. There is a greater risk of death driving to the grocery store than losing this bet. The Bayesian analysis is just a formal way of saying that, once one has accounted honestly for the possible

Table 10.4 Comparison of treatments A and B based on counts of the four elementary toxicity and response outcomes. Posterior credible intervals for the four elementary outcome probabilities within each treatment arm were computed assuming a non-informative Dirichlet (0.25, 0.25, 0.25, 0.25) prior on the probabilities of the four elementary outcomes

	Elementary outcomes			
	Response and no toxicity	Early toxicity	Late toxicity	No response and no toxicity
Arm A	40/100	20/100	8/100	32/100
95% posterior ci	0.31 − 0.50	0.13 − 0.28	0.04 − 0.14	0.23 − 0.41
Arm B	9/100	90/100	1/100	0/100
95% posterior ci	0.04 − 0.16	0.83 − 0.95	0.00 − 0.04	0.00 − 0.02

elementary outcomes, the data tell us we can be virtually certain that (1) A has a much higher probability of the event Best = [response without toxicity] than B, and that (2) B has a much higher probability of toxicity than A. This is the opposite of the earlier conclusion based on the "evaluable" patients.

The key difference between the two analyses of the data is very simple. Table 10.4 does not play The Evaluability Game by excluding patients who discontinued treatment before computing response percentages. Given the summary in Table 10.4, the observation that patients with toxicity discontinued their therapy after one cycle says that B is so toxic that most patients simply were either unable or unwilling to receive enough of it for it to have a substantive anti-disease effect. This example illustrates a central issue in oncology, that many treatments have harmful effects, and ultimately any treatment evaluation must consider risk–benefit trade-offs. A similar, slightly more complicated version of this example would allow a patent to have both toxicity and response, but the main points would not change.

While this example may seem to be extreme, it illustrates a very common mistake in treatment evaluation. In any setting where dropout is not some completely random event but actually is due to adverse treatment effects, discarding dropouts as being inevaluable misrepresents the trial results. What nominally may be called a "dropout," or a patient being considered "inevaluable" because some predetermined amount of treatment was not given, actually is an outcome. If one removes the dropouts before computing the rates of desirable clinical outcomes, such as response or survival time, using the rationale that they are "inevaluable," the results can be very misleading. While, by definition, the response rate was 10% among nominally evaluable patients, this must be accompanied by a transparent account of what actually happened, as shown in the table with all four possible outcomes. If one were to define the nominal outcome as a binary Response variable, but count a patient as "inevaluable" if RRT = regimen-related toxicity occurs, and Response and RRT are disjoint events, then even in this simpler case, there are three possible elementary outcomes: response, RRT, and neither response nor RRT.

No rational physician would want to give a treatment with a 90% toxicity rate to their patients, unless perhaps "toxicity" is an unpleasant but transient event and response is a necessary precursor to long-term survival. In contrast, if a patient never received any therapy at all for some reason, then that is a different matter.

A closely related version of this problem is the causal relationship between treatment, the time when treatment is discontinued, and response. To think about this, change "response" to survival time, and consider how treatment may play out over time. For example, consider a randomized trial in allogeneic stem cell transplantation (allosct) in which patients in the experimental arm are scheduled to receive a particular agent, A, following allosct each month for 12 months, while patients in the control arm receive standard of care during follow-up. The motivating idea is that A given post-transplant may decrease the disease recurrence rate, and thus improve survival time. Suppose that only 30% of the patients in the experimental arm complete the planned 12 monthly administrations (cycles) of A, and the number of cycles received varies substantially on the set $\{1, 2, \ldots, 12\}$. A key question is whether adverse events, or some other variable occurring over time post-allosct, are associated with whether and when a patient in the experimental arm discontinues therapy. For example, if toxicity is positively associated with the number of cycles of A received, but more A also is associated with longer survival time, then it is not obvious what is meant by "the effect of A." So this turns out to be a nontrivial problem in causal inference.

Takeaway Messages About Evaluability

1. In any clinical trial protocol, or data analysis, look carefully at how "evaluable" is defined and, given that, how patient outcomes are defined and counted.

2. Do not allow yourself to be misled by a relabeling of toxicity or dropout as "inevaluable." Toxicities and dropouts are outcomes, not incidental complications to be swept aside.

3. Safety is never a secondary consideration in a clinical trial.

4. To figure out what is going on in a dataset, if possible, write down all of the possible elementary clinical outcomes.

10.4   The Fox, the Farmer, and the Chickens

A farmer, after discovering that several of his chickens had been killed and eaten by an unknown predator, decided to do something about it. After careful thought and a detailed analysis of this dire situation, the farmer decided to hire a fox to guard his chickens. His reasoning was that nobody knows more than a fox about killing chickens, so his newly hired guard fox should do an excellent job. Running a clinical trial is not the same thing as running a farm, but there are some similarities.

Recall that safety is never a secondary consideration in any clinical trial. What this means in practical terms is that, if toxicity is observed at an unacceptably high rate, there must be explicit rules in place to stop an individual patient's treatment until the toxicity is resolved, de-escalate dose, suspend accrual, or stop the trial entirely. Also recall that what is at stake in a given trial may be very different for a patient enrolled in the trial, future patients treated after the trial, the trial's PI, a federal regulatory agency, or the pharmaceutical company that provided an investigational drug and funding for the trial. They all have different utilities, and thus different reasons for being involved in the trial. But any clinical trial protocol should contain explicit rules for protecting patient safety.

The following example illustrates an increasingly common practice with regard to interim monitoring, for either safety or efficacy, in clinical trial protocols. Consider a design for a phase II-III clinical trial to treat patients with advanced liposarcoma that cannot be resected, that is, surgical removal of the tumor is not possible. The primary endpoint in both phases is progression-free survival (PFS) time. In phase II, a total of 50 patients will be randomized fairly between a new agent, A, and placebo, P. Physicians and patients will be blinded to the randomization, and patients will be stratified by number of prior therapies (1 versus ≥ 2). No interim futility or safety rules for stopping the phase II trial early are given. At the discretion of the treating physician, a patient on the P arm may be crossed over to receive A once their disease has progressed. Without suspending enrollment, phase III will immediately follow phase II, with 130 patients randomized to A and 65 to P, again with no interim decision or stopping rules. At the end of phase III, a one-sided, 0.025-level test of whether A has longer expected progression-free survival time than P will be performed, using all of the phase II and phase III data, based on a fitted Cox model.

Problems with This Design

1. The use of a placebo arm implies that no treatment with substantive anti-disease effect exists for this disease. However, no explanation or rationale for this is provided.
2. Since there are no interim decisions or stopping rules in either phase, it raises the question of whether it will be ethical to continue randomization if a very large superiority of A over P in terms of PFS time is seen based on the interim data. The same issue pertains to possible adverse effects of A.
3. What if the PFS data do not satisfy the proportional hazards (PH) assumption that underlies the validity of the Cox model? Will this be checked using a goodness-of-fit analysis? If so, and if the data show that the PH assumption is violated, will a different test that is not based on the Cox model be performed?
4. What if the P arm turns out to have substantially longer PFS, on average, than the A Arm? The one-sided test ignores this possibility.

With regard to point 2, above, the protocol does include the following interesting statement:

"Accrual to the trial may be stopped early at the sole discretion of the Sponsor for any reason, including medical or ethical reasons that affect the continued performance of the study, or difficulties in the recruitment of patients."

Notice the vagueness of this general statement. Since the primary motivation for a pharmaceutical company to fund a clinical trial is to get their new drug approved and market it, one may ask who will protect the patients enrolled in the trial if A has much worse toxicity, or much shorter PFS, compared to P. The above statement provides no guarantee, much less specific decision rules, for doing this.

The last I heard about the guard fox, when asked one day by a fox friend what he planned to have for supper, he replied "A delicious chicken. It's a perk of my job."

10.5 Planned Confounding

Never try to teach a pig to sing. It wastes your time, and it annoys the pig.
Robert Heinlein

The following example may be considered a toy version of many actual clinical trials that involve multiple experimental treatments for a given disease. The example is purely hypothetical. You may decide whether it resembles any real trials that you may have seen, or in which you may have been involved. The hypothetical Principal Investigator is Doctor Happy. The hypothetical trial enrolls patients with advanced or metastatic non-squamous non-small cell lung cancer (NSCLC) who have progressive disease after receiving prior therapy. The aim of the trial is to evaluate the effects of adding each of three exciting new targeted agents, which I will denote by A_1, A_2, A_3, to a standard "normal" agent, denoted by N. The trial is designed to study the three combinations $N + A_1$, $N + A_2$, and $N + A_3$ in terms of their probabilities of tumor response. Patients are stratified into two subgroups, determined by whether they previously received another particular type of targeted therapy or not, denoted by G and $\overline{G}$, respectively. The trial is conducted in three hypothetical medical centers, M_1, M_2, and M_3.

The hypothetical design, called a "parallel phase II trial," specifies that $N + A_1$ will be studied in M_1, $N + A_2$ will be studied in M_2, and $N + A_3$ will be studied in M_3. That is, it is really three separate trials conducted at the same time in three different medical centers. Within each center, at the start, two single-arm phase II trials of the center's specified combination are conducted, one trial in the patients in subgroup G who previously received the targeted agent, and a second trial in the subgroup $\overline{G}$ of patients who previously did not receive the targeted agent. Within each center, the two subgroup-specific trials will be conducted independently of each other, so that the data from one subgroup's trial are not used in any way to affect or modify decisions in the other subgroup's trial. For each of the six (Center/Treatment, Subgroup) combinations, the plan is to start by conducting a 17-patient single-arm trial, but allow the possibility of treating an additional 83 patients as an "expansion cohort," with the decision of whether to do this or not made by each center's Principal Investigator, based entirely on their personal judgment. No explicit rules or criteria are given for deciding whether or not to treat an expansion cohort, or whether to stop any trial early for either futility or safety. No mention is made of randomization, or of any comparisons between the three treatment combinations.

To summarize, there are three treatment combinations, three medical centers, and two subgroups. The single agent N is not studied in the trial, despite the fact that this has been used as a standard therapy in this NSCLC patient group. To keep track of these three variables, denote the indices $j = 1, 2, 3$ for N with each of the added target agents, where $j = 0$ identifies the single agent N, $k = 1, 2, 3$ for the medical centers, with $g = 1$ for subgroup G and $g = 2$ for subgroup $\overline{G}$. Denote the response probabilities

$$\pi_{j,g,k} = \Pr(\text{Response} \mid N + A_j, \ g, \ k)$$

for a patient in subgroup g treated with the combination $N + A_j$ in center M_k, where $N + A_0 \equiv N$. Thus, $\pi_{0,s,j}$ corresponds to treatment with the single agent N in subgroup g and center M_j, although this is not included in the parallel phase II trial for any (Subgroup, Center) combination (g, j). But I will include an N only arm in the alternative designs that I will describe in the following text, which decouple treatments and centers, where I will consider $4 \times 3 \times 2 = 24$ possible combinations of (Treatment, Subgroup, Center). The parallel phase II trial as actually designed will study only six of these 24 combinations, without any randomization, and a vague account of how each of these six single-arm trials actually will be conducted. If one wished to simulate this trial design to investigate its properties under various possible scenarios, it would be impossible, since no rules are given either for determining sample size or making interim safety, futility, or superiority decisions, and no comparisons between treatments are planned.

If you think about it, nobody really is interested in the 24 response probabilities $\{\pi_{j,g,k}\}$. The probabilities of interest only depend on the combination $N + A_j$ for $j = 0, 1, 2, 3$, and subgroup $g = 1$ or 2. Scientifically, the center effects are a nuisance since, if one wishes to estimate and/or compare treatment effects, either overall or within subgroups, all that the centers do is increase variability due to between-center differences. Confounding treatments and centers, by design, renders such center effects indistinguishable from treatment effects.

To see what the parallel phase II design does and does not do, as a start, refer to Table 10.5. This table gives a simple cross-tabulation of centers, treatments, and subgroups, including a column for N given alone as a treatment that is not included in the trial. The purpose of this table is to identify all of the (Treatment, Subgroup, Center) combinations that are, and are not, included in the trial.

Worse than Useless

The following discussion is motivated by the simple considerations that the treatment effects are what are of interest, allowing the possibility that they may differ within the prognostic subgroups, but the center effects are nothing more than a nuisance. Moreover, inevitably, an important goal is to compare treatments, within subgroups if there are treatment–subgroup interactions, and overall. Recall that the simulation study of Wathen and Thall (2017) showed that a randomized study of several experimental treatments without a control arm is useless, and never should be conducted. Consequently, because the parallel phases II design of $N + A_1$, $N + A_2$, and $N + A_3$ does not include an N arm, and does not even randomize, it is worse than useless.

Table 10.5 Combinations of centers, treatments, and subgroups in the parallel phase II trial in non-small cell lung cancer. The symbol 'X' in a cell identifies a (Center, Treatment, Subgroup) combination included in the trial, where patients were treated in a single-arm sub-trial. Empty cells represent combinations that were not studied. No patients were randomized

Centers	Treatments in subgroup G			
	N	$N + A_1$	$N + A_2$	$N + A_3$
M_1		X		
M_2			X	
M_3				X
Centers	Treatments in subgroup $\overline{G}$			
	N	$N + A_1$	$N + A_2$	$N + A_3$
M_1		X		
M_2			X	
M_3				X

Problems with the Parallel Phase II Design

1. The parallel phase II design completely ignores the fundamental question "Compared to what?" Because patients are not randomized, and treatment-center confounding is intrinsic to the design, it is not possible to obtain unbiased comparative between-treatment estimates. For example, an unbiased estimate of the $(N + A_1)$ versus $(N + A_2)$ effect cannot be obtained, either within each prognostic subgroup $g = 1, 2$ or overall, since this effect is confounded with the M_1 versus M_2 between-center effect.

2. Because each $N + A_j$ is studied in M_j only, the effect of $N + A_j$ is confounded with the effect of center M_j. Thus, in terms of the response probability, only a confounded treatment-center effect, of $N + A_j$ in M_j, can be estimated within each prognostic subgroup $g = 1$ or 2.

3. Because N alone is not studied in the trial, and moreover patients are not randomized, it is not possible to estimate the effects of adding A_1, A_2, or A_3 to N on the probability of response, or any other outcome, either within subgroups or overall. For example, if adding A_1 to N is completely ineffective, this can only be determined by randomizing and estimating the $N + A_1$ versus N effect on the probability of response, overall or within each subgroup.

4. There is no provision for borrowing strength between the sub-trials of the two subgroups, either within each center-treatment combination or overall. The design implicitly assumes that the response rate in one subgroup has no relationship with the response rate in the other subgroup.

To summarize things more briefly, the flaws of this parallel phase II design arise from the following facts:

• Patients are not randomized between treatments.

• N is not included as a control arm to evaluate the effects of adding the targeted agents.

• Each of the three treatment combinations is perfectly confounded with a center.

The hypothetical Doctor Happy and the other investigators were quite pleased with this parallel phase II design, however, and in the hypothetical protocol they made the following hypothetical statement:

"To control bias in assignment of individual patients to treatment arms and to mitigate the risk of medication errors with the three investigational study treatments administered in specific regimens, study sites will be aligned with one specified treatment combination."

If this statement were not hypothetical, it would be an astonishingly egregious misuse of the word "bias," since bias actually is built into the design by treatment-center confounding. To anyone with an understanding of elementary experimental design, this statement is akin to saying

"Up is down, black is white, and there are fairies living in the bottom of my garden who come out and sing at night."

Here is the actual definition of bias.

The Actual Definition of Bias. A statistic $\hat{\theta}$ is an *unbiased* estimator of a parameter θ if the distribution of $\hat{\theta}$ has mean θ.

This is written formally as $E(\hat{\theta}) = \theta$. The whole point of randomization is to obtain unbiased estimators of between-treatment effects. If one simply wants to estimate the six (Treatment, Subgroup) response probabilities, $\{\pi_{j,g} : j = 1, 2, 3, g = 1, 2\}$, this also is impossible with the parallel phase II design, because each combination $N + A_j$ is confounded with the center M_j in which it is tested. The only sample proportions that can be estimated are the confounded treatment-center response probabilities $\pi_{j,g,j} = \Pr(\text{Response with } N + A_j \text{ in center } M_j \text{ for subgroup } g)$. Due to the designed confounding, it is impossible to estimate and remove the center effects, so an estimator of $\pi_{j,g}$, the response probability of $N + A_j$ in subgroup g without the effect of center M_j, cannot be computed. In turn, unbiased estimators of between-treatment comparisons cannot be computed, either overall or within the subgroups.

For example, the response data on $N + A_1$ pertain to how this combination works in center M_1. If there is substantial between-center variation, say with a lower response rate in M_1, compared to other centers, on average, for any treatment of NSCLC, then the data may misleadingly make $N + A_2$ and $N + A_3$ appear to be superior treatments compared to $N + A_1$. That is, *what actually are center effects may be seen as treatment effects, with no way to disentangle the two.* Center effects can be quite large, and typically are a result of some combination of differences in supportive care, additional treatments such as antibiotics, patient selection that may be either intentional or not, physician effects, nurse effects, and unknown latent variables affecting outcome that are unbalanced between centers purely due to the play of chance.

As a simple numerical example of how treatment-center confounding may work in a parallel phase II trial, for simplicity consider a single prognostic subgroup and suppress the subgroup index g. Denote $\pi_{j,j} = \Pr(\text{Response with } N + A_j \text{ in center } M_j)$, where $j = 1, 2,$ or 3 due to designed confounding of center with agent. Suppose that, in fact, the three treatment effects are identical, that is, $N + A_1$, $N + A_2$, and

$N + A_3$, have exactly the same true response probability, $\pi = e^\mu/(1 + e^\mu)$, but the (Treatment, Center) response probabilities follow a logistic regression model of the form

$$\text{logit}(\pi_{j,j}) = \log\left(\frac{\pi_{j,j}}{1 - \pi_{j,j}}\right) = \mu + \alpha_j, \quad \text{for } j = 1, 2, 3.$$

Thus, μ is the baseline effect of all three treatments, and α_j is the additional effect of center M_j, for $j = 1, 2, 3$. Suppose that the true (Treatment, Center) response probabilities are $\pi_{1,1} = 0.10$, $\pi_{2,2} = 0.20$, and $\pi_{3,3} = 0.40$. If the data-based estimates reflect these values, then it may appear that, aside from variability in the data, $N + A_3$ has twice the response probability of $N + A_2$, which in turn has twice the response probability of $N + A_1$. Solving the above equations shows that, assuming $\alpha_1 = 0$ for simplicity, $\mu = \text{logit}(0.10) = -2.197$, and the real-valued M_2 versus M_1 between-center effect is $\alpha_2 = 0.811$, which is $0.20 - 0.10 = 0.10$ on the probability domain. Similarly, the real-valued M_3 versus M_1 between-center effect is $\alpha_3 = 1.792$, which is $0.40 - 0.10 = 0.30$ on the probability domain. So, in terms of response probabilities, what may appear to be large differences in treatment effects, reflected by the response probabilities 0.10, 0.20, and 0.40, actually are due entirely to between-center effects. Consequently, response probabilities very different from these values will be seen in subsequent studies conducted in other medical centers, or in clinical practice. This sort of flawed design is a prominent reason that results of such clinical trials cannot be replicated in future studies.

Simply put, studying three treatments in separate single-arm phase II trials, each conducted within a different medical center, yields data that are worse than useless. This trial may be considered a scientific disaster before it is run. In terms of future patient benefit, the data resulting from this parallel phase II trial, or any similarly designed trial, is very misleading and potentially harmful to future patients. The existence of center effects, which may be as large as, or larger than, treatment effects, is not a new discovery. For many years, statisticians involved in clinical trial design have worked to develop methods that control for center effects in multicenter studies, so that one may obtain unbiased, reliable estimates of treatment effects. Reviews of methods that adjust for center effects in multicenter randomized clinical trials are given, for example, by Localio et al. (2001) and Zheng and Zelen (2008).

A fundamental method for dealing with nuisance factors, like the three centers in this hypothetical trial, is called "blocking," which has been around for a century, and is discussed in any elementary textbook on experimental design. But blocking is not used in this design, which does not even include randomization. Early on, blocking was a basic tool in agricultural experiments that involved nuisance factors, such as different plots of land, when comparing different types of seeds or fertilizers in terms of yield per acre. Blocking on plots of land used in the studies was done to remove plot effects, and thus avoid mistaking plot effects for treatment effects. In any multicenter clinical trial, the medical centers should be used as blocks, since they are sources of variability and their effects are not of interest. Any reasonable design should involve randomization of patients among the treatments within each medical center. If this is not done and, say, center M_3 has a comparatively positive effect on response probability compared to centers M_1 and M_2 and, due to the play of chance, M_3 receives a disproportionately larger number of $N + A_3$ patients in an unblocked randomization, then it may appear, as in the numerical example above, that $N + A_3$ is superior, when in fact observed differences in responses rates are due mainly to the superiority of center M_3. One may regard a parallel phase II trial as the opposite of blocking, since it ensures the most extreme imbalance possible, namely, that all $N + A_j$ patients are treated in M_j for each j.

In summary, here are three key elements of any reasonable design in this sort of setting:

1. Patients should be randomized between treatment arms to obtain unbiased comparisons.

2. Given what Wathen and Thall (2017) learned about the uselessness of multi-arm trials of several experimental treatments without a control arm, the trial should include a control arm in which N alone is administered.

3. The randomization should block on medical centers.

As discussed in Chap. 6, there are numerous possible ways to design a clinical trial to evaluate multiple treatments. Obviously, Dr. Happy's "parallel phase II" design is not one that should be used. Ever. How lucky we all are that this trial, and the patients enrolled in it, are purely hypothetical.

10.6   Select-and-Test Designs

Ellenberg and Eisenberger (1985) addressed the problem of bias that results from the conventional practice, most commonly seen in oncology, of conducting a single-arm trial of an experimental agent E to obtain an estimate, $\hat{\pi}_E$ of its response rate, and then using this as a basis for deciding whether to conduct a large randomized trial of E versus C. They noted the problem that, given some criterion used to evaluate whether $\hat{\pi}_E$ is large enough for E to be considered promising based on a single-arm study, E often performs less well in a subsequent phase III trial. To avoid bias in any comparison of E to C, they proposed a two-stage phase II-III design that randomizes between E and C throughout. The design includes an interim rule to stop for futility after stage 1 based on the unbiased stage 1 estimator of the E versus C effect, and otherwise the randomization is continued in stage 2, with a conventional comparative test done at the end. At the time, the idea of randomizing in phase II was quite iconoclastic. Unfortunately, to this day single-arm phase II trials still are conducted quite commonly, especially in oncology. The parallel phase II design described above just does three of them at the same time.

A harder, more realistic problem is that of evaluating and screening several experimental treatments, $E_1, \ldots, E_k$ for $k \geq 2$, and comparing one or more of the best of them to a control treatment, C. Conducting a $k + 1$ arm confirmatory randomized phase III trial usually is unrealistic, because it would require an extremely large sample size to make so many between-treatment comparisons reliably. A common strategy is to first screen the E_j's. But if one does a preliminary randomized trial and selects the E_j having largest empirical response probability, $\hat{\pi}_{[k]}$, then this procedure will suffer from selection bias. Recall that, even if the true response probabilities are identical, that is, $\pi_1 = \pi_2 = \cdots = \pi_k = \pi$, then the distribution of the selected maximum estimator $\hat{\pi}_{[k]}$ will have a mean larger than π. For example, if 90 patients are randomized to E_1, E_2, E_3 with samples of $n = 30$ each, and in fact $\pi_1 = \pi_2 = \pi_3 = 0.20$, then the mean of $\hat{\pi}_{[3]}$ is 0.26, not 0.20. Unavoidably, selection of the best of several treatments based on empirical response rates, mean or median survival times, etc., overstates the mean of the selected treatment in this way.

The following two-stage phase II-III design, proposed by Thall et al. (1988), solves this problem. The design compares experimental treatments $E_1, \ldots, E_k$ and control treatment C based on binary outcomes, assuming that patients are homogeneous. I will refer to this as the "TSE" design. It was the first of a family of what came to be known as "select-and-test" designs. After establishing the TSE design, I will explain how it may be extended and modified to accommodate patient subgroups, or use a time-to-event rather than a binary outcome. The TSE design randomizes throughout, does treatment selection in stage 1 and, if the trial is not stopped early due to the decision that no E_j is promising compared to C, it does a two-arm comparison in stage 2 based on the pooled data from both stages. The design is constructed to control both the overall false positive and false negative error probabilities. In stage 1, $(k + 1)n_1$ patients are randomized fairly among $E_1, \ldots, E_k$ and C. The response probabilities are denoted by $\pi_1, \ldots, \pi_K$ and π_0, with $j = 0$ indexing the control C. To utilize established ranking and selection theory, as given by Bechhofer et al. (1995), Thall et al. (1988) based the design on a specified fixed $\delta_1 > 0$ defined to be a small, clinically insignificant improvement, and $\delta_2 > 0$ a larger value such that $\pi_0 + \delta_2$ is considered a clinically meaningful improvement over π_0. The design is based on a two-stage group sequential testing procedure of the global null hypothesis $H_0 : \pi_1 = \cdots = \pi_k = \pi_0$ versus the one-sided alternative hypothesis that at least one $\pi_j > \pi_0$, for $j = 1, \ldots, k$.

To define the test statistics used at each of the two stages, first let $\hat{\pi}_{j,s}$ denote the usual sample proportion in the jth treatment arm used to estimate π_j based on the data from stage $s = 1$ or 2. Denote the variance stabilizing transformation $\phi(\pi) = sin^{-1}\sqrt{\pi}$, which is used to obtain improved normal approximations for the test statistic distributions. Denote $Z_{j,s} = (4n_s)^{1/2}\phi(\hat{\pi}_{j,s})$ for each $j = 0, 1, \ldots, k$, and $s = 1, 2$, and let $[k]$ represent the index of the experimental treatment having the largest estimated stage 1 response proportion $\hat{\pi}_{j,1}$. That is, $\hat{\pi}_{[k],1}$ is the maximum of $\{\hat{\pi}_{1,1}, \ldots, \hat{\pi}_{k,1}\}$. The stage 1 test statistic is

$$T_1 = \frac{1}{\sqrt{2}} \max_{1 \le j \le k} (Z_{j,1} - Z_{j,0}).$$

The stage 2 test statistic is the weighted average

$$T_2 = \frac{1}{\sqrt{2}} \left\{ \frac{n_1}{n_1 + n_2}(Z_{[k],1} - Z_{0,1}) + \frac{n_2}{n_1 + n_2}(Z_{[k],2} - Z_{0,2}) \right\},$$

which is based on the pooled $E_{[k]}$ and C data from both stages. Let t_1 and t_2 denote the stage 1 and stage 2 test cutoffs. At the end of stage 1, if $T_1 \leq t_1$ then the trial is terminated and the global null hypothesis H_0 is accepted, with no E_j selected. If $T_1 > t_1$, then an additional $2n_2$ patients are randomized fairly between $E_{[k]}$ and C in stage 2, the test statistic T_2 is computed, and a final test is performed with H_0 is rejected in favor of $\pi_{[k]} > \pi_0$ if $Z_2 > t_2$, and H_0 is accepted if $T_2 \leq t_2$.

To account for the hybrid "select-and-test" nature of this two-stage design, the usual definition of power must be extended to that of *Generalized Power* (GP), as follows. Assume for simplicity that $\pi_1 \leq \pi_2 \leq \cdots \leq \pi_k$, and denote $\pi = (\pi_0, \pi_1, \ldots, \pi_k)$. First, the *least favorable configuration* (LFC) of π is defined as follows. Suppose that (1) at least one $\pi_j \geq \pi_0 + \delta_2$, that is, at least one E_j is "superior" in that π_j provides at least the desirable targeted improvement δ_2 over π_0, and (2) no π_j is between $\pi_0 + \delta_1$ and $\pi_0 + \delta_2$. This is motivated by the idea that it is impossible for any statistical method to reliably distinguish between arbitrarily close parameter values. The *least favorable configuration* (LFC) π^* is defined as the parameter vector where $\pi_1 = \cdots = \pi_{k-1} = \pi_0 + \delta_1$ and $\pi_k = \pi_0 + \delta_2$. The GP of the two-stage select-and-test design is defined to be the probability, $1 - \beta(\pi)$, that the design rejects H_0 in favor of a truly superior E_j under the LFC. This requires two actions, that a superior E_j is selected in stage 1, and H_0 is rejected in stage 2. So, it is a much more demanding requirement than that of usual power, since it includes the requirement that the stage 1 selection is correct. It is easy to prove that, under assumptions (1) and (2), the LFC minimizes the GP. Since the sample size is either $n = (k + 1)n_1$ or $n = (k + 1)n_1 + 2n_2$, Thall et al. (1988) derived values of the four design parameters $\{t_1, t_2, n_1, n_2\}$, for given overall Type I error and GP, to minimize the expected sample size $E(n) = (k + 1)n_1 + 2n_2 \Pr(T_1 > t_1)$, computed as the equally weighted average $(1/2)E(n \mid H_0) + (1/2)E(n \mid \pi^*)$ between $E(n)$ under H_0 and under the LFC. Figure 10.1 gives a schematic of the select-and-test decision-making process of the TSE phase II-III design.

To apply the TSE design within each of the G and $\overline{G}$ subgroups of the parallel phase II trial example, $k = 3$, $E_j = N + A_j$ for $j = 1, 2, 3$ and $C = N$. Assuming $\pi_0 = 0.20$, Type I error 0.05, $\delta_1 = 0.05$, and $\delta_2 = 0.20$, a design with GP $= 0.80$ would

Fig. 10.1 Illustration of the randomization and decision-making process of the TSE phase II-III select-and-test clinical trial design, with three experimental treatments

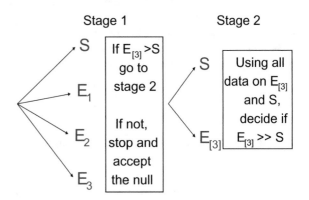

require $n_1 = 48, n_2 = 57$, with test cutoffs $t_1 = 0.762$ and $t_2 = 1.902$. The total stage 1 sample size would be $4n_1 = 192$, the stage 2 sample size would be $2n_2 = 114$, the maximum overall sample size would be $192 + 114 = 306$, and the expected sample size would be 267. One may compare these values to the completely arbitrary maximum sample size of $3(17 + 83) = 300$ of the parallel phase II design conducted in one of the two subgroups.

An elaboration of the TSE design that accommodates the two subgroups G and $\overline{G}$ in a single trial while borrowing strength between them, rather than conducting two separate trials will be given in Chap. 11. For settings where the primary outcome is an event time, Schaid, Wieand, and Therneau (1990), (SWT) provided a very useful two-stage phase II-III design that is logically similar to the TSE design. The major difference in terms of possible decisions is that the SWT design allows more than one of the E_j's to be selected for study in the second stage. To deal with event times, SWT make a proportional hazards assumption, with the design's decisions based on the hazard ratios λ between the E_j's and C. Fixed decision times $t_1 < t_a < t_2$ and decision cutoffs $x_1 < x_2$ and x_3 must be specified, along with the sample sizes n_1 and n_2. As in the TSE design, $(k + 1)n_1$ patients are randomized fairly among $E_1, \ldots, E_k$ and C in stage 1. Denote the log-rank statistic comparing E_j to C at study time u by $T_j(u)$. The trial is conducted as follows.

Stage 1
1. If $\max\{T_1(t_1), \ldots, T_k(x_1)\} < x_1$ then the trial is terminated with all E_j's declared not promising compared to C.
2. If any $T_j(t_1) > x_2$ then the trial is terminated early for superiority, with the conclusion that all such treatments provide a substantive survival improvement over C.
3. Otherwise, all $k_2 \le k$ treatments for which $x_1 \le T_j(t_1) \le x_2$ are moved forward to stage 2.

Stage 2
1. Randomize $(k_2 + 1)n_2$ patients fairly among the selected treatments and C.
2. Terminate accrual at time t_a.
3. Make final pairwise comparisons at time t_2, with the conclusion that E_j is superior to C if $T_j(t_2) > x_3$.

SWT derived design parameters to control the pairwise comparison false positive probability to be $\le \alpha$ and the pairwise power to be $\ge 1 - \beta$, while minimizing the null expected total sample size for given α, $1 - \beta$, targeted pairwise hazard ratio λ, follow-up duration $t_2 - t_a$, and accrual rate, assuming uniform accrual during $[0, t_a]$ and an exponential event time distribution. For example, if the pairwise hazard ratio $\lambda = 1.5$ is targeted, $k = 2, 3$, or 4, and the approximate overall Type I error $k\alpha = 0.05$ and pairwise power $= 0.80$, for accrual rate $= 50$, the null expected total sample size ranges from 233 to 488. For the larger targeted hazard ratio 2.0, the expected sample size ranges from 109 to 188. An advantage of this more flexible approach with time-to-event outcomes is that, since survival differences may not be seen until the second stage, it provides protection against false negative decisions in the stage 1 screening. The SWT design has the desirable properties that the stage 1 data are incorporated into

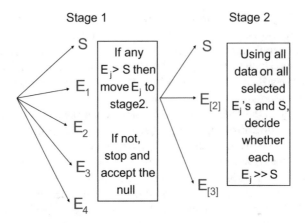

Fig. 10.2 Illustration of the randomization and decision-making process of the SWT phase II-III clinical trial design, with four experimental treatments, where two experimental treatments are selected for further study in stage 2 of the trial

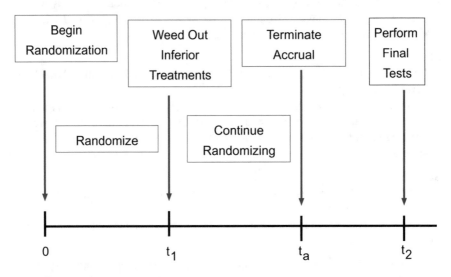

Fig. 10.3 Illustration of how the randomization and decision-making of a trial run using the SWT phase II-III clinical trial design play out over time

the stage 2 tests, it randomizes patients throughout to obtain unbiased comparisons, and it controls the pairwise false positive and false negative error rates. Figures 10.2 and 10.3 give schematics of the select-and-test decision-making process of the SWT phase II-III design.

Since publication of the TSE and SWT designs, numerous elaborations of the two-stage phase II-III select-and-test paradigm have been proposed. These include designs allowing interim hypothesis selection, as described by Bretz et al. (2006), or more than two stages with group sequential decisions. Some references are Bauer and Kieser (1999), Royson et al. (2003), Stallard and Todd (2003), Kelly et al. (2005), Schmidli et al. (2006), Jennison and Turnbull (2007), and Hampson and Jennison (2015), among many others.

Takeaway Messages About Select-and-Test Designs

1. They do away with all of the problems caused by a parallel phase II design, including treatment-center confounding.
2. They provide a very efficient way to screen multiple experimental treatments based on unbiased comparisons to standard treatment.
3. They control overall false positive and false negative decision probabilities.
4. They require about the same maximum sample size, but provide a smaller expected sample size, compared to a parallel phase II design.
5. They impose the additional burden that patients must be randomized while treating medical centers as blocks.

Chapter 11
Getting Personal

It is far more important to know what person the disease has than what disease the person has.
Hippocrates

Contents

Abstract In day-to-day practice, a physician uses each patient's individual characteristics, and possibly diagnostic test results, to make a diagnosis and choose a course of treatment. Because the best treatment choice often varies from patient to patient due to their differing characteristics, routine medical practice involves personalized treatment decisions. The advent of sophisticated machines that provide high-dimensional genetic, proteomic, or other biological data for use in this process has made it much more complex, and this often is called "precision" or "personalized" medicine. In this chapter, I will discuss the simple version of personalized medicine in which one or two patient covariates or subgroups may interact with treatment. Examples will include (1) a biomarker that interacts qualitatively with two treatments, (2) an illustration of why the routine practice of averaging over prognostic subgroups when comparing treatments can lead to erroneous conclusions within subgroups, (3) a randomized trial design that makes within-subgroup decisions, (4) a phase II–III select-and-test design that makes within-subgroup decisions, and (5) a Bayesian nonparametric regression survival analysis that identifies optimal dosing intervals defined in terms of the patient's age and disease status.

© Springer Nature Switzerland AG 2020
P. F. Thall, *Statistical Remedies for Medical Researchers*, Springer Series
in Pharmaceutical Statistics, https://doi.org/10.1007/978-3-030-43714-5_11

11.1   From Bench to Bedside

Physicians have been practicing precision medicine for thousands of years. What the phrase "precision medicine" actually means is that a physician considers all of a patient's known covariates, including the observed signs and symptoms of their disease, and their personal characteristics, such as age and general health status, when choosing a treatment. Simply put, proper medical practice has always been precision medicine. Using the new phrase "precision medicine" has become very fashionable in recent years, in part because a great deal of new biological information has become more readily available from various technologies, such as microarrays, and it sounds very appealing. In terms of medical practice, the issues are whether such modern individual covariate information is of any use, and if so how it can be applied reliably to choose a better treatment for each patient. These are statistical problems that can be solved only if high-quality data including covariates, treatments, and outcomes on a reasonable number of patients are available. Again, no matter how promising or exciting preclinical data may be, the only way to find out how a treatment, or treatments, work in humans with a given disease is to administer it to humans with the disease and see what happens. The scientific experiments for doing this are called clinical trials.

Somehow, these basic facts seem to be ignored by many laboratory-based researchers. It appears likely that they, and many people who see news reports of breakthroughs in preclinical laboratory experiments, believe that a positive treatment effect seen in cells or xenografted mice is sufficient evidence that substantial beneficial clinical effects will be seen in humans with the disease being studied.

Here is an example. A researcher may repeat a laboratory experiment until they get data showing that an experimental agent, E, kills a higher percentage of "biomarker +" cancer cells than "biomarker −" cancer cells. This often is based on bioinformatics-based technology and may motivate the following two clinical trials. In biomarker + patients, first, a $3 + 3$ algorithm is used to run a very small phase I trial to find a putatively "safe" dose, the MTD, of E, which in fact is likely to be unsafe, ineffective, or both. Next, a single-arm phase II trial, or possibly an "expansion cohort" without any interim monitoring rules, is run, again enrolling only in biomarker + patients, all treated at the MTD. Any existing standard therapy is ignored. One then may record numerous outcome variables, such as disease status (CR, PR, stable disease, progressive disease) at one or more times, numerous biological variables, and the rates of numerous subgroup-specific outcomes, each accompanied by a p-value. Any outcome–subgroup combinations with p-values <0.05 are reported, and the treatment then is advertised as "precision medicine" for biomarker + patients in the subgroups having "significant" p-values.

Here are some things that are very wrong with this sort of approach. I have discussed all of these errors in earlier chapters. I just didn't call them "precision medicine."

1. Because biomarker − patients were not treated with E, whether the effect of E on any given outcome differs between biomarker + and biomarker − patients cannot be estimated.

2. Because patients were not randomized between E and a standard therapy S, the E versus S effect on any given outcome cannot be estimated within biomarker + patients, within biomarker − patients, or overall.

3. Cherry-picking subgroups with "significant" p-values risks mistaking random variation for an actual subgroup effect.

So, with this common approach to "precision medicine" research, you cannot determine whether the biomarker actually matters in terms of treatment effect, or whether E actually provides a treatment advance. You also have a risk of making false positive discoveries of "biomarker sensitive" subgroups, and this risk increases with the number of subgroups examined. Here are two obvious questions:

Obvious Question 1: What should one do if being biomarker + does not matter, because (a) E misses its target or (b) E hits the biomarker target, but this does not translate to therapeutic benefit?

Obvious Question 2: What should one do if E is (a) less efficacious or (b) more toxic compared to S? Is it ethical to continue a phase II trial, or to treat a large expansion cohort, if the data show that one or both of these possibilities is likely?

A pediatric oncologist once asked me to help him design a clinical trial to study the effects of targeted agents on brain tumors in children. He had constructed an extremely detailed table with columns labeled "Pathway," "Genomics Biomarker," and "Agent." In each row of the table, the entries in the "Pathway" column were things like Akt, growth factor TRK, DNA repair, and Angiogenesis. Each "Genomics Biomarker" entry was a long list of genes or proteins with alphanumeric names. Each entry in the "Agent" column included one to three targeted agents. The table had 15 rows. It was obvious that he had invested quite a lot of thought and time to construct the table. Here is what he proposed to do. A study committee comprised of two neuro-oncologists, two neuropathologists, a neurosurgeon, a neuroradiologist, and a pharmacist would examine the "Pathway" and "Genomics Biomarker" data for each newly enrolled child once the child's pathway and genomics data had been evaluated. Based on this, the committee then would choose a combination of one to three agents as treatment for the child. A requirement was that each drug, given as a single agent, should have an MTD previously determined in a phase I trial. He desired to assess the outcomes toxicity, response, and the times to tumor progression and death, as functions of pathway, genomics biomarker, and the combination of agents given. He planned to treat a maximum of 30 patients, which would take about 2 years to accrue, and then follow each patient for up to one additional year. This may be considered an example of "Bench to Bedside" or "precision medicine" research. Unfortunately, this trial was so ambitious, and the sample size was so small, that it was impossible to conduct it in any way that would provide useful inferences that might benefit future patients.

Here are some problems with this proposed clinical trial design:

1. No explicit algorithm or rules for mapping PGB = (Pathway, Genomics Biomarker) to a combination of agents was given. Rather, it was to be "Treatment Selection by Committee." So, if the trial were run, the questions then would be how to use the resulting data to choose treatment combinations for future patients, or how to design a subsequent clinical trial. If, for some magical reason, the committee's choices were wildly successful and all 30 children were cured, how could the trial results be used to benefit future patients? The simple point is that any decision algorithm must be made explicit, so that others may follow it, if they choose.

2. In terms of the 15 PGB combination rows, the 30 children would be a tiny subset, with on average two children per combination. In other words, the population of patients was so heterogeneous, in terms of PGB alone, that it would be impossible to make any sort of even minimally reliable inferences. For example, if there were only three PBG pairs and three treatment combinations, there still would be 27 possible mappings. Certainly, some mappings might not make sense biologically. But with 15^{15} possibilities, which comes out to over 400 quadrillion, an important first question is, what subsets of these treatment selection mappings might be reasonable to consider?

3. If the treatment combination chosen for a given child by the committee was, in fact, either greatly suboptimal or optimal, there would be no way to determine this statistically, since there would be a subsample of size one. Of course, in the happy but extremely unlikely case where a child's tumor disappeared completely, of course the (PGB, combination treatment) pairing would undergo intense subsequent evaluation. At this point, statistical strategies would become irrelevant. But, again, while optimism is a pleasant mindset, it is not a useful strategy.

4. While single-agent MTDs had been established, there was no knowledge of how any combinations of two or three agents would act. So, it would be necessary to guess the doses of the agents in any given combination.

I certainly admired this oncologist's goals. I just had no idea how to assist him in achieving them.

11.2   Age Discrimination

Suppose that you wish to compare the response probabilities, π_A and π_B, of treatments A and B, and that the data are identical for the two treatments. For example, suppose $X_A = X_B = 52$ responses out of the same sample sizes $n_A = n_B = 100$. Any reasonable statistical analysis would conclude that the parameters π_A and π_B are likely to be similar. If one takes a Bayesian approach and assumes that π_A and π_B start with independent beta (0.50, 0.50) priors then, given the data, there is about an 85% chance that $|\pi_A - \pi_B| \leq 0.10$. Any reasonable frequentist test of the hypothesis $H_0 : \pi_A = \pi_B$ would accept H_0 based on these data.

But suppose that you discover that there is additional information in the form of a Magic Biomarker that gives the refinement of the data shown in Table 11.1. Since this provides a basis for discriminating between two different types of patients, one

Table 11.1 Response rates cross-classified by biomarker ± subgroup and treatment. Each entry is [# responses]/[#patients] followed by the empirical probability in parentheses

| | Treatment | | A versus B difference |
	A	B	
Biomarker +	33/60 (0.550)	24/60 (0.400)	0.150
Biomarker −	19/40 (0.475)	28/40 (0.700)	−0.225

may follow the advice of Hippocrates. If we denote the four (Treatment, Biomarker) response probabilities by $\pi_{A,+}$, $\pi_{A,-}$, $\pi_{B,+}$, $\pi_{B,-}$ and assume, taking a Bayesian approach as above, that they all begin with non-informative beta(0.50, 0.50) priors then, a posteriori, for the biomarker + patients

$$\Pr(\pi_{A,+} > \pi_{B,+} \mid data) = 0.95.$$

But for the biomarker − patients, the reverse inequality is very likely, with

$$\Pr(\pi_{A,-} < \pi_{B,-} \mid data) = 0.98.$$

So, based on the observed data, it appears very likely that A is better than B in biomarker + patients, and that B is better than A in biomarker − patients. This is called a *treatment–biomarker interaction*. It is the Holy Grail for laboratory researchers in pursuit of grant support, fame, and a valid excuse to wear a tuxedo to something other than a wedding. The message from this dataset is very simple. Ignoring known heterogeneity, or not knowing about the biomarker, can lead to the completely incorrect conclusion that the two treatments have the identical, or at least very similar, response rates. *Notice that ignorance of the biomarker gives the wrong conclusion in both subgroups.* It is true that the two agents have nearly identical overall response rates. *But in this example, once you know about the biomarker, the overall response rates are not what matter, and they actually are misleading.* If you do not know about the biomarker, then unfortunately you are stuck. But if you do know about it, then obviously you should account for treatment effects within the two biomarker subgroups. In general, for this sort of effect to occur, the biomarker could be replaced by any variable that dichotomizes patients, or separates patients into several subgroups and possibly interacts with treatment. It could, for example, be a well-known classification of patients into good and poor prognosis. This leads to what should be an obvious question.

An Obvious Question: Why do medical researchers design so many clinical trials assuming that patients are as alike as peas in a pod?

Here is an example of how ignoring patient heterogeneity in clinical trial design can lead to a severe error. Suppose that a randomized clinical trial of standard chemotherapy for poor prognosis untreated AML/MDS patients, consisting of cytarabine + daunorubicin, ± the new targeted agent G (chemo ±G) is conducted. Suppose that, from historical experience, the mean overall survival (OS) in older patients (Age ≥ 60 years) is $\mu_{old} = 15$ months and mean OS in younger patients (Age < 60 years) is $\mu_{young} = 24$ months, and furthermore 2/3 of the patients are older, while 1/3 are younger. Suppose that, for simplicity, one constructs a group sequential design in terms of the average mean OS, computed as $\mu_{ave} = (2/3)\mu_{old} + (1/3)\mu_{young}$. This has null historical value $(2/3)15 + (1/3)24 = 18$ months. If one constructs a group sequential test-based design with a targeted improvement of 33% in this average mean OS, then the targeted alternative is $\mu_{ave}^a = 1.33 \times 18 = 24$ months. Equivalently, this is a targeted Hazard Ratio 0.75 in favor of chemo + G over chemo alone. But notice that the targeted value $\mu_{ave}^a = 24$ months for chemo + G equals the null mean with chemo alone in younger patients, which may suggest that something is not quite right about using the average mean survival time to construct a "one size fits all" design.

To see what may go wrong, suppose that, in fact, adding G to chemo decreases mean OS slightly in older patients, by 10%, to 13.5 months, but adding G increases mean OS by 30%, to 31.2 months, in younger patients. That is, there is a fairly substantial treatment–age subgroup interaction. The overall mean OS with chemo + G, averaging older and younger patients, is $(2/3)13.5 + (1/3)(31.2) = 19.4$ months, which is well below the targeted average mean OS of 24 months and much closer to the null value 18 months. If the trial is designed based on the overall average mean OS, the group sequential procedure is unlikely to conclude that adding G to chemo improves OS. But this would be a false negative conclusion in younger patients, where G actually provides a 30% improvement in mean OS, from 24 to 31.2 months. Suppose that one completes the trial and the group sequential test fails to reject the null hypothesis, but a later data analyses accounting for age show that there is an age–treatment interaction, with G + chemo superior to chemo in younger patients. Someone reporting this may be accused of data dredging, however, that is, of searching the dataset for subgroups where the G effect is "significant." After all, if one looks at enough subgroups, it is likely that a p-value ≤ 0.05 will be obtained for some subgroup, even if there is no actual G effect in any of the subgroups examined.

What has gone wrong is very simple. The design failed to perform separate tests of $C + G$ = chemo + G versus chemo within the two age subgroups that would allow different within-subgroup conclusions to be made. This might be accomplished very simply by conducting simultaneous but separate trials, each based on a group sequential log-rank testing procedure, within the two age subgroups. But this would be inefficient, because it would not borrow strength between the subgroups.

As an alternative approach, rather than doing two separate trials one could do a single trial, but allow different conclusions within the two prognostic subgroups, and assume a model that borrows strength between subgroups. The design and comparative tests could be based on, for example, a piecewise exponential (PE) regression model with linear component of the form

$$\eta(\beta) = \beta_1 \text{ Age } +$$
$$\left(\beta_G + \beta_{G,Old}\, I[Age \geq 60] + \beta_{G,Young}\, I[Age < 60]\right)I[C+G].$$

In the PE model, $\exp\{\eta(\beta)\}$ would multiply the baseline piecewise hazard function. For example, after adjusting for the effect of Age, the parameter $\beta_G + \beta_{G,Old}$ is the total effect on the risk of death of adding G to chemo for older patients. The treatment–subgroup interaction parameter $\beta_{G,Old}$ is the additional effect of G, beyond the baseline effect β_G, in older patients. Similarly, $\beta_G + \beta_{G,Young}$ is the total effect of adding G to chemo for younger patients, and $\beta_{G,Young}$ is the additional effect of G, beyond the baseline effect β_G, in younger patients. One may use estimates of these parameters to make separate decisions of whether adding G to chemo is advantageous within each age subgroup. For example, it might turn out that adding G to chemo is beneficial for younger patients but not for older patients, as in the numerical illustration above. The baseline parameter β_G allows the model to borrow strength between the two subgroups. If desired, other prognostic covariates known to affect OS also could be included in the linear component $\eta(\beta)$.

In a case where the subgroups Young and Old are replaced by some other complementary subgroups, say patients who do or do not have a particular gene signature, $+$, and $-$, then one possible alternative model and parametrization would be to do away with β_G, but assume a hierarchical Bayesian structure that induces association between $\beta_{G,+}$ and $\beta_{G,-}$ through a hyperprior. This sort of approach is appropriate if, before any clinical data are observed, the two biomarker subgroups are exchangeable in terms of possible treatment effects. That is, while one expects younger patients to have better prognosis than older patients, there may be no prior reason to favor one biomarker subgroup over the other. For example, one might assume that $\beta_{G,+}$ and $\beta_{G,-}$ follow a $N(\mu_G, \sigma^2)$ prior with the level 2 priors (hyperpriors) $\mu_G \sim N(\mu, 10)$ and $\sigma^2 \sim Gamma(10, 10)$, where a fixed value of μ might be estimated from historical data. With this parameterization, one could construct Bayesian posterior decision rules, of the general form

$$\Pr\{e^{\beta_{G,-}} > 1.33 \mid data\} > c_1$$

$$\Pr\{e^{\beta_{G,-}} < 0.75 \mid data\} > c_1.$$

Since a larger value of $e^{\beta_{G,-}}$ corresponds to a higher risk of death, in words the above two inequalities say, respectively, that adding G to chemo is likely to increase or decrease the death rate in biomarker $-$ patients. The similar rules for biomarker $+$ patients are

$$\Pr\{e^{\beta_{G,+}} > 1.33 \mid data\} > c_2$$

$$\Pr\{e^{\beta_{G,+}} < 0.75 \mid data\} > c_2.$$

These could be used as comparative testing criteria for each of $\beta_{G,-}$ and $\beta_{G,+}$. The operating characteristics of the procedure, including overall Type I and Type II error probabilities, could be established by simulation, with this used as a tool to derive decision cutoffs, c_1 and c_2, and calibrate prior hyperparameter values.

11.3 Comparing Treatments Precisely

Murray et al. (2018b) describe a Bayesian design for a randomized comparative group sequential trial to compare nutritional prehabilitation (N) to standard of care (C) for esophageal cancer patients undergoing chemoradiation and surgery. The trial includes both primary patients (P) who have not been treated previously, and salvage patients (S) whose disease has recurred after previous therapy. The outcome is post-operative morbidity (POM), scored within 30 days of surgery, which Clavien et al. (1992) and Dindo et al. (2004) defined as the six-level ordinal variable having possible values 0 if normal recovery, 1 if minor complication, 2 if complication requiring pharmaceutical intervention, 3 if complication requiring surgical, endoscopic or radiological intervention, 4 if life-threatening complication requiring intensive care, and 5 if death. A central difficulty in constructing a test is comparing different vectors of probabilities for the five outcomes. To see why this has no obvious solution, consider a simpler example where the three possible outcomes are (Alive with Response, Alive without Response, Dead). If one wishes to compare treatments A and B using their outcome probability vectors, $\pi_A = (0.40, 0.10, 0.50)$ and $\pi_B = (0.20, 0.50, 0.30)$, then it is unclear which is better, since A has double the response probability of B, but A also has a higher probability of death. One may solve this problem by assigning utilities to the three outcomes, say (100, 80, 0). This gives mean utilities $\overline{U}_A = 48$ and $\overline{U}_B = 60$, so B is preferred over A. If, instead, one assigns the utilities (100, 30, 0) then $\overline{U}_A = 43$ and $\overline{U}_B = 35$, so A is preferred over B. The fact that utilities are subjective is a desirable property, rather than a disadvantage, because they explicitly reveal the basis for decision-making. In practice, eliciting numerical utilities of possible clinical outcomes from practicing physicians is straightforward, since utilities are easy to understand, physicians have a very good idea of the relative importance and desirability of different clinical outcomes and, if necessary, reaching a consensus among a group of physicians can be done quickly and easily.

The nutritional prehabilitation trial design is based on the elicited numerical utilities of the possible POM outcomes given in the upper portion of Table 11.2. These numerical utilities were determined by first fixing the two extreme outcome utilities to be $U(0) = 100$ and $U(5) = 0$, and eliciting the four intermediate values from the trial's Principal Investigator. The numerical utilities show that $Y \leq 2$ is much more desirable than $Y \geq 3$, but dichotomizing POM score in this way would throw away a great deal of information. Moreover, the differences in utilities between the successive levels are far from equal, so using the integer-valued scores 0, ..., 5 would be a poor way to quantify POM outcome.

It is useful to examine how the numerical utilities in Table 11.2 turn a vector $\pi = (\pi_0, \pi_1, \ldots, \pi_5)$ of POM score probabilities into a single mean utility $\overline{U}$ that quantifies the desirability of π. This is at the heart of how the utility-based design works. In the lower portion of Table 11.2, the probability vector π_1 may be considered a null distribution and π_2 a targeted alternative distribution for salvage patients. Going from π_1 to π_2 reduces the probability $\Pr(Y \geq 3)$ of an undesirably high POM score from 0.35 to 0.14, and increases $\overline{U}$ from 60.0 to 81.4. Similarly, π_3 may be

Table 11.2 Elicited utilities of Clavien-Dindo post-operative morbidity (POM) scores, and four examples of POM probability distributions and mean utilities. Each π_j is vector of POM score probabilities

POM score	0	1	2	3	4	5	
Elicited utility	100	80	65	25	10	0	
	POM score probabilities						$\overline{U}$
π_1	0.30	0.25	0.10	0.10	0.10	0.15	60.0
π_2	0.60	0.14	0.12	0.08	0.04	0.02	81.4
π_3	0.50	0.20	0.10	0.10	0.05	0.05	75.5
π_4	0.73	0.14	0.05	0.05	0.03	0	89.0

considered a null distribution and π_4 a targeted alternative distribution for primary patients. Going from π_3 to π_4 reduces the probability $\Pr(Y \geq 3)$ of a high POM score from 0.20 to 0.08, and increases $\overline{U}$ from 75.5 to 89.0.

An important property of the design is that it accounts for patient heterogeneity by allowing possibly different conclusions within the two subgroups P and S when comparing treatment N and S. A group sequential testing procedure is based on the posterior utilities for the POM score computed under the following Bayesian probability model. For $Y =$ POM score, a nonproportional odds (NPO) model is assumed in which the linear term is

$$\eta_{NPO}(y, X, Z) = \text{logit}\{\Pr(Y \leq y \mid X, Z)\}$$
$$= \alpha_y + \gamma_{1,y} X + \gamma_{2,y} Z + \gamma_{3,y} X Z,$$

for $y = 0, 1, 2, 3, 4$, where the subgroup covariate is defined as $X = -0.5$ for P and $+0.5$ for S, and the treatment variable is defined similarly as $Z = -0.5$ for C and $+0.5$ for N. The subgroup and treatment variables are defined in this way numerically, rather than as more traditional 0/1 variables, in order to balance the variances under the Bayesian model. The probability $\Pr(Y \leq y \mid X, Z)$ of the ordinal POM variable is increasing in y due to a constraint in the prior of the parameters (α, γ). A hierarchical model is assumed, to borrow strength between subgroups while allowing shrinkage, with location parameters determined from elicited prior information on POM score. Details are given in Murray et al. (2018b).

The NPO model differs from the usual proportional odds (PO) model of McCullagh (1980) in that all elements of the treatment and subgroup effect parameter vector $\gamma_y = (\gamma_{1,y}, \gamma_{2,y}, \gamma_{3,y})$ vary with the level y of POM score. The PO model would have the simpler linear term

$$\eta_{PO}(y, X, Z) = \alpha_y + \gamma_1 X + \gamma_2 Z + \gamma_3 X Z. \tag{11.1}$$

Under the PO model, only the intercept parameters $\alpha = (\alpha_0, \alpha_1, \alpha_2, \alpha_3, \alpha_4)$ account for the levels of Y, but the treatment effects do not vary with y. Thus, the NPO model is much more flexible, since it accounts for treatment and subgroup effects that vary with the level of POM score, in addition to allowing treatment–subgroup interaction.

Denoting

$$\pi_y(X, Z \mid \alpha, \gamma) = \Pr(Y = y \mid X, Z, \alpha, \gamma),$$

the mean utility of treatment Z in subgroup X is defined as

$$\overline{U}(X, Z, \alpha, \gamma) = \sum_{y=0}^{5} U(y)\pi_y(X, Z \mid \alpha, \gamma). \tag{11.2}$$

The trial has a maximum of 200 patients, with interim and final tests in each subgroup $X = P$ and S performed at $n = 100$ and 200 patients, based on the posterior probability decision rules

$$\Pr\{\overline{U}(X, z, \alpha, \gamma) > \overline{U}(X, z', \alpha, \gamma) \mid data\} > p_{cut}, \tag{11.3}$$

for treatments $(z, z') = (N, C)$ or (C, N). The numerical value of p_{cut} was determined to control the two subgroup-specific Type I error probabilities to be ≤ 0.025. In subgroup X, a large value of (11.3) for $(z, z') = (N, C)$ corresponds to superiority of N over C. A large value of (11.3) for $(z, z') = (C, N)$ corresponds to superiority of C over N, equivalently inferiority of NuPrehab to standard.

The power figures of the tests were determined to detect the ratio

$$\frac{\Pr(Y \geq 3 \mid X, N)}{\Pr(Y \geq 3 \mid X, C)} = 0.60$$

for each subgroup $X = P$ and $X = S$. This translates into a reduction in $\Pr(Y \geq 3 \mid P)$ from 0.20 to 0.08 for primary patients, and a reduction in $\Pr(Y \geq 3 \mid S)$ from 0.35 to 0.14 for salvage patients. This, in turn, determines targeted mean utility increases with N compared to C from 75.5 with C to 89.0 with N for $X = P$ patients, and from 60.0 with C to 81.3 with N for $X = S$ patients.

As a comparator, one may consider a conventional PO model-based design that ignores subgroups and makes "one size fits all" treatment comparisons. For this conventional design, the linear term in the PO model is

$$\eta_{PO,conv}(y, Z) = \alpha_y + \gamma_1 X + \gamma_2 Z$$

for $y = 0, 1, 2, 3, 4$, so γ_2 is the only treatment effect parameter. Thus, the posterior decision criterion is simply $\Pr(\gamma_2 > 0 \mid data) \geq p_{cut,\gamma}$, where the cutoff $p_{cut,\gamma}$ is calibrated to ensure an type overall I error ≤ 0.05. Both designs were simulated under each of the four scenarios given in Table 11.3. For example, under Scenario 3, NuPrehab achieves an improvement of $89.0 - 75.5 = 13.5$ in POM score mean

Table 11.3 Simulation scenarios for the nutritional prehabilitation trial designs, in terms of the true mean utility of the POM score for each treatment–subgroup combination

Scenario	Treatment	Subgroup	Mean utility
1 (Null/Null)	Control	Primary	75.5
	NuPrehab	Primary	75.5
	Control	Salvage	60.0
	NuPrehab	Salvage	60.0
2 (Null/Alt)	Control	Primary	75.5
	NuPrehab	Primary	75.5
	Control	Salvage	60.0
	NuPrehab	Salvage	81.3
3 (Alt/Null)	Control	Primary	75.5
	NuPrehab	Primary	89.0
	Control	Salvage	60.0
	NuPrehab	Salvage	60.0
4 (Alt/Alt)	Control	Primary	75.5
	NuPrehab	Primary	89.0
	Control	Salvage	60.0
	NuPrehab	Salvage	81.3

utility in the primary patients, but there is no difference between the two treatments in the salvage patients. In other words, there is a treatment–subgroup interaction in Scenario 3. With either design, to ensure balance and also comparability, patients are randomized using stratified block randomization with blocks of size four, so that for each block of four patients within each prognostic subgroup, two patients receive NuPrehab and two receive the control.

After the paper by Murray et al. (2018) was accepted for publication, the trial's Principal Investigator recruited other medical centers to participate, so it then became feasible to double the maximum sample size, from 100 to 200. Consequently, the following numerical simulation results are different from those given by Murray et al. (2018), which were based on a maximum sample size of 100. Table 11.4 summarizes the simulation results. It shows that both the conventional design that makes the same decision for both the Primary and Salvage subgroups because it ignores them, and the design with subgroup-specific decisions have overall Type I error probabilities very close to 0.05, due to the way they are designed. Scenarios 2 and 3 are treatment–subgroup interaction cases in that one treatment is superior to the other in one subgroup but the two treatments are equivalent in the other. These scenarios show the great advantage of making subgroup-specific decisions. For example, in Scenario 2, N is truly superior to C in Salvage patients, since N gives an increase in mean POM utility of $81.3 - 60.0 = 21.3$, but the two treatments are equivalent in the Primary patient subgroup. For this scenario, the subgroup-specific design has power 0.808 in Salvage patients and Type I error probability 0.029 in Primary patients.

Table 11.4 Simulation results for the two nutritional prehabilitation trial designs. In each scenario, each cell gives the proportion of simulated trials that declared the NuPrehab superior or inferior compared to the control within a subgroup. Each correct superiority decision probability is given in a box. $\bar{n}$ = mean sample size. Each result is based on 2000 computer simulations

Scenario (Prim/Salv)	Prim	Salv	Prim	Salv	$\bar{n}$
	N superior		N inferior		
Precision design accounting for subgroups					
1 (Null/Null)	0.023	0.025	0.027	0.031	199.2
2 (Null/Alt)	0.029	**0.808**	0.022	0.000	189.6
3 (Alt/Null)	**0.777**	0.036	0.000	0.023	187.0
4 (Alt/Alt)	**0.823**	**0.845**	0.000	0.000	172.4
Conventional design that ignores subgroups					
1 (Null/Null)	0.024	0.024	0.028	0.028	199.4
2 (Null/Alt)	0.440	**0.440**	0.000	0.000	193.0
3 (Alt/Null)	**0.562**	0.562	0.000	0.000	189.6
4 (Alt/Alt)	**0.977**	**0.977**	0.000	0.000	145.1

Comparisons of the subgroup-specific design to the conventional design that assumes homogeneity in Scenarios 2 and 3 may be considered shocking. In Scenario 2, the conventional design that ignores subgroups has power 0.440, equivalently, Type II error probability $1 - 0.44 = 0.56$, in Salvage patients, and Type I error probability 0.44 in Primary patients. The same sort of thing is seen in Scenario 3. These huge within-subgroup false positive and false negative rates with the conventional design in Scenarios 2 and 3, where there are treatment–subgroup interactions, suggest that one could do as well by not running a trial at all, and instead deciding which treatment is better by flipping a coin. The conventional design does have substantially larger power than the subgroup-specific design in Scenario 4, where N provides an improvement over C in both subgroups, that is, if there are improvement and no treatment–subgroup interaction. Thus, the loss of power in Scenario 4 is the price that must be paid if one chooses to use a comparative design that allows different decisions to be made within subgroups. This should be considered along with the Type I and II error probabilities 0.440 and 0.560 using the conventional design in the mixed Scenarios 2 and 3. These simulation results suggest that the choice to use the conventional methodology, which may have within-subgroup Type I and Type II error probabilities in the range 0.46 to 0.56, should reflect a strong belief that any N versus C effect will be the same in the two prognostic subgroups.

More generally, the properties of the two designs for this particular trial suggest something much more general that is rather disturbing. For decades, large-scale randomized clinical trials have been designed to make overall decisions that ignore well-known prognostic subgroups. They compare treatments in terms of a single overall effect, such as the difference or ratio of mean survival times in two treatment arms. But if, in fact, treatment effects are substantively different between prognostic subgroups due to treatment–subgroup interactions, then estimating one overall effect essentially averages the different, subgroup-specific effects. As illustrated above, this easily can result in huge probabilities of false positive or false negative decisions within subgroups. Thus, conventional clinical trials that make decisions based on the assumption that patient are homogeneous with regard to treatment effects may have been a multi-decade disaster in the medical research community. The common practice of searching for "significant" treatment–subgroup interactions, not specified on the design, after a trial is completed may be regarded as data dredging or cherry-picking. Designing a trial from the start to make reliable within-subgroup decisions certainly requires a larger sample size, but not doing so may produce very misleading conclusions. It appears that it may be time to make a radical change in the conventions that govern how randomized comparative clinical trials are designed.

11.4 A Subgroup-Specific Phase II–III Design

It is worthwhile to return to the phase II–III select-and-test design in order to describe how it may be extended to accommodate subgroups. The following model-based Bayesian design allows one to make different decisions for each of the subgroups, while borrowing strength between subgroups. Using the hypothetical worse-than-useless parallel phase II trial as an illustration, recall that the two subgroups were G and $\overline{G}$, and the four phase II–III treatment arms that actually should be considered are $N + A_1$, $N + A_2$, or $N + A_3$ and N. So, a design with subgroup-specific decisions must account for eight (Treatment, Subgroup) combinations. The main inferential issues are whether any of the three combinations $N + A_1$, $N + A_2$, or $N + A_3$ is likely to provide a substantive improvement over N alone within either G or $\overline{G}$.

The following design mimics the select-and-test structure of the TSE design, with the two main elaborations being that it makes within-subgroup decisions, and it is based on an assumed Bayesian regression model. Denote the eight subgroup-specific response probabilities by

$$\pi_{j,g} = \Pr(\text{response} \mid j, g),$$

for treatment arm index $j = 0, 1, 2, 3$, and subgroup $g = 1$ or 2. Define the indicator variable $W_g = 1$ if $g = 1$ and $W_g = -1$ for if $g = 2$, and the linear terms

$$\eta_{j,g} = \text{logit}\{\pi_{j,g}\} = \mu + \alpha_j + \gamma W_g + \xi_{j,g}$$

for $j = 0, 1, 2, 3$, and $g = 1, 2$, with $\alpha_0 \equiv 0$ for the control arm. The baseline linear terms for C are $\mu + \gamma$ for subgroup G and $\mu - \gamma$ for $\overline{G}$, α_j is the additional effect of adding A_j to N, and $\{\xi_{j,g}\}$ are additional treatment–subgroup interactions, with $\xi_{0,g} = 0$ for $g = 1, 2$. There are various ways to model the parameter priors. One possibility is to assume two identical hierarchical structures, one for each subgroup, with $\xi_{1,g}, \xi_{2,g}, \xi_{3,g} \sim$ iid $N(0, \sigma_g^2)$ for each subgroup g, and inverse gamma or some other hyperpriors on σ_1^2 and σ_2^2. A simpler approach is to fix these prior variances at a large value such as 10. With either approach, the data would determine whether the $\xi_{j,g}$'s are far enough from 0 to infer that there are treatment–subgroup interactions, along with the posteriors on the main experimental treatment effects $\alpha_1, \alpha_2, \alpha_3$. The priors on the model parameters $\mu, \alpha_1, \alpha_2, \alpha_3, \gamma$ and the $\xi_{j,g}$'s would need to be balanced among $j = 1, 2, 3$ and sufficiently non-informative to ensure that the data dominate all decisions.

The trial may be conducted in two stages, to mimic the Thall et al. (1988) design, but allowing different decisions to be made within the subgroups. The decision rules would be formulated using Bayesian posterior probabilities as decision criteria, similar to the nutritional prehabilitation trial design. In stage 1, within each medical center and each subgroup, at the time of accrual patients are randomized fairly among the four treatment arms, but restricted so that successive groups of size b_1 have the same number of patients per arm within each group G and $\overline{G}$ in each medical center. For example, $b_1 = 16$ would give 4 patients per treatment arm, with the stage 1 group size b_1 a multiple of 4, chosen depending on the overall sample size requirements. The idea is to balance the samples of the treatment arms within each (Prognostic Subgroup, Medical Center). Since the accrual rates likely will differ between prognostic subgroups, and between medical centers, an overall stage 1 sample size $4n_1$ should be specified, so n_1 is the stage 1 sample size of each treatment arm. The group size b_2 in stage 2 may be any appropriate multiple of 2, and the stage 2 sample size is $2n_2$. At the end of stage 1, for each $g = 1, 2$, denote the index of the experimental treatment having maximum empirical estimate $\hat{\pi}_{j,g}$ in prognostic subgroup g by $[k(g)]$. For each stage $s = 1, 2$ and each prognostic subgroup $g = 1, 2$, let $\pi_{s,g}^*$ be a fixed targeted prognostic subgroup-specific probability value larger than the corresponding subgroup-specific prior means with N, formally $\pi_{s,g}^* > E\{\pi_{0,g}\}$ for each $g = 1, 2$. Denote the stage 1 data by $data_1$ and the combined stage 1 and stage 2 data by $data_{1,2}$.

STAGE 1: Randomize patients fairly among the four treatment arms within each prognostic subgroup and each medical center in groups of size b_1. For each subgroup $g = 1$ or 2, if

$$\Pr(\pi_{[k(g)]} > \pi_{1,g}^* \mid data_1) > c_{1,g}$$

then continue to stage 2 in prognostic subgroup g with treatments $N + A_{[k(g)]}$ and N. Otherwise, terminate accrual in subgroup g with no $N + A_j$ selected in that subgroup. If both prognostic subgroups are terminated, the trial is stopped and stage 2 is not conducted.

STAGE 2: Randomize patients fairly, within each prognostic subgroup g that has been continued and each medical center in groups of size b_2, between $N + A_{[k(g)]}$ and N. For each $g = 1$ or 2, if

$$\Pr(\pi_{[k(g)]} > \pi_{2,g}^* \mid data_{1,2}) > c_{2,g}$$

then conclude that treatment $N + A_{k(g)}$ is superior to N, and otherwise conclude that no $N + A_j$ is superior to N in that subgroup.

Table 11.5 Combinations of centers, treatments, and subgroups in the randomized phase II–III trial in which non-small cell lung cancer patients are treated. The symbol "**X**" in a cell identifies a (Center, Treatment, Subgroup) combination where a subsample is included in the trial

Centers	Treatments in subgroup G			
	N	$N + A_1$	$N + A_2$	$N + A_3$
M_1	X	X	X	X
M_2	X	X	X	X
M_3	X	X	X	X
Centers	Treatments in subgroup $\overline{G}$			
	N	$N + A_1$	$N + A_2$	$N + A_3$
M_1	X	X	X	X
M_2	X	X	X	X
M_3	X	X	X	X

While the four target response probability values $\{\pi^*_{s,g}\}$ may be chosen by the physicians planning the trial to be meaningful improvements over what has been seen historically with N, one must choose the design parameters n_1, n_2, b_1, b_2, $c_{1,g}$, and $c_{2,g}$, say to control overall within-subgroup Type I error and generalized power. As usual with highly structured Bayesian designs, this is done by conducting preliminary computer simulations to determine design parameters that give the desired properties.

At the risk of beating a dead horse, I have included Table 11.5, which shows the combinations of (Treatment Arm, Medical Center, Prognostic Subgroup) that are included in the above randomized phase II–III design that accounts for subgroups. This is an example of a balanced complete block design. Table 11.5 should be compared to Table 10.5, which shows what the parallel phase II trial evaluates. When looking at Table 11.5, one also should keep in mind the important additional facts that, with this design, patients are randomized in such a way to provide balance among the prognostic subgroups and medical centers, and that the design includes a hybrid select-and-test decision procedure based on comparisons that are unbiased due to randomization.

11.5 Precision Pharmacokinetic Dosing

This next example requires some background first to be established in two areas, one medical and the other statistical. The example explains how a particular pharmacokinetically (PK) guided dosing strategy for intravenous busulfan (IV Bu) in the preparative regimen for allosct was developed over many years. The statistical methodology is the family of Bayesian nonparametric models (BNPs).

IV Bu was established as a very desirable component of the preparative regimen for allosct. A key variable for predicting survival time is the patient's systemic busulfan exposure, represented by the area under the plasma concentration versus time curve, AUC, which quantifies the actual delivered dose. The earlier practice of giving busulfan orally, rather than IV, results in 10 to 20 times the variability in AUC (Andersson et al., 2000), since with oral administration the delivered busulfan dose cannot be controlled. As shown by Bredeson et al. (2013), IV administration has improved patient survival substantially, since administering it IV rather than orally greatly improves its bioavailability and delivered dosing accuracy. With IV Bu, Andersson et al. (2002) estimated an optimal AUC range by fitting a Cox proportional hazards model for overall survival time and constructing a smoothed martingale residual plot, which provided a graphical representation of the risk of death as a function of AUC. The plot showed that the hazard of death is a "U-shaped" function of AUC, which implies that there is an optimal intermediate interval for the AUC of IV Bu that gives the lowest hazard of death, equivalently the longest mean survival time. This is because higher rates of life-threatening adverse events are associated with an AUC that is either too high or too low. The optimal interval of IV Bu dose currently is considered to be AUC values approximately 950 to 1520 μMol-min. This discovery resulted in a change in therapeutic practice, with a preliminary test dose first used to determine each patient's pharmacokinetics, so that a therapeutic dose of IV Bu can be determined that targets the optimal AUC window. Bartelink et al. (2016) reported that, when treating children and young adults with allosct, the optimal daily AUC range in a 4-day busulfan-based preparative regimen was 78 to 101 mg*h/L, corresponding to a total course AUC of about 19,100 to 21,200 μMol-min. The optimal interval for younger patients thus appears to be much higher than for older patients.

Motivated by these new results, and based on the fact that Age and whether the patient is in complete remission (CR) or has active disease (No CR) at the time of allosct both are very important prognostic covariates for survival, Xu et al. (2019) addressed the problem of determining optimal patient-specific AUC intervals that are functions of these two covariates. The goal was to develop a "personalized" or "precision medicine" method to determine an optimal targeted AUC interval tailored to each patient's prognosis. To do this, they analyzed a dataset of 151 patients who had undergone allosct for AML or MDS without using targeted AUC to determine IV Bu dose. The goal was to fit a regression model accounting for the joint impact of patient Age, CR status, and AUC on patient survival, for subsequent use to predict survival of future patients and thus choose an optimal (Age, CR status)-specific delivered dose interval to target.

To avoid the strong, restrictive parametric assumptions made by Andersson et al. (2002) or Bartelink et al. (2016) for their data analyses, Xu et al. (2019) fit a BNP survival time regression model to this dataset. In general, the family of BNP models can accommodate a broad range of data structures and, due to their robustness and flexibility, they often can overcome limitations of conventional statistical models. BNP models can be applied to problems in density estimation, regression, survival analysis, graphical modeling, neural networks, classification, clustering, population models, forecasting and prediction, spatiotemporal models, and causal inference. A major advantage of BNP models is that they can accurately approximate any distribution or function, a property known as "full support." In many cases, this allows one to avoid making invalid inferences due to assuming an incorrect or overly restrictive model. Due to their flexibility, BNP models often identify unexpected structures in a dataset that cannot be seen from fitted conventional statistical models. Examples include identifying patient clusters, treatment–subgroup interactions, and complex patterns of biomarker change over time. Reviews of BNP models and methods are given by De Iorio et al. (2009), Müller and Mitra (2013), and Thall et al. (2017), and many examples of applications are given in the books by Müller and Rodriguez (2013) and Mitra and Mueller (2015).

BNP models began with Ferguson et al. (1973), who proposed assuming a Dirichlet process (DP) prior on an arbitrary probability distribution function F. The DP has been used widely in BNP analyses as a general prior model for random unknown probability distributions, due largely to the fact that modern MCMC methods for computing posteriors have made it practical to implement BNP models. A $DP(\alpha, G)$ distribution is characterized by a fixed scaling parameter $\alpha > 0$ and a base probability measure G. Sethuraman (1994) provided the so-called stick-breaking definition for constructing a $DP(\alpha, G)$, as a weighted average of random distributions $\theta_1, \theta_2, \ldots$, given by

$$G = \sum_{h=1}^{\infty} w_h \delta_{\theta_h}, \quad \text{with } \theta_h \sim \text{iid } G_0 \text{ and } w_h = v_h \prod_{l<h} (1 - v_l) \text{ for } v_h \sim \text{beta}(1, \alpha).$$

The symbol $\delta_{\theta_h}(\cdot)$ denotes the Dirac delta function, which assigns probability 1 to θ_h and is equal to 0 everywhere else. The weights $w_1, w_2, \ldots$ are obtained as functions of iid beta$(1, \alpha)$ random variables. The idea is to obtain a sample $X_1, X_2, \ldots, X_n$, in order, from a distribution F following a $DP(\alpha, G)$ prior by successively sampling, with $X_n \mid X_1, \ldots, X_{n-1}$ equal any X_i for $i = 1, \ldots, n-1$ with probability $1/(n-1+\alpha)$ and equal to a new draw from G with probability $\alpha/(n-1+\alpha)$. In practice, the sum in h actually does not go to ∞, but stops at some finite number depending on the data being analyzed.

MacEachern (1999) extended the DP model to include regression on covariates, $X = (X_1, \ldots, X_p)$, by assuming that each component distribution in the sum is a normal distribution with mean function $\theta_h(X)$, which yields the regression model

$$F(y \mid X) = \sum_{h=1}^{\infty} w_h \, N(y; \theta_h(X), \sigma^2).$$

This model is completed by specifying a stochastic process as the prior for $\{\theta_h(X)\}$. This BNP regression model, called a "dependent Dirichlet process" (DDP), allows one to represent any distribution as a mixture of an arbitrary number of normal distributions, each having a mean that is a random function of the covariates X. Since virtually any regression distribution can be closely approximated by such a mixture, one is virtually guaranteed that the DDP model will provide a good fit to one's data. Barrientos et al. (2012) provide a formal explanation of the full support properties of the DDP. MacEachern (1999) further proposed assuming a Gaussian process (GP) prior for $\{\theta_h(X)\}$, which is called the DDP-GP model. Suppressing the subscript h, a GP prior is characterized by the marginal distribution for any n-tuple $(\theta(X_1), \ldots, \theta(X_n))$ being a multivariate normal distribution with mean vector $(\mu(X_1), \ldots, \mu(X_n))$ and $(n \times n)$ variance–covariance matrix V with (i, j) element $V(X_i, X_j)$, for any set of covariate vectors $X_1, \ldots, X_n$. This GP model is denoted by $\theta(X) \sim GP(\mu, V)$.

For the AUC data, denote $Y = \log$ time to death, $C =$ censoring on the log(time) domain, $T = \min\{Y, C\}$, and $\delta = I(Y \leq C)$, so the observed data for patient $i = 1, \ldots, 151$ are the outcomes (T_i, δ_i), and the three covariates are $X_i = (\text{Age}_i, \text{CR}_i, \text{AUC}_i)$. Xu et al. (2019) constructed a DDP-GP model for the survival distribution $F(Y \mid X)$ by starting with a model for a discrete random distribution $G(\cdot)$, using a Gaussian kernel to extend this to a prior for a continuous random distribution and, following the general approach of MacEachern and Müller (1998), determined a regression structure by defining a prior on $\{F(Y \mid X)\}$. Xu et al. (2019) applied an extended version of the DDP-GP model given by Xu et al. (2016), obtained by including a scale parameter, λ_d, for each covariate X_d, and an overall multiplicative scale parameter σ_0^2, in the covariance function,

$$Cov(X_i, X_\ell) = \sigma_0^2 \exp\left\{-\sum_{j=1}^{p} \frac{(X_{ij} - X_{\ell j})^2}{\lambda_j^2}\right\} + \delta_{i\ell} J^2. \tag{11.4}$$

The scale parameter σ_0^2 accounts for variability in the data beyond that explained by the variance σ^2 of the normal distributions, and this model provides a more robust fit to one's data. Denoting the vector of all model parameters by θ and the data by $\mathcal{D}_n = \{T_i, \delta_i, X_i\}_{i=1}^n$, the likelihood function for the survival data is

$$L(\theta \mid \mathcal{D}_n) = \prod_{i=1}^{n} f_{X_i}(y_i \mid \theta)^{\delta_i} \{1 - F_{X_i}(y_i \mid \theta)\}^{1-\delta_i},$$

where $f_X(\cdot)$ and $F_X(\cdot)$ denote the density and cumulative distribution function of Y for a patient with covariates X. Given the assumed DDP-GP prior on $F_X(\cdot)$, the model is completed by assuming the priors

$$\beta_h \sim \text{iid } N(\beta_0, \Sigma_0), \ 1/\sigma^2 \sim \text{Gamma}(a_1, b_1), \ \alpha \sim \text{Gamma}(a_2, b_2),$$

$\sigma_0 \sim N(0, \tau_\sigma^2)$, and the covariate scale parameters $\lambda_j \sim \text{iid } N(0, \tau^2)$, for $j = 1, \ldots, p$.

To obtain posterior inferences from a fitted DDP-GP survival regression model, the full model is marginalized analytically with respect to the random probability measures $F_X(\cdot)$ by expressing the model equivalently as a hierarchical model with a set of new latent indicator variables $\{\gamma_i\}$, assuming that

$$Y_i \mid \gamma_i = h, X_i \sim N(\theta_h(X_i), \sigma^2) \text{ and } p(\gamma_i = h) = w_h, \qquad (11.5)$$

for $i = 1, \ldots, n$. The fitted BNP model is used to derive the posterior predictive distribution $p(Y_{n+1} \mid X_{n+1}, \mathcal{D}_n)$ of the survival time Y_{n+1} of a future $n + 1$st patient with covariates X_{n+1}. To implement the DDP-GP model for a robust survival regression analysis, the computations can be performed using the R package, DDPGPSurv, which can be downloaded from CRAN.

For the IV busulfan allosct data, X includes the key treatment variable AUC that quantifies the patient's delivered dose of IV busulfan. Based on the fit of the IV busulfan data using the DDP-GP model, for a future patient $n + 1$, the posterior predictive distribution of Y_{n+1} may be used to compute the optimal AUC, given the patient's Age and CR status, to maximize the patient's expected log survival time, as

$$\widehat{\text{AUC}}_{n+1} = \text{argmax}_{\text{AUC}} E(Y_{n+1} \mid X_{n+1}, \mathcal{D}_n). \qquad (11.6)$$

Since laboratory-based PK evaluation methods for determining a median daily Bu-SE may have about a 3–6% error, this motivated the optimal AUC $\pm 10\%$ as a reasonable interval to be targeted. Thus, the optimal AUC interval for future patient $n + 1$ with covariates X_{n+1} is defined to be

$$\left[0.9 \times \widehat{\text{AUC}}_{n+1}(X_{n+1}), \quad 1.1 \times \widehat{\text{AUC}}_{n+1}(X_{n+1}) \right].$$

The analyses of Xu et al. (2019) identified, for each combination of CR status and Age, an optimal AUC range that yields higher expected survival times compared to an AUC that is either below or above the optimal range. Figure 11.1, which is similar to portions of Fig. 6 in Xu et al. (2019), gives predicted posterior mean survival time as a function of AUC for each of the four combinations Age = 30 or 60 and CR status = Yes or No. A key point is that these inferred optimal AUC ranges differ substantially between the (Age, CR status) combinations. This has extremely important implications for a physician choosing an individual allosct patient's personalized targeted AUC based on their Age and CR status. For example, the optimal AUC interval for a 50-year-old patient not in CR is $4.7 \pm 0.47 = [4.23, 5.17]$, while the optimal interval is $5.8 \pm 0.58 = [5.22, 6.38]$ for a 40-year-old patient in CR. Because these intervals do not overlap, these two patients should have very different targeted AUC values to maximize their expected survival times.

Figure 11.2, which is similar to Fig. 7 in Xu et al. (2019), gives a striking illustration of the optimal targeted AUC intervals as functions of Age and CR status, based on predictions obtained from the fitted DDP-GP regression model. The figure shows the negative association between optimal AUC and Age. Moreover, while CR status has no effect on the optimal AUC interval for young patients with Age ≤ 28, the optimal AUC interval for patients in CR at transplant increases with Age, with the optimal intervals for CR = Yes versus CR = No becoming completely disjoint for patients with Age ≥ 55 years. The lower portions of the curves for CR and No CR

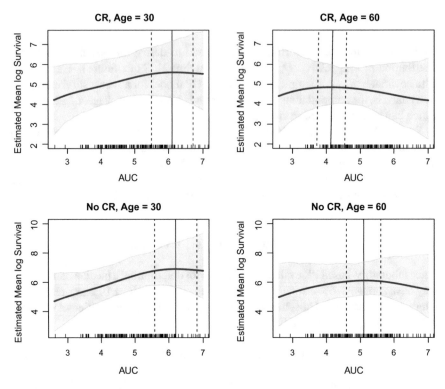

Fig. 11.1 Estimated log mean survival time for a patient either in complete remission (CR) or with active disease (No CR) at transplant, Age = 30 or 60 years, as a function of AUC, with the optimal targeted AUC interval for each subgroup

Fig. 11.2 Estimated optimal targeted AUC interval for intravenous busulfan, in the allogeneic stem cell transplant conditioning regimen, as a function of Age and disease status

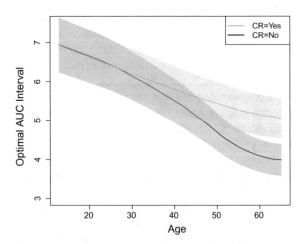

in Fig. 11.2, for Age $\leq$ 28, agree with the conclusion of Bartelink et al. (2016) for pediatric and adolescent patients, while the higher portions for Age > 28, show that CR status matters a great deal for older patients. This provides a concrete basis for planning a targeted AUC for an AML/MDS patient undergoing allosct based on their Age and CR status, that is, for doing precision medicine.

Chapter 12
Multistage Treatment Regimes

> *Life can only be understood backwards; but it must be lived forwards.*
> Soren Kierkegaard

Contents

Abstract For each patient, treatment for a disease often is a multistage process involving an alternating sequence of observations and therapeutic decisions, with the physician's decision at each stage based on the patient's entire history up to that stage. This chapter begins with discussion of a simple two-stage version of this process, in which a Frontline treatment is given initially and, if and when the patient's disease worsens, i.e., progresses, a second, Salvage treatment is given, so the two-stage regime is (Frontline, Salvage). The discussion of this case will include examples where, if one only accounts for the effects of Frontline and Salvage separately in each stage, the effect of the entire regime on survival time may not be obvious. Discussions and illustrations will be given of the general paradigms of dynamic treatment regimes (DTRs) and sequential multiple assignment randomized trials (SMARTs). Several statistical analyses of data from a prostate cancer trial designed by Thall et al. (2000) to evaluate multiple DTRs then will be discussed in detail. As a final example, several statistical analyses of observational data from a semi-SMART design of DTRs for acute leukemia, given by Estey et al. (1999), Wahed and Thall (2013), and Xu et al. (2016), will be discussed.

© Springer Nature Switzerland AG 2020   247
P. F. Thall, *Statistical Remedies for Medical Researchers*, Springer Series
in Pharmaceutical Statistics, https://doi.org/10.1007/978-3-030-43714-5_12

12.1   The Triangle of Death

Survival time sometimes is referred to as "The Gold Standard" when considering different possible outcomes to use as a basis for evaluating and comparing treatments for life-threatening diseases. This is biased on the ideas that, in this type of medical setting, how long a patient lives from the start of treatment is what matters most and, while other, intermediate events may be important, they do not matter as much as survival time. Some examples of intermediate nonfatal events measured after the start of Frontline treatment are disease progression or severe regimen-related toxicity in oncology, worsening pulmonary function for patients with severe respiratory disease, severe sepsis in patients who have been admitted to an intensive care unit, myocardial infarction for a patient with acute coronary disease, or hospitalization for a variety of different diseases. There is an ongoing controversy in the medical research and regulatory communities regarding whether, in a given setting, it is appropriate to use a particular intermediate outcome as a surrogate for survival time, or possibly for some other later event that is considered to be of primary importance. This is a central issue in the FDA's Accelerated Approval Program, and in "fast tracking" the development and review of new drugs. Because this is a complex issue with medical, economic, political, and ethical aspects, I will not discuss it here per se, although most of the examples that follow are closely related to the issue of surrogacy.

This chapter will focus on examining some statistical aspects of the practicing physician's medical decision-making process of repeatedly choosing treatments and observing clinical outcomes for each patient. To start, consider an oncology setting where a Frontline anticancer treatment is given first, after which the patient may either experience disease progression or not before death. At the time of cancer progression, patients typically are given some form of salvage treatment. Figure 12.1 gives a schematic of the two possible pathways that each patient may follow, with the transition times T_1 = time from the start of Frontline therapy to progression, T_2 = time from progression to death, and T_0 = time from the start of Frontline therapy to

Fig. 12.1 The Triangle of Death. Possible outcomes for cancer patients who may either die without disease progression or have disease progression and then receive Salvage treatment. The two-stage treatment regime is (Frontline, Salvage)

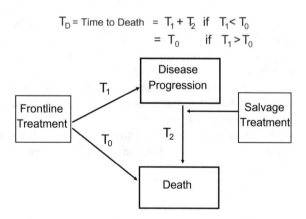

death without progression. One of my colleagues once referred to this basic graphical representation as "The Triangle of Death."

The therapeutic goal of salvage treatment is to make the subsequent survival time T_2 longer. In the happy case where $T_0 = 20$ years, then the Frontline treatment obviously was successful. If $T_0 = 3$ months then the outcome often is called "regimen-related death." This schematic is a simplification of the actual therapeutic process, which usually includes many additional things, such as prophylactic treatments given with the Frontline therapy, regimen-related adverse events such as toxicity, impaired organ function, or infection, and additional treatments or actions to deal with these events. Moreover, depending on the disease and treatment setting, a patient may experience more than one repetition of the sequence (Salvage Treatment, Disease Remission, Disease Progression) before death. So, Fig. 12.1 may be considered the simplest case, with numerous possible additional structures ignored at this point for simplicity. This paradigm is not limited to cancer therapy. It may be considered to be a prototype example of the therapeutic process for any fatal disease where an intermediate event that characterizes disease worsening may occur prior to death, including the examples given above, and a second treatment may be given if and when the disease worsening occurs.

An interesting aspect of this schematic representation is that all three transition times are potential outcomes. Aside from the possibility of administrative censoring at the time when follow-up ends, the observed outcomes for any patient are either (T_1, T_2) or T_0, since it is impossible to observe both for any patient. Each patient's survival time is either $T_D = T_1 + T_2$ if $T_1 < T_0$, that is, if the patient experiences progressive disease, or $T_D = T_0$ if $T_0 < T_1$, that is, if the patient dies without first experiencing progressive disease. The inequality $T_1 < T_0$ is interesting in that it only makes sense if you think of T_1 and T_0 as potential outcomes, only one of which can be observed. It implies that, if the patient suffers disease progression, then T_0 is a counterfactual outcome that is not observed. Another way of putting this is that progression and death without progression are *competing risks*. The Frontline treatment may have effects on all three transition times T_1, T_2, T_0, but Salvage treatment can only affect T_2. The physician often chooses which Salvage treatment to give a patient adaptively, based on the patient's Frontline treatment and T_1. Thus, a patient's overall therapy consists of the two-stage adaptive treatment strategy denoted by (Frontline, Salvage). To elaborate this structure so that it represents medical practice somewhat more accurately, denote the patient's covariates at the start of therapy by X_0, and updated covariate values at the time of progression by X_1. The vectors X_0 and X_1 may include age, performance status, indices of organ function such as bilirubin and creatinine levels, disease subtype, disease severity, number of prior therapies, genomic or molecular biomarkers thought to be related to the disease, and so on. These are the variables that a physician uses to make a diagnosis and choose a first treatment, that is, the physician's choice of Frontline treatment depends on X_0. But the choice of Salvage treatment may depend on a lot more data, including (Frontline, T_1, X_0, X_1). Thinking of the multistage process, the treatment at each stage typically is chosen adaptively based on the patient's current history of covariates, previous treatments, and outcomes. A very important point is that the current data used for choosing a Salvage treatment may include the Frontline treatment that the patient was given. If so, the two treatment choices are linked to each other. In general, (Frontline, Salvage) actually is a pair of consecutive decision rules. These may be represented as the following alternating sequences, where each arrow represents either "Observe" or "Decide/Treat":

Two-Stage Adaptive Treatment Strategy for Fatal Diseases

1. $X_0 \Rightarrow$ Frontline $\Rightarrow T_1$ or T_0.
2. If $T_1 < T_0$ then $(X_0, X_1,$ Frontline, $T_1) \Rightarrow$ Salvage $\Rightarrow T_2$.

Below, I will argue by example that, and if you are analyzing a dataset with different (Frontline, Salvage) combinations given to a sample of patients, your inferences should focus on how to choose a best (Frontline, Salvage) pair together at the start of a patient's therapy, rather than choosing each separately in time. This is a central issue when considering two-stage or multistage treatment regimes, either statistically or in medical practice.

The pair (Frontline, Salvage) is a two-stage example of a dynamic treatment regime (DTR). This is a statistical formalism for what physicians actually do when they practice medicine. In general, a DTR may be represented operationally as the sequence

[observe baseline variables] $\rightarrow$ [make a treatment decision] $\rightarrow$
[observe outcome variables] $\rightarrow$ [make a second treatment decision] $\rightarrow$
[observe updated outcome variables] $\rightarrow \cdots$

Despite the fact that DTRs are ubiquitous in medical practice, they usually are ignored in clinical trial designs because most investigators feel that it is too difficult to account for multistage treatment strategies, rather than individual treatments. I will discuss DTRs and give some examples in this chapter.

Consider the common problem of comparing two Frontline treatments, A and B, for a particular cancer, and suppose that you want to design a clinical trial that sticks to The Gold Standard by using survival time as the primary outcome. In this case, I have some Very Bad News and some Very Good News. The Very Bad News is that, if you stick to The Gold Standard, your clinical trial may reach a ridiculous, obviously incorrect, and possibly embarrassing conclusion. The Very Good News is that there is a wonderful new area of statistics that can rescue you from this terrible possibility. This new area of statistics has several different names, including DTRs, adaptive treatment strategies, treatment policies, or multistage treatment strategies. But whatever you call them, the statistical methods provide remedies for inferential and decision-making problems that may arise in medical settings, like that described above, where giving an initial Frontline treatment is not the only thing that the physician may do.

Here are some examples of how sticking to The Gold Standard when evaluating Frontline treatments can lead to serious inferential problems if one is not careful. First, suppose that A and B are treatments for a rapidly fatal disease, with the same mean time to disease progression of $\mu_{A,1} = \mu_{B,1} = 14$ months, and patients are randomized fairly between them. Suppose that, per protocol, all patients in arm A receive Salvage therapy S_1, which gives an additional mean survival time of $\mu_{S_1,2} = 10$ more months, and all patients in arm B receive Salvage therapy S_2, which gives an additional mean survival time of $\mu_{S_2,2} = 2$ more months. For simplicity, suppose temporarily that no patients die prior to disease progression, so we do not need to consider T_0. The strategy to start with A and then give S_1 as Salvage at disease progression, written as (A, S_1), gives total expected survival time $\mu_{A,1} + \mu_{S_1,2} = 14 + 10 = 24$ months, while the strategy (B, S_2) gives total expected survival time $\mu_{B,1} + \mu_{S_2,2} = 14 + 2 = 16$ months. If A and B are compared in terms of mean survival time while ignoring the salvage therapies, then it may appear that A provides a 50% improvement in mean survival compared to B. But the difference is due entirely to the superiority of salvage S_1 over S_2. Of course, if it were known from the start that S_1 is greatly superior to S_2 then no sensible physician would use the strategy (B, S_2). But in practice, such things often are not known.

This simple example may be elaborated in numerous ways. In many clinical trials, per protocol, Salvage treatment is chosen by the attending physician. Either a particular Salvage treatment is not specified in the protocol, or a set of possible salvage treatments may be listed. Suppose that A and B are two different remission induction chemotherapies for relapsed/refractory AML, and in fact $\mu_{1,A} = 15$ months and $\mu_{1,B} = 10$ months. Suppose that the attending physicians give $S = $ Allosct as salvage for patients who have Poor prognosis at progression, defined in terms of worse X_1 and shorter T_1. But if a patient has Good prognosis at the time of progression, then the physician's selected S will be one of a set of investigational targeted agents, TA. I will simplify things by lumping all the TAs together. Formally, dichotomizing the data (X_1, T_1) observed at the time of disease progression as either *Good* or *Poor* prognosis, the deterministic Salvage treatment decision rule is $S(Good) = TA$ and $S(Poor) = $ Allosct. This rule for choosing S as either Allosct or a TA may be based on a variety of rationales. Many hematologists who use chemotherapeutic or targeted agents regard Allosct as a such an aggressive treatment that it should be used mainly as Salvage, as the last hope for *Poor* prognosis patients at the time of progression. There invariably is optimism about a new TA based on preclinical results, and there may be a desire to give a new TA a better chance of success by administering it only to *Good* prognosis patients. Moreover, clinicians often ignore the earlier history $(X_0, Frontline)$ and only use the more recent data (X_1, T_1) to choose a Salvage treatment.

Suppose that the mean residual survival times after S are $E(T_2 \mid Allosct) = \mu_{Allosct,2} = 24$ months and $E(T_2 \mid TA) = \mu_{TA,2} = 6$ months, regardless of T_1 and Frontline. Finally, suppose that the probability of *Good* prognosis at the time of progression after Frontline A is $\Pr(Good \mid A) = 0.80$, but Frontline B is a more toxic chemotherapy that is more likely to harm the patient's immune system and organ function, so $\Pr(Good \mid B) = 0.20$. It follows that the mean survival time for the two-stage strategy (A, S) is computed as

$$\mu(A, S) = \mu_1(A) + \mu_2\{A, S(Good \mid A)\} \times \Pr(Good \mid A)$$
$$+\mu_2\{A, S(Poor \mid A)\} \times \Pr(Poor \mid A)$$
$$= \mu_1(A) + \mu_2\{TA\} \times \Pr(Good \mid A)) + \mu_2\{Allosct\} \times \Pr(Poor \mid A)]$$
$$= 15 + (6 \times 0.80) + (24 \times 0.20) = 24.6.$$

Similarly, the mean survival time for the two-stage strategy (B, S) is

$$\mu(B, S) = \mu_1(B) + \mu_2\{B, S(Good \mid B)\} \times \Pr((Good \mid B)$$
$$+\mu_2\{B, S(Poor \mid B)\} \times \Pr(Poor \mid B)$$
$$= \mu_1(B) + \mu_2\{TA\} \times \Pr(Good \mid B)) + \mu_2\{Allosct\} \times \Pr(Poor \mid A)$$
$$= 10 + (6 \times 0.20) + (24 \times 0.80) = 30.4.$$

So, based on overall survival time while ignoring Salvage treatment effects, it appears that B is superior to A because B has a 5.8 month (24%) longer mean survival time. But notice that patients who have *Poor* prognosis at progression actually do better subsequently than those with *Good* prognosis at progression, due to the salvage therapy selection rule. The apparent superiority of B over A actually is due to (1) the superiority of Allosct over TA as Salvage, (2) the fact that B is much more likely to have *Poor* prognosis patients at progression than A, and (3) the fact that all of the attending physicians choose to give Allosct as salvage rather than an investigational TA to *Poor* prognosis patients. This toy example was constructed to be fairly simple to show how salvage treatment selection can affect overall survival time. The example can be elaborated by making the salvage selection rule a more complex function of the patient's entire history, (Frontline, T_1, X_0, X_1). This example also is similar to what was seen in the analyses of actual data reported by Hernan et al. (2000) to assess the effect of AZT on survival time in HIV positive men, described in Sect. 6 of Chap. 6, where the physicians were more likely to give AZT to patients with worse prognosis.

In the (Frontline, Salvage) oncology setting, researchers often ignore salvage therapies in clinical trial designs based on the assumption that the effects of different salvage treatments will somehow "average out" between two Frontline treatments A to B, which are compared based on mean survival time. In the above example, it then would appear that B is superior to A, since the naive statistical estimate of $\mu(B)$ will vary around $\mu(B, S) = 30.4$, while the naive estimate of $\mu(A)$ will vary around $\mu(A, S) = 24.6$. But the apparent advantage of B as Frontline treatment actually is entirely due to the superiority of Allosct over TA as Salvage treatment. To summarize, what is going on in this example is the following:

1. A is superior to B as a Frontline treatment in terms of time to progression, since $\mu_1(A) = 15$ compared to $\mu_1(B) = 10$.
2. Patients treated with A initially are much more likely to have *Good* prognosis at the time of progression compared to B given initially.
3. Taking the interim data into account when choosing Salvage, attending physicians give all *Good* prognosis patients TA and give all *Poor* prognosis patients Allosct.

4. Allosct provides a much longer mean residual survival time after progression than a TA.

This example illustrates the point that there is nothing wrong with sticking to The Gold Standard by using mean survival time as the criterion for treatment evaluation. The rationale for using mean survival time is perfectly valid. But, given this, one must evaluate and compare each entire DTR, which in this example are (A, S) and (B, S), and not just the Frontline treatments A and B, for the simple reason that different Salvage treatments may affect post-progression survival time T_2 differently. Also, it is important to bear in mind that S actually is an adaptive rule for selecting a salvage treatment based on the patient's history at the time of progression.

To elaborate this example further by including the possibility of death before progression, which is the event $T_0 < T_1$, suppose that $\Pr(T_0 < T_1 \mid B) = \Pr(T_0 < T_1 \mid A) = 0.10$. If mean time to death without progression is the same for the two Frontline treatments, say $\mu(0, A) = \mu(0, B) = 3$ months, then the mean overall survival times of the two DTRs are

$$\mu_D(A, S) = (24.6 \times 0.90) + (3 \times 0.10) = 22.44$$

and

$$\mu_D(B, S) = (30.4 \times 0.90) + (3 \times 0.10) = 27.66.$$

Again, although A is a better Frontline treatment than B in terms of mean time to progression, the two-stage DTR (B, S) is superior to (A, S) in terms of mean overall survival time, although the expected difference is only about 5 months. Again, this example shows that it is better to optimize the entire two-stage regime, rather than to first optimize Frontline therapy and then optimize Salvage.

An important, closely related practical consideration in this example is that progression-free survival (PFS) time often is used in place of survival time as the primary endpoint. Formally, this is the time $T_{PFS} = \min\{T_0, T_1\}$ to progression or death, whichever occurs first. One may call T_{PFS} a "surrogate" for T_D. One common argument for using T_{PFS} rather than T_D for treatment evaluation is that salvage therapies given at progression may muddy the A versus B comparison. Since PFS ignores T_2 and thus the effects of salvage therapy, it provides a perfectly valid comparison of the Frontline treatments. But this comparison may result in a misleading conclusion. Comparing the Frontline treatments in the above example based on their mean PFS times gives

$$\mu_{PFS}(A) = \mu(0, A) \times \Pr(T_0 < T_1 \mid A) + \mu(1, A) \times \Pr(T_0 > T_1 \mid A)$$
$$= (3 \times 0.10) + (15 \times 0.90) = 13.8$$

and

$$\mu_{PFS}(B) = \mu(0, B) \times \Pr(T_0 < T_1 \mid B) + \mu(1, B) \times \Pr(T_0 > T_1 \mid B)$$
$$= (3 \times 0.10) + (10 \times 0.90) = 9.3.$$

Table 12.1 Obtaining a response in one or two stages of therapy

Treatment	Response probabilities		
	Frontline	Salvage with C	Overall in 1 or 2 stages
A	0.40	0.10	0.46
B	0.20	0.50	0.60

So, in terms of mean PFS time, A is 48% better than B, although the mean difference of $13.8 - 9.3 = 4.5$ months is a small improvement in terms of actual patient benefit. In this setting, a clinical trial based on PFS certainly would be completed much more quickly, which is attractive from a practical viewpoint. But this approach and its conclusion are myopic. These patients are very likely to suffer disease progression, and the attending physicians almost certainly will choose each patient's salvage therapy based on their data (Frontline, T_1, X_0, X_1), or maybe just (T_1, X_1), at disease progression. To say "A is superior to B in terms of mean PFS" ignores this simple reality. Additionally, using T_{PFS} treats progression and death as "failure" events that are perfectly equivalent, which is complete nonsense. Using PFS, if the conclusion that A is better than B is published and on that basis clinicians use A rather than B as Frontline, then what they actually will do is apply the DTR (A, S). To quantify the consequence of this myopic inference based on an A versus B comparison of PFS, if (A, S) rather than (B, S) is used for 1000 patients, then the total expected loss in survival time would be $\{(27.66 - 22.44)/12\} \times 1000 = 435$ years.

Here is a very simple example of different two-stage DTRs evaluated in terms of a binary response variable in each of two discrete stages of therapy. It illustrates the idea that an aggressive Frontline treatment may maximize the probability of response at the start of therapy, but if it fails then the chance of subsequent response may be greatly reduced. A common example in oncology is an aggressive chemo that maximizes the stage 1 response probability, but if it fails then the patient's immune system and health are so damaged by the chemo that any stage 2 chemo must be given at a reduced dose, in order to control the likelihood of further damage. But this also makes a stage 2 response unlikely. Suppose that the goal is to achieve a response in at most two stages of therapy, and stage 2 treatment is given only if response is not achieved in stage 1. Suppose that either A or B is given in stage 1, and if a response is not achieved then C is given in stage 2. Table 12.1 gives an example where A has twice the stage 1 response probability of B, 0.40 versus 0.20, but the overall probabilities of response in one or two stages are $0.40 + (1 - 0.40) \times 0.10 = 0.46$ for two-stage strategy (A, C) and $0.20 + (1 - 0.20) \times 0.50 = 0.60$ for two-stage strategy (B, C). So, if achieving a response within at most two stages of therapy is the goal, then the myopic approach of starting with A based in its larger stage 1 success probability is a mistake.

This example may be elaborated to a process where the goal is to achieve a response within three stages. Suppose salvage with D is given in stage 3 if the treatments given in stages 1 and 2 both fail. If response with D in stage 3 has probability 0.05, then the overall three-stage response probability starting with A is $0.40 + 0.60 \times \{0.10 + (0.90 \times 0.05)\} = 0.487$. The overall three-stage response probability starting with B is $0.20 + 0.80 \times \{0.50 + (0.50 \times 0.05)\} = 0.620$. The point is that the damage due to an aggressive stage 1 treatment that fails to achieve a response propagates through all subsequent stages, so such a Frontline therapy actually may lock one into a strategy that is far from optimal.

Takeaway Messages

1. If one wishes to use mean survival time to evaluate treatments, then each entire DTR, including Frontline and one or more subsequent Salvage treatment decisions, should be evaluated, not just Frontline treatments.

2. Using PFS time to evaluate and compare Frontline treatments is convenient, but it treats death and progressive disease as equivalent failure events. Because a DTR, and not just Frontline treatment, will actually be administered, using PFS may lead to erroneous conclusions and bad recommendations if the Frontline treatment effects on the PFS time distribution differ substantially from the (Frontline, Salvage) DTR effects on the survival time distribution.

3. In any medical setting that involves multiple stages of therapy, choosing treatments myopically by optimizing the first stage of therapy may be a mistake. The first action of the optimal DTR may be very different from the myopic action that optimizes only the first stage outcome. Ideally, one should determine an optimal multistage DTR.

12.2   Observe, Act, Repeat

In everyday life, each action that you take has consequences that may be good or bad, large or small. Later on, when you see what happens after some action you have taken, you may take another action to remedy what has gone wrong, or to reinforce what has gone right. Of course, you never know for sure that a given action *caused* what happened later, since your action actually may have been an Innocent Bystander while some other, unseen forces were at work. Life goes on like this for us all, with all the consequences of all the actions that we take rippling forward in time. Looking back, you might see that some of your actions may have been the right choices because they were followed by good outcomes, and see that other decisions you made and things you did had bad consequences. But you can't go back in time and do things differently.

Imagine that you could, somehow, stand outside of time and watch all of the things that you have done, or will ever do, and all their consequences. The question is, even if you knew everything that has happened or will happen in the course of your life, could you figure out the best sequence of actions to take? In each medical example given in this chapter, a physician repeatedly gives treatments to a patient, observes the clinical outcomes that follow, and decides what to do next. Treating a patient must be done moving forward, the same way that life is lived. But these days, clinical practice data are being collected and stored on a massive scale. So now we can look at data from what many physicians have done over time to treat a particular disease and approach the problem statistically. The above question about how best to live a single life may be rephrased as two statistical questions about how to determine optimal multistage adaptive treatment strategies, also called DTRs, that physicians may use.

Question 1: If you have an observational dataset from patients treated repeatedly over time for a particular disease, including each patient's entire history of covariates, treatments, and clinical outcomes, can you identify an optimal DTR statistically?

Question 2: In the absence of such observational data, can you design a clinical trial to define and study a reasonable set of DTRs and reliably determine which is optimal?

So, while no one can go back and relive their life making smarter decisions, ideally, medical researchers can analyze observational data or design clinical trials to learn what DTR is likely to be best for a given disease, and communicate this to the community of practicing physicians to help them improve therapy for future patients. But this sort of data analysis and clinical trial design is quite modern, and far from straightforward. They require a new way of thinking about statistics and medicine. The simple examples of this given in Sect. 12.1 may help to motivate what follows.

   Here is a general description of a multistage treatment process. What a physician does with each patient can be described as an alternating sequence of observations and actions, with some final payoff, Y, such as resolution of an infection, the time until an alcoholic undergoing treatment takes a drink during a specified follow-up period, or survival time for a fatal disease. The observation at each stage of the process includes the most recent clinical outcomes and possibly updated baseline covariates, and each action is a treatment or possibly a decision to either delay or terminate treatment. Notation for a DTR can get very complicated, so here is a fairly simple representation. Indexing the stages of therapy by the integer k, with $k = 0$ identifying the baseline before any action is taken, the sequence of observations and actions for a three-stage DTR can be written as

$$O_o \quad \rightarrow \quad A_1 \quad \rightarrow \quad O_1 \quad \rightarrow \quad A_2 \quad \rightarrow \quad O_2 \quad \rightarrow \quad A_3 \quad \rightarrow \quad Y.$$

It might seem that the payoff Y should be written as $Y = O_3$ instead, but if $Y =$ survival time then this would not really be correct, since a patient might die at the end of stage 1 or 2 before all three actions can be taken. This process may be carried out up to a predetermined maximum number of stages, as in the examples given above and in Sect. 12.1, or continued until the disease is cured, the physician decides to stop because either it is futile to continue or the treatment produces unacceptable toxicity that outweighs its potential benefits, the patient decides to discontinue therapy, or the patient dies. The first, baseline observation O_o includes all of the patient's initial prognostic covariates, diagnosis, and disease severity. A_1 is the initial, Frontline treatment given, and O_1 includes the patient's initial clinical outcomes, such as response or toxicity, and possibly updated values of some of the baseline covariates included in O_0. Each action A_k is not necessarily a treatment, since it may be things like "Do not treat" or "Delay treatment until toxicity is resolved." In some settings, such as treatment of acute or rapidly fatal diseases, there are only a small number of stages. In treatment of chronic diseases, the alternating sequence usually has many stages that go on for a long time. For example, a nephrologist may repeatedly adjust the dose of an immunosuppressive agent given to a kidney transplant patient to prevent rejection over a long period of time, based on repeatedly observed variables like creatinine level and glomerular filtration rate to quantify kidney function, as well as adverse events associated with the agent. Similar repeated treatment decisions, which usually are a dose adjustment, augmentation by adding something new to previous therapy, or a switch to a new drug, are made in long-term therapeutic management of chronic diseases such as high blood pressure, epilepsy, or diabetes.

Since the physician decides the actions $A_1, A_2, \ldots$, and observes the baseline data and outcomes $O_0, O_1, O_2, \ldots$, the next point is central to thinking about how these actions are chosen. Consider a two-stage process. The Frontline therapy typically is chosen based on the baseline information, O_0, which we may write as $A_1 = A_1(O_0)$. The complete history for the patient when the second therapeutic action is taken is (O_o, A_1, O_1), which includes the physician's first action. So, if the physician wishes to use all of this history to choose A_2, then we may write this as $A_2 = A_2(O_o, A_1, O_1)$. In words, this says that the physician's second action depends on not only the patient's covariates and stage 1 outcomes, but also what the physician gave as Frontline treatment. That is, the two-stage DTR $= (A_1, A_2)$ is a pair of decision rules, where A_2 depends on A_1. In general, the action sequence $A_1, A_2, A_3, \ldots$, is embedded in the alternating sequence $O_0, A_1, O_1, A_2, O_2, A_3, \ldots$. When I describe the DTR paradigm to physicians with whom I collaborate, invariably they say something like, "Yes, that is what I do."

Given the payoff variable, Y, the problem of estimating an optimal DTR may be formulated with the goal to maximize $E(Y \mid DTR)$, the expected payoff if a given DTR is used. It turns out that, in general, this is quite a difficult and complicated statistical problem. Recall the examples where first myopically choosing the Frontline treatment $A_1 = A_1(O_0)$ to optimize the expected outcome in the first stage, $E(O_1 \mid A_1, O_0)$, may be the wrong thing to do, since it may not give the largest expected final payoff, $E(Y \mid A_1, A_2) = E(Y \mid DTR)$.

Since the G-estimation approach of Robins (1986) will be applied in some of the examples given below, here is a general explanation of the G-computation algorithm, for a three-stage DTR. Denote the patient's entire history of observed covariate and outcome data through k stages by $\overline{O}_k$, and all of the actions taken by the physician through k stages by $\overline{A}_k$. Suppressing notation for model parameters, the joint distribution of all observations and actions and the final outcome $(O_0, A_1, O_1, A_2, O_2, A_3, Y)$ for a three-stage DTR is

$$p(\overline{O}_2, \overline{A}_3, Y) = p(O_0) \times p(A_1 \mid O_0) \times p(O_1 \mid O_0, A_1) \times p(A_2 \mid \overline{O}_1, A_1)$$
$$\times p(O_2 \mid \overline{O}_1, \overline{A}_2) \times p(A_3 \mid \overline{O}_2, \overline{A}_2) \times p(Y \mid \overline{O}_2, \overline{A}_3).$$

Notice that defining a conditional distribution for each action A_k treats it as random, which accounts for things like randomization or observational data where a given treatment decision at a given stage is a random function of the patient's current history. Also, $\overline{A}_3 = (A_1, A_2, A_3)$ is the three-stage DTR that we wish to learn about. G-estimation is carried out by considering only the conditional distributions of the observed outcomes while ignoring the treatment assignment distributions and, under some assumed model, fitting the likelihood function

$$\mathcal{L}(\overline{O}_2, \overline{A}_3, Y) = p(O_0) \times p(O_1 \mid O_0, A_1) \times p(O_2 \mid \overline{O}_1, \overline{A}_2) \times p(Y \mid \overline{O}_2, \overline{A}_3)$$

to the data. Averaging over $\overline{O}_2 = (O_0, O_1, O_2)$ then gives $p(Y \mid A_1, A_2, A_3)$, so the expected payoff $E(Y \mid A_1, A_2, A_3) = E(Y \mid DTR)$ can be computed. This says that, to do G-estimation, one needs to model the conditional distribution of each outcome, fit the resulting likelihood function, and integrate or average out all of the observations except the final Y. In practice, O_0 usually includes patient baseline covariates, X, and a sensible approach is simply to average over the sample distribution of X.

There is an extensive, constantly growing published literature on methods for constructing and evaluating DTRs, in numerous fields, including observational medical data analysis, causal inference, sequential analysis, control theory, computer science, treatment of chronic substance abuse, and clinical trial design. I will not attempt to review this literature here, but rather will just give some examples. Papers on models and methods for analysis of DTRs are Robins (1986, 1997, 1998), Murphy et al. (2001), Murphy (2003), Moodie et al. (2007), and Almirall et al. (2010), among many others. A trial of DTRs for treating mental health disorders is described by Rush et al. (2003), and other clinical trial designs or analytic methods are described by Lavori and Dawson (2004), Lunceford et al. (2002), and Wahed and Tsiatis (2004). Q-learning methods were proposed by Watkins (1989), and extended by Murphy (2005b), and many others. Other approaches are the outcome weighted learning methods of Zhou et al. (2015), and regression-based Bayesian machine learning proposed by Murray et al. (2018a), implemented using Bayesian adaptive regression trees, Chipman et al. (1998). Two recent books on DTRs are a formal treatment by Chakraborty and Moodie (2013), and a collection of chapters on a wide array of topics edited by Kosorok and Moodie (2015).

12.3 SMART Designs

Most commonly, the DTRs that physicians actually use are determined by come combination of their own clinical practice experience, what they learned in medical school or more recently from the published medical literature, advice from colleagues, intuition, and trying out different DTRs that are motivated by some combination of these elements. Constructing or identifying optimal DTRs empirically is a very new process, since it shares the data from many physicians either observationally or through a clinical trial, and it is not yet well established in the medical community.

After researchers identified the general DTR paradigm and began to construct models and methods for analyzing longitudinal data involving DTRs, they came up with a general strategy for designing clinical trials to evaluate and compare DTRs fairly. Thus, the sequential multiple assignment randomized trial (SMART) was born. A SMART has two key motivations. The first is that, because actual medical therapy usually involves multiple stages, we should be evaluating DTRs, not just the treatments given in one stage of the process. The great importance of this was illustrated by the examples given above in Sect. 12.1. The second motivation is the scientific goal to obtain an unbiased estimator of the mean of some final payoff variable for each DTR, so that one can do unbiased comparisons. Combining these two ideas motivates randomizing patients at more than one stage of the regime, with the randomizations done after the first stage often depending on the patient's history of treatments and outcomes. The goal of a SMART is to identify an optimal DTR from the set of DTRs being studied. Murphy (2005a) presented a general treatment of SMARTs, and introductions to the basic ideas are given by Lei et al. (2012), and in each of the books by Chakraborty and Moodie (2013) (Chap. 2) and Kosorok and Moodie (2015) (Chaps. 2, 3, 4, and 5). Most SMARTs have been conducted to develop and evaluate DTRs for chronic diseases, including trials discussed by Almirall et al. (2014) in weight loss research, by Collins et al. (2005) for optimizing a DTR for behavioral interventions, by Breslin et al. (1999) for alcohol abuse, and by Murphy et al. (2005) for treatment of substance abuse. In oncology, SMARTs have been designed by Thall et al. (2000) for prostate cancer, which I will discuss below in Sect. 12.4, and by Thall et al. (2002) for acute leukemia.

The first step in constructing a SMART design is to determine the DTRs to study. This often is a difficult exercise, and it must involve close collaboration between the statistician and the research scientists planning the trial. Each DTR must be viable, as described by Wang et al. (2012). That is, it must be a sequence of actions that the medical professionals involved in the trial actually will take. While there are general classes of DTRs, those studied in a particular SMART will reflect the nature of the disease, clinical practice, observable patient outcomes, and resource limitations.

Figure 12.2 gives a schematic representation of a SMART to evaluate eight DTRs for treatment of a severe hematologic disease, such as AML, ALL, or non-Hodgkin's lymphoma. All patients first are randomized fairly between two Frontline induction treatments, F_1 and F_2. Patients for whom an early response, $O_1 = R$, is achieved are re-randomized between two maintenance treatments M_1 and M_2, while patients for whom early response is not achieved, $O_1 = NR$, are re-randomized between two salvage treatments S_1 and S_2. In conventional DTR terminology, the first binary O_1 outcome is used as a "tailoring variable" to decide whether the second round of treatment is maintenance or salvage. The four DTRs starting with F_1 are (F_1, M_1, S_1), (F_1, M_1, S_2), (F_1, M_2, S_1), (F_1, M_2, S_2), with similar structures for patients starting with F_2, for a total of eight DTRs. Notice that, since the first outcome is the binary $O_1 = R$ (response) versus $O_1 = NR$ (Resistant Disease), tailored by O_1, a patient may receive either maintenance or salvage as second stage therapy, but not both. The final "payoff" outcome O_2 may be survival time.

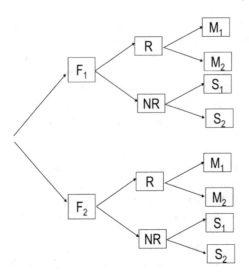

Fig. 12.2 Schematic of a SMART to evaluate eight DTRs for treatment of acute hematologic diseases. F_1 and F_2 denote Frontline remission induction treatments, M_1 and M_2 denote remission maintenance treatments, and S_1 and S_2 denote salvage treatments

This basic structure may be elaborated in numerous ways. For example, if F_1 and F_2 differ qualitatively, the two salvage treatments following (F_1, NR) may be different from the two salvage treatments following (F_2, NR). If a patient whose Frontline therapy achieves R suffers severe toxicity, then maintenance therapy may either be delayed until the toxicity is resolved, or maintenance may be given at a reduced dose, according to an adaptive rule depending on the patient's white blood cell count and indices of organ function such as bilirubin and creatinine levels. So, the first outcome is the more specific bivariate $O_1 = $ (Response, Toxicity), rather than only response. The point is that the eight DTRs represented in Fig. 12.2 may be simplified versions of the actual DTRs.

Still, if the goal is to choose a DTR to maximize mean survival time, then in addition to accrual rate and follow-up needed to obtain a reasonable number of deaths to make reliable inferences, depending on the response rates for F_1 and F_2, the sample size of each DTR will be limited. The expected subsample sizes of the eight DTRs are not equal, but rather they depend on the distribution of the tailoring variable, O_1, for each Frontline treatment. For example, if $\Pr(O_1 = R \mid F_1) = 0.70$ and $\Pr(O_1 = R \mid F_2) = 0.60$, and 1000 patients are enrolled, then the expected number of patients whose treatment–outcome sequence is (F_1, R, M_j) is 175 for each $j = 1$ or 2, while the expected number of patients whose treatment–outcome sequence is (F_1, NR, S_j) is 75 for $j = 1$ or 2. The corresponding subsample sizes are 150 for each (F_2, R, M_j) and 100 for each (F_2, NR, S_j). If the eight DTRs are refined further, as described above, the subsample sizes decrease further. If a patient who has received F_1 as induction dies before response can be evaluated, then that patient's short survival time is consistent with four DTRs, so that patent's data can be used to estimate all of their mean survival times. One simplified version of the design illustrated in Fig. 12.2 would give the same maintenance treatment, M, to all responders, so only nonresponders would be re-randomized, and the trial would study the four DTRs (F_1, S_1), (F_1, S_2), (F_2, S_1), (F_2, S_2), with the within-patient rules (1) initially, randomize fairly between F_1 and F_2, (2) if $O_1 = \text{R}$ then treat with M, and if $O_1 = \text{NR}$ then randomize fairly between S_1 and S_2.

It is important to think about what is gained by running a SMART rather than separate trials for frontline, maintenance, and salvage treatments. Suppose that, instead of conducting the SMART illustrated in Fig. 12.2, one conducts (1) a randomized trial of F_1 versus F_2 based on O_1 or $Y = $ survival time, (2) a randomized trial of S_1 versus S_2 in patients for whom $O_1 = \text{NR}$, and (3) a randomized trial of M_1 versus M_2 in patients for whom $O_1 = \text{R}$.

1. If Trial 1 is based on Y, then it ignores the effects of Maintenance or Salvage treatments on Y, as well as the way that a given (F, S) or (F, M) combination works together.

2. If Trial 1 is based on $O_1 = $ response, then it completely ignores the relationship between (F, O_1) and the rest of the therapeutic process that follows.

3. Trials 2 and 3 each ignore the effects of the Frontline treatments on Y, as well as the way that a given (F, S) or (F, M) combination works together.

4. Three trials are being run, instead of one trial, and no information on the DTRs is obtained, which is a waste of time and resources.

Next, consider a two-stage SMART trial to study DTRs for opioid addiction. In stage 1, patients are randomized between an opioid antagonist, A, given at an initial dose d_1, denoted by A_{d_1}, and behavioral intervention, BI. The initial response is defined as the patient not using an opioid during the following 2-month period, which is stage 1. Responders to either A_{d_1} or BI are randomized between a 12-step program, $P12$, and telephone disease management, TDM. Nonresponders to A_{d_1} are randomized between A_{d_2}, where $d_2 > d_1$ is an increased dose of the opioid agonist, and $A_{d_1} + BI$, that is, the opioid agonist given again at the initial dose but augmented by the addition of BI. Nonresponders to BI are randomized between switching to A_{d_1} and $BI + A_{d_1}$, an augmentation. The final outcome is obtained after an additional 2 months, with stage 2 response defined as the absence of opioid use during the second 2-month period. Thus, the DTR denoted as $(A_{d_1}, P12, A_{d_2})$ says to start with the agonist at dose d_1, and then treat with the 12-step program if the patient responds in stage 1 or with a higher dose of the agonist if the patient does not respond. The other three DTRs starting with A_{d_1} may be denoted as (A_{d_1}, TDM, A_{d_2}), $(A_{d_1}, P12, A_{d_1} + BI)$, and $(A_{d_1}, TDM, A_{d_1} + B)$. The four DTRs starting with BI may be denoted similarly by $(BI, P12, A_{d_1})$, (BI, TDM, A_{d_1}), $(BI, P12, A_{d_1} + BI)$, $(BI, TDM, A_{d_1} + BI)$. The SMART is illustrated by Fig. 12.3. A bit of thought shows that Figs. 12.2 and 12.3 actually have the same structure. That is, treating the stage 1 responders in the opioid addiction trial with P12 or TDM is like treating the stage 1 responders in the cancer trial with the maintenance therapies M_1 or M_2. The treatments given to the opioid addiction patients who are nonresponders to the stage 1 treatment are analogous to the salvage treatments for a cancer that is resistant to Frontline treatment in Fig. 12.2. The main structural difference between the two SMARTs is that the stage 2 treatments

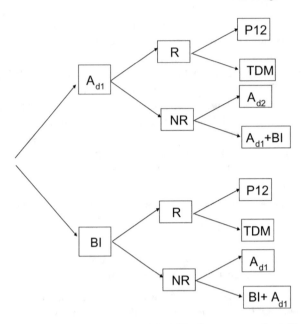

Fig. 12.3 Schematic of a SMART to evaluate eight DTRs for treatment of opioid addiction

for nonresponders in the opioid addiction trial are tailored to the treatment given in stage 1. The analogous structure in the cancer trial would be if the salvage therapies for nonresponders differed depending on which drug was given as Frontline. Thus, while the opioid addiction trial deals with a chronic disease, rather than cancer, and accordingly the treatments at each stage are completely different from those used for hematologic malignancies, the SMART structures are very similar. Still, noncompliance and dropouts are much more likely in an opioid addiction trial than a trial of cancer treatments, where an apparent noncompliance likely would be due to intercurrent toxicity, which actually is an outcome. So the different natures of the diseases necessarily would make conduct of these two SMART trials very different, as would be the case for conventionally designed trials.

In a SMART, to obtain unbiased comparisons all randomization probabilities are fixed at the start. Recall the example in Sect. 12.1 of DTRs for hematologic malignancies with Frontline A or B and Salvage Allosct or a TA. The salvage therapies were chosen by a deterministic rule that assigned all patients with *Good* prognosis at progression a TA as salvage, and all patients with *Poor* prognosis at progression Allosct as salvage. But many conventional practices in multistage adaptive treatment turn out to be suboptimal. An important question is whether the underlying assumption that Allosct is best for *Poor* prognosis patients and TA is best for *Good* prognosis patients is right or wrong, or possibly something in between. If one wishes to determine empirically how well Allosct and a TA actually compare, following either A or B, within each prognostic group, then one may use a SMART design that randomizes patients fairly between A and B as Frontline and then, regardless of whether their prognosis is *Good* or *Poor* at progression, re-randomizes each patient fairly between Allosct and TA as salvage. This SMART design fairly compares the four possible two-stage strategies for Frontline and Salvage, $(A, Allo), (A, TA), (B, Allo), (B, TA)$. For example, "$(A, Allo)$" means to first give A and then, regardless of prognosis at progression, do an allogeneic stem cell transplant. Given the previous assumed true numerical values, the overall mean survival times of these four two-stage regimes would be

$$\mu(A, Allo) = 15 + 24 = 39 \quad \mu(A, TA) = 15 + 6 = 21$$
$$\mu(B, Allo) = 10 + 24 = 34 \quad \mu(B, TA) = 10 + 6 = 16.$$

This is a simple toy example, but the SMART design that randomizes fairly at both stages would give unbiased estimators of the overall survival times of these four DTRs, and unbiased comparisons between their mean survival times. Thus, the SMART would provide useful data for inferring, ideally, that the regime $(A, Allo)$ is optimal.

For treating patients with AML, a similar but different SMART may be constructed by using prognosis at progression as a tailoring variable. This design might randomize patients with *Poor* prognosis at progression between Allosct and TA as salvage, but treat all patients with *Good* prognosis using a Salvage chemo combination that contains the agent cytarabine, which is well established for treating AML. The two salvage rules thus would be either $S_1(Poor) = $ Allosct, $S_1(Good) = $ cytarabine, or $S_2(Poor) = TA$, $S_2(Good) = $ cytarabine, with both using post-progression prognosis as a tailoring variable. The four DTRs evaluated by this SMART would be $(A, S_1), (A, S_2), (B, S_1), (B, S_2)$. Since there are numerous TAs available for AML at any given point in time, as reviewed by Shafer and Grant (2016), it is worth mentioning that either of these SMART designs could be elaborated by breaking out the TAs into different subtypes, say TA_1 and TA_2, such as epigenetic modifiers or signaling pathway inhibitors. This would increase the number of DTRs, however, which would be an extremely important practical consideration. Another approach might be to replace TA in the above DTRs with a combination of different specific types of TA for *Poor* prognosis patients. Yet another possibility, which avoids the somewhat controversial comparison of TA to Allosct, would be to replace $(TA, Allosct)$ with (TA_1, TA_2). It should be obvious, at this point, that a SMART taking any of these forms could only be constructed by close collaboration between a statistician and physicians familiar with the biology of the disease and currently available TAs.

Any DTR that physicians will not follow, especially one that is unethical, never should be included in a clinical trial design. That is, only viable DTRs should be studied. Thus, when constructing a SMART, each randomization must be acceptable in this sense. While there may be a gray area regarding what is or is not ethical, in some cases it should be obvious that a particular DTR is not viable. Interactions between the statistician and the investigators planning the START are essential in making this determination, since a statistician might not understand why a particular DTR is or is not viable. For example, suppose that, in a trial to evaluate DTRs for chronic pain, e.g., in patients with migraine headaches, rheumatoid arthritis, or sickle cell anemia, each patient first is randomized between two nonsteroidal anti-inflammatory drugs (NSAIDs), N_1 and N_2, and pain subsequently is categorized ordinally as $O_1 =$ Higher, Same, or Lower, compared to the patient's baseline pain level at enrollment. Here are some adaptive treatment rules for stage 2:

1. If $O_1 =$ Higher, then a more aggressive treatment is given by randomizing the patient fairly among three opioids: oxycodone, codeine, and morphine.

2. If $O_1 =$ Same, then the patient is switched to the other NSAID not given in stage 1.

3. If $O_1 =$ Lower, then the initial NSAID is repeated, since it worked.

But what if, instead of using rule 3, a statistician plans to switch or randomize a patient for whom $O_1 =$ Lower to some other drug or drugs? From the patient's viewpoint, this makes no sense, since they would be switched away from a medication that reduced their pain. To put this another way, such a rule would be unethical.

Methods for computing an overall sample size, N, when planning a SMART have been given by several authors, including Feng and Wahed (2009), Dawson and Lavori (2010), and Li and Murphy (2011). Formulas and simulation results for achieving a given power in each pairwise comparison of DTRs that start with different stage 1 treatments were provided by Murphy (2005a). Almirall et al. (2012) proposed sizing pilot SMARTs as feasibility studies having a specified minimum probability of enrolling at least some specified number of patients for each possible subgroup of nonresponders. Alternatively, one may determine a trial's sample size by first eliciting an anticipated accrual rate, individual patient follow-up time, and the maximum trial duration that the investigators planning the trial would consider. This may be used to determine a range of feasible values of N. Using as the criterion reliability of estimation of the payoff for each DTR, or the probability of selecting an optimal DTR, computer simulation may be used to determine how the SMART performs for each value of N considered. This exercise may motivate either reducing the complexity of the design if a simpler trial is feasible and still worthwhile, or concluding that a multicenter trial will be needed to accrue enough patients to obtain useful results.

In most applications, designing a SMART is more challenging and time-consuming than constructing a conventional RCT. This is because DTRs are inherently more complex than single-stage treatments, statistical modeling of the sequences of treatments and outcomes is required, and the properties of the trial must be validated by computer simulation, so computer software must be developed for this purpose. In contrast, the actual conduct of a trial to evaluate and compare DTRs actually is very similar to that of a conventional trial that includes randomization and within-patient adaptive rules. So, a SMART trial is fairly routine to conduct, and it only requires harder work at the design and data analysis stages.

12.4   Repeat or Switch Away

Physicians know that, for most diseases, no perfect treatment exists. Moreover, because patients are heterogeneous, a given treatment that works in one patient may not work in another, and often it is unclear why this is so. Chapter 11 gave some examples of how, in some cases, one may learn which treatment is best for a patient based on some of their characteristics. Since therapy of many diseases usually consists of multiple stages, the way that many physicians deal with patient heterogeneity in practice is to proceed sequentially with each patient by trying a sequence of possibly different treatments, and using the patient as their own control. A general algorithm used by many physicians to construct the DTR that they actually use is called the "Repeat a winner, switch away from a loser (RWSL)" rule, which may be described as follows.

The RWSL Rule: Try an initial treatment, possibly chosen based on the patient's baseline characteristics. If the initial treatment works, then keep giving it until either the disease is cured or the treatment can no longer be administered to the patient due to toxicity or treatment failure. If the treatment does not work, then switch to a different treatment. Repeat this until either the disease is cured, it is futile to continue therapy, the patient decides to discontinue therapy, or the patient dies.

The RWSL rule is motivated by the idea that whether a given treatment succeeds or fails in a given patient may be affected by between-patient heterogeneity, treatment–covariate interactions that are not fully understood, and random variation. There are many versions of the RWSL algorithm. Most start by using conventional prognostic covariates, such as disease type and severity, how many times the patient has been treated previously for their disease, age, and performance status. In multistage therapy, the RWSL algorithm is a very commonly used form of personalized medicine.

Thall et al. (2000) described a design to study 12 DTRs for treatment of advanced prostate cancer. Each DTR was determined by a Frontline chemo and a Salvage chemo combination, administered using a tailored version of the RWSL rule. Each patient first was randomized fairly among four Frontline chemo combinations, and later re-randomized if their Frontline chemo did not achieve overall success. This was one of the first SMART trials in oncology. At the time, we had no idea that we were being smart. Re-randomization just seemed like a sensible thing to do. The four chemos evaluated in the trial were denoted by CVD, KA/VE, TEC, and TEE. These four combinations were determined by the trial's PI, Dr. Randall Millikan. They were given within an adaptive multistage regime, according to the following set of RWSL rules. Rather than conducting a conventional four-arm randomized trial of these agents as Frontline therapy, we designed the trial to mimic the way that oncologists who treat prostate cancer actually behave. At enrollment, each patient's baseline disease, prostate-specific antigen (PSA) level, and other prognostic covariates were evaluated, and the patient then was randomized fairly among the four chemos. The patient's disease and PSA level were reevaluated at the end of each of up to four successive 8-week treatment courses, or stages. A distinction was made between success for the chemo administered in a given course, called "per-course success," and overall success of the entire multistage regime. For a chemo used for the first time in any of courses $k = 1$, 2, or 3, initial per-course success was defined as a drop of at least 40% in PSA compared to baseline and absence of advanced disease (AD). If this occurred, then that chemo was repeated for that patient in the next course; if not, then that patient's chemo in the next course, $k + 1$, was chosen

Table 12.2 The seven possible per-course outcomes with the Adaptive Treatment Strategy (C, B) using the RWSL algorithm, and the overall outcome, in the prostate cancer trial. S = overall success = two consecutive successful courses, and F = overall failure = a total of two unsuccessful courses

Per-course outcomes	Chemos	Overall outcome	# courses
$(Y_1, Y_2) = (1, 1)$	$(A_1, A_2) = (C, C)$	S	2
$(Y_1, Y_2, Y_3) = (0, 1, 1)$	$(A_1, A_2, A_3) = (C, B, B)$	S	3
$(Y_1, Y_2, Y_3, Y_4) = (1, 0, 1, 1)$	$(A_1, A_2, A_3, A_4) =$ (C, C, B, B)	S	4
$(Y_1, Y_2) = (0, 0)$	$(A_1, A_2) = (C, B)$	F	2
$(Y_1, Y_2, Y_3) = (1, 0, 0)$	$(A_1, A_2, A_3) = (C, C, B)$	F	3
$(Y_1, Y_2, Y_3) = (0, 1, 0)$	$(A_1, A_2, A_3) = (C, B, B)$	F	3
$(Y_1, Y_2, Y_3, Y_4) = (1, 0, 1, 0)$	$(A_1, A_2, A_3, A_4) =$ (C, C, B, B)	F	4

by re-randomizing fairly among the three chemos not given initially to that patient. Success in course $k = 2, 3,$ or 4 following a success in the previous course $k - 1$ with the same chemo was defined in a more demanding way, as a drop of at least 80% in PSA level compared to baseline and absence of AD. Overall success was defined as two consecutive successful courses. Under this particular RWSL algorithm, this could be achieved only by using the same chemo in both courses. Overall failure was defined as a total of two courses without per-course success, with the patient's therapy terminated if this occurred. There were a total of 12 possible two-stage regimes, say (C, B), where C and B were chosen from the set of four chemos given above, with the requirement that C and B had to be different. Dr. Millikan and I constructed the design for this trial over the course of many meetings that took quite some time. Christopher Logothetis, the Chair of the Genitourinary Oncology Department at M.D. Anderson, was so bemused by the complexity of this clinical trial that he dubbed it "The Scramble." The seven possible outcomes generated by the RWSL algorithm are summarized in Table 12.2.

The trial enrolled 150 patients, but 47 (31%) dropped out before completing their regime per protocol. Thall et al. (2007) presented the first analysis, focusing on the per-course response probabilities, in addition to summary statistics and Kaplan–Meier survival plots. They assumed a conditional logistic regression model in which each per-course success outcome probability was defined, given the patient's current history, as a function of the previous course outcome, a smoothed average number of failures through the most recent course that was a failure, and an indicator of low versus high disease volume. In this model, the unit of observation was one course of a patient's therapy, not a patient. Substantive conclusions from the fitted model were that (1) TEC was the best stage one treatment, (2) CVD had the highest overall (two-stage) success rate, and (3) KA/VE had the highest response rate as Salvage following a failure with a different chemo, that is, the lowest cross-resistance. Bembom and van der Laan (2007) provided two alternative analyses of this dataset that focused on estimating the overall success probability of each of the 12 two-stage DTRs.

They applied both the G-computation algorithm of Robins (1986) and IPTW, and also fit an additive model to estimate odds ratios for pairwise comparisons of the four chemos in terms of both their Frontline and Salvage success probabilities. They implemented IPTW, following Robins and Rotnitzky (1992), by (1) identifying all patients whose observed treatment history at each stage was compatible with the DTR being considered, (2) weighting each of these patients' per-stage outcomes by the inverse of the probability of having been assigned to the treatment history he actually received, and (3) averaging the re-weighted outcomes to estimate the overall success probability of the DTR.

Wang et al. (2012) provided a third analysis of The Scramble dataset, motivated by the desire to account for the fact that 31% of the 150 patients randomized in the trial dropped out. Both Thall et al. (2007) and Bembom and van der Laan (2007) assumed that the dropouts were non-informative. However, it turned out that most of the "dropouts" were not really dropouts. At the behest of Xihong Lin, I asked Randy Millikan the reason for each dropout in the trial. He explained that most of the patients classified as "dropouts" actually had experienced either toxicity or progressive disease (PD) so severe that the protocol's RWSL algorithm simply could not be completed. Instead, these patients were taken off protocol and given palliative care or some other salvage, at their attending physician's discretion. When we designed the trial, Dr. Millikan considered this practice to be so routine that he did not bother to include it in the RWSL algorithm. After some further discussion, we decided to elaborate the definition of per-course outcome to reflect what actually had been done in the trial. The aim was to account for severe toxicity and severe PD as the outcomes that they actually were, rather than incorrectly calling the patients with such outcomes "dropouts." The new per-course outcomes went far beyond the binary response/failure variables that had been defined in the design and used for previous statistical analyses. We first defined Toxicity as an ordinal outcome with the following three possible values:

T0 = No Toxicity.

T1 = Toxicity occurring at a level of severity that precluded further therapy but allowed efficacy to be evaluated.

T2 = Toxicity so severe that therapy was stopped and efficacy could not be evaluated.

We defined Efficacy as a variable with the following four possible values:

E0 = Favorable per-course response to a chemo.

E1 = Non-favorable per-course response to chemo but no PD.

E2 = PD.

E3 = Response was inevaluable due to severe toxicity.

It turned out that only seven of the 12 combinations of the newly defined two-dimensional (Efficacy, Toxicity) variable actually were possible in each stage. To deal with this in a practical way, I elicited numerical outcome desirability scores from

Table 12.3 Expert desirability scores for the seven possible combinations of Toxicity and Efficacy outcomes in each course of therapy

Toxicity	Efficacy			
	E0	E1	E2	E3
T0	1.0	0.5	0.1	–
T1	0.8	0.3	0	–
T2	–	–	–	0

Dr. Millikan, given in Table 12.3. If these elicited "expert" scores look suspiciously like utilities, it's because that is exactly what they are.

This new definition of per-course outcome reduced the number of dropouts from 47 (31%) to 12 (8%), and effectively expanded the dataset's information. This allowed Wang et al. (2012) to carry out better informed IPTW-based analyses, to estimate the benefits of the 12 different DTRs in reducing disease burden over four 8-week courses of therapy (32 weeks). The analyses considered three different sets of numerical scores, including the expert scores in Table 12.3, a binary indicator scoring 1 for two consecutive responses, that is, for overall success as defined in the trial's RWSL algorithm, and 0 otherwise, and ordinal scores with possible values $\{1, 0.5, 0\}$. Wang et al. (2012) focused on "viable switch rules," defined as the per-protocol RWSL rules, but with the important additional provision that patients who develop either severe toxicity or severe PD will be switched to a non-prespecified therapeutic or palliative strategy, determined by the attending physician. Based on a detailed construction of potential outcomes in each cycle for the viable DTRs that included toxicity and PD as outcomes, the resulting estimates identified three promising regimes. (TEC, CVD) had the highest mean expert score of 0.78, (TEC, KA/VE) and (TEC, TEE) had respective mean expert scores 0.73 and 0.74, and (CVD, TEE) had the lowest mean expert score of 0.56. IPTW estimates of mean log survival time showed very small differences between the 12 DTRs. However, for the analyses of both overall 32-week success and survival time, the estimated 95% confidence intervals for the mean outcomes of the 12 DTRs overlapped substantially, so any comparative inference was at best suggestive.

Focusing on the chemo combinations studied in The Scramble, Armstrong et al. (2008) wrote a letter that severely criticized the trial, primarily because estramustine was included in each of the chemo combinations KA/VE, TEC, and TEE. Citing the TAX327 trial of Tannock et al. (2004), they stated

"The additional cardiovascular, thrombotic, and gastrointestinal toxic effects associated with estramustine treatment have led to the conclusion that this agent likely has a minimal, if any, additive benefit in the first-line treatment of patients with metastatic castration-resistant prostate cancer. ... Because there was no docetaxel single-agent comparator arm in the study by Thall et al. (2007), it is difficult to judge the merits of this aggressive and toxic approach."

To respond to these criticisms based on actual data, Millikan et al. (2008) first obtained the fitted survival time regression model used to analyze the TAX327 study data cited by Armstrong et al. (2008). To estimate how long each patient in The Scramble would have survived if he had received docetaxel + prednisone every 3 weeks,

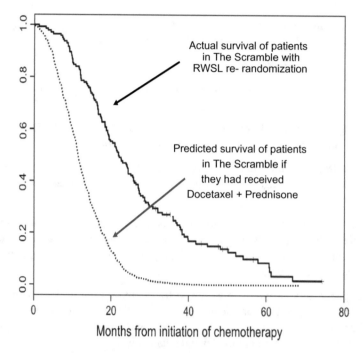

Fig. 12.4 Comparison of the Kaplan–Meier survival probability estimate based on the actual survival data from The Scramble (solid line), and the predicted survival probability curve computed from the fitted model of Armstrong et al. (2008) if all patients received docetaxel + prednisone every 3 weeks, as in the superior arm of the TAX327 study (dotted line)

as in the superior arm of the TAX327 study, they substituted the baseline covariates of each Scramble patient into this fitted prognostic model, and computed the covariate-specific survival time estimate. The covariates in the fitted TAX327 study model included liver involvement, number of metastatic sites, analgesic-requiring pain medication, Karnofsky performance status, Gleason score, measurable disease, PSA, alkaline phosphatase, progression in bone, hemoglobin, and PSA doubling time. Figure 12.4 gives the resulting estimated survival probability curve (dotted line), along with the Kaplan–Meier curve estimated from the actual survival time data of the patients in The Scramble as a comparator (solid line). Although this graphical comparison is informal, and no correction for bias due to between-trial effects or other possible confounding factors was made, the large separation between the two survival probability curves strongly suggests that the RWSL algorithm, applied with the four chemos in The Scramble, probably provided much longer survival for the patients enrolled in this trial compared to how long they would have survived if treated with docetaxel + prednisone, as in the superior arm of TAX327 study. One may hypothesize how much of the putative benefit implied in Fig. 12.4 may have been due to the "aggressive and toxic" chemo combinations studied in The Scramble, how much might be attributed to the RWSL algorithm, or how much might have been due to superior supportive care at M.D. Anderson Cancer Center.

Takeaway Messages from The Scramble

1. One cannot design a complex experiment optimally until after it already has been carried out.

2. Applying a reasonable RWSL rule as part of a DTR likely will provide greater patient benefit than giving one treatment without switching.

3. Because dropouts and other deviations from protocol designs occur quite commonly, analysis of data from a randomized clinical trial often is similar to analysis of observational data.

4. A medical statistician who wants to do a good job of study design or data analysis should talk to their doctor. This process may play out over months or years, during which the statistician will learn about medical practice and the physician will learn about statistical science.

5. What appear to be dropouts in a clinical trial may turn out to be patients whose treatment was stopped by a physician applying an adaptive decision rule as part of medical practice. Such a rule, once recognized and defined, should be incorporated into the DTR and the statistical model and data analysis.

6. If you want to study and compare 12 DTRs reliably, then you will need a sample size a lot larger than 150.

12.5   A Semi-SMART Design

The last example of unanticipated DTRs arising from medical practice is a randomized trial of four combination chemotherapies (chemos) given as Frontline treatments for patients with poor prognosis AML or MDS. Chemotherapy of acute AML/MDS begins with remission induction, aimed at achieving a complete remission (CR), which is defined as the patient having less than 5% blast cells, a platelet count greater than $10^5/mm^3$, and white blood cell count greater than $10^3/mm^3$, based on a bone marrow biopsy. Unfortunately, the phrase "Complete Remission" is extremely well established, despite the fact that it is one of the worst misnomers in medicine. All that a CR does is give an AML/MDS patient some hope of not dying very soon. Achieving a CR is well known to be necessary for long-term survival in AML/MDS, but most patients suffer a disease recurrence, at which point they have a much higher risk of death than if their CR had persisted. Consequently, if the induction chemo does not achieve a CR, or if a CR is achieved but the patient suffers a relapse, then some form of Salvage chemo is given in a second attempt to achieve a CR.

The AML/MDS trial, described by Estey et al. (1999), randomized patients fairly among the four chemo combinations FAI = fludarabine + cytarabine (ara-C) +

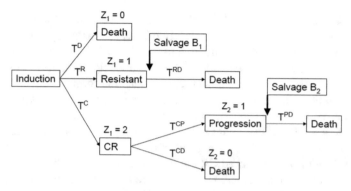

Fig. 12.5 Schematic of the adaptive treatment strategy (Induction, Salvage B_1, Salvage B_2) for acute myelogenous leukemia, with the seven possible transition times between states

idarubicin (n = 54), FAI + ATRA = all-trans-retinoic acid (n = 54), FAI + GCSF = granulocyte colony-stimulating factor (n = 52), and FAI + ATRA + GCSF (n = 50). The primary aim was to estimate the effects of adding ATRA, GCSF, or both to FAI as Frontline chemo combinations on the probability of treatment success, defined as the patient being alive and in CR at 6 months. Estey et al. (1999) analyzed the data using conventional logistic regression, Kaplan–Meier estimates, and Cox model regression, assuming that the only relevant treatments were the frontline therapies. They concluded that FAI $\pm$ GCSF $\pm$ ATRA improved the CR rate versus FAI $\pm$ ATRA, but that adding ATRA, GCSF, or both to FAI had no significant effect on survival.

Unfortunately, like most randomized trials that focus only on Frontline treatments, the AML/MDS trial was only semi-SMART. The design started out very smartly indeed, by randomizing patients fairly among the four Frontline chemo combinations using a 2×2 factorial design. But after that, following common oncology practice, the design implicitly assumed that Salvage chemo combinations had no effect on survival time, or that their different effects would somehow "average out." The trial design allowed Salvage chemos to be chosen for each patient by their attending physician, rather than re-randomizing patients among Salvage chemos, as would be done in a SMART design. Figure 12.5, which is a simplified version of Fig. 1 in Wahed and Thall (2013), shows the actual treatment and outcome pathways that an AML/MDS patient undergoing chemotherapies may follow. The three possible outcomes for Remission Induction chemo are CR, Resistant Disease, which means that the Induction chemo failed to bring the patient into CR, or Death during Remission Induction. The transition times for these events are given in the figure as T^C = time to CR, T^R = time to Resistant Disease, and T^D = time to Death. Patients with resistant disease are given some form of Salvage treatment, called B_1, and T^{RD} = the transition time from Resistant Disease to Death. For patients who are lucky enough to have their Induction chemo achieve CR, the two possible events that come next are disease Progression or Death, with T^{CP} = the transition time from CR to progressive disease (PD), and T^{CD} = the transition time from CR to Death. For the latter, a patient may die in CR for a variety of reasons, including organ failure or infection due to the Induction chemo. A patient with PD following CR is given some form of Salvage treatment, B_2, with T^{PD} = the transition time from post-CR PD to Death. Thus, the possible treatment regimes take the form (F, B_1, B_2) where F denotes Frontline induction therapy, B_1 denotes salvage therapy for patients whose disease was resistant to induction, and B_2 denotes salvage therapy for patients with PD following a CR achieved with induction. For regime (F, B_1, B_2) each patient received F alone if they died during induction, (F, B_1) if they had resistant disease, or (F, B_2) if they achieved CR but later had progressive disease. That is, no patient could receive both B_1 and B_2. A closely related point is that all of the transition times in Fig. 12.5 are potential outcomes. For example, if a patient's induction chemo achieves a CR, then T^C is observed but the potential times T^R and T^F are not observed. Figure 12.5 shows that a patient could follow one of four different paths to Death as the final event. In terms of the transition times between successive states, each patient's survival time could take one of four different potential forms:

$$T^D \text{ if death during induction therapy}$$
$$T^R + T^{RD} \text{ if death after salvage for resistant disease}$$
$$T^C + T^{CD} \text{ if death in CR}$$
$$T^C + T^{CP} + T^{PD} \text{ if death after salvage for PD after CR.}$$

Wahed and Thall (2013) estimated mean survival time for patients in the AML/MDS trial by first defining the possible DTRs as follows. Since many different Salvage treatments were given, in order to make any progress estimating overall survival time as a function of DTR, Wahed and Thall (2013) categorized each Salvage as either containing high dose ara-C (HDAC) or not (OTHER). A rationale for this is that HDAC is known to be among the most effective chemo agents for treating AML/MDS. This gave a total of $4 \times 2 \times 2 = 16$ possible DTRs. Since the design was only half-SMART due to subjective selection of Salvage chemos, it was necessary to use some form of bias correction to estimate mean overall survival, $\mu(F, B_1, B_2)$, for each DTR (F, B_1, B_2), where the salvage chemos B_1 and B_2 each are either HDAC or OTHER.

Since ignoring the salvage chemos may lead to biased estimation of induction chemo effects on survival, Wahed and Thall (2013) applied both the likelihood-based G-estimation approach of Robins (1986) and IPTW to estimate mean survival as a function of DTR. For G-estimation, they assumed that the conditional distribution of each transition time $\{T^C, T^R, T^D, T^{CP}, T^{CD}, T^{RD}, T^{PD}\}$, given the patient's current history of treatments and previous transition times, followed an accelerated failure time distribution (exponential, Weibull, log-logistic, or log-normal). They chose a specific AFT distribution for each transition time based on Bayes information criterion in preliminary goodness-of-fit analyses. A final estimate of each mean survival time $\mu(F, B_1, B_2)$ was obtained by averaging over the empirical distribution of the covariates.

Xu et al. (2016) reanalyzed the AML/MDS trial dataset by taking a similar G-estimation approach, but modeling the conditional distribution of each transition time very differently. They took a Bayesian approach and assumed that each conditional transition time distribution followed a Bayesian DDP-GP structure. This is the same class of Bayesian nonparametric survival time regression models used for analysis of survival as a function of (AUC, CR, Age) in the allogeneic transplant data, described earlier in Chap. 11. This much more flexible model obviated the need to perform goodness-of-fit analyses of different candidate distributions for each transition time. The DDP-GP also addresses the possible problem that none of the parametric models considered by Wahed and Thall (2013) may fit the data well, for example, if some transition time has a multimodal distribution. Wahed and Thall (2013) were concerned about extreme values that they labeled as "outliers," and fit the model both including and excluding the putative outliers. In contrast, the mixture structure of the DDP-GP model accommodates such extreme values quite naturally. To establish a prior, Xu et al. (2016) performed a preliminary fit by assuming that each transition time T^j approximately followed a lognormal distribution $\log(T^j \mid X) \sim N(X\beta^j, \sigma_0^2)$, fit this model, and used the estimated β^j values as the prior means of β, with prior dispersion hyperparameters calibrated to ensure a non-informative prior. For their analyses, the goal was to estimate the posterior mean survival time of each DTR. The posterior estimates, with accompanying 90% credible intervals, are given in Fig. 12.6, which is similar to Fig. 8 in Xu et al. (2016).

In addition to the DDP-GP-based reanalysis of the leukemia data, to see how well this Bayesian nonparametric method for estimating mean survival with DTRs compared to IPTW and AIPTW, Xu et al. (2016) performed a computer simulation study. It was designed to mimic the AML/MDS trial, but as a simplified version. Samples of size 200 were simulated, with patient baseline blood glucose X_g generated from a N(100, 10^2) distribution. Patients initially were randomized equally between two Frontline induction chemos, with the two possible stage 1 outcomes complete response (C) or resistant disease (R), with C followed by progressive disease P, then salvage treatment B_P, and then death (D), or R followed by salvage treatment B_R, and then D. Thus, the five transition times were T^C, T^R, T^{CP}, T^{PD}, and T^{RD}, and survival time was one of the two potential sums $T^R + T^{RD}$ or $T^C + T^{CP} + T^{PD}$. The simulation study assumed two possible induction chemos $\{a_1, a_2\}$, two possible salvage chemos $\{b_{R,1}, b_{R,2}\}$ following resistant disease, and two possible salvage chemos $\{b_{P,1}, b_{P,2}\}$ following progressive disease after C. To reflect clinical practice, the salvage treatments were simulated as random decisions based on X_g, with $\Pr(B_R = b_{R,1} \mid X_g) = 0.80$ if $X_g < 100$ and $= 0.20$ if $X_g \geq 100$. That is, if the simulated $X_g < 100$, then the salvage treatment following resistant disease was simulated as $B_R = b_{R,1}$ with probability 0.80 and $B_R = b_{R,2}$ with probability 0.20, with these two probabilities reversed if $X_g \geq 100$. Similarly, the salvage treatment following PD was simulated with $\Pr(B_P = b_{P,1} \mid X_g) = 0.20$ if $X_g < 100$ and $= 0.85$ if $X_g \geq 100$. The eight possible DTRs in the simulations thus were $(a_1, b_{R,1}, b_{P,1})$, $(a_2, b_{R,1}, b_{P,1})$, $(a_1, b_{R,2}, b_{P,1})$, and so on.

Figure 12.7, which is given as Fig. 4b in Xu et al. (2016), provides a graphical comparison of the estimated mean survival times of the eight DTRs using G-estimation

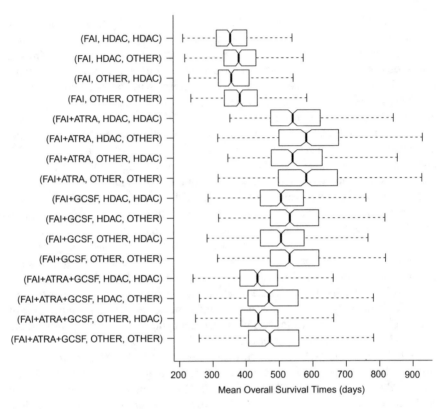

Fig. 12.6 Estimated posterior mean survival and 90% credible interval for each of the 12 DTRs in the AML/MDS chemotherapy trial

with the Bayesian DDP-GP to estimate posterior means, IPTW, and AIPTW. Recall that AIPTW requires at least one of the models for the propensity score (estimated treatment probabilities at each stage) or outcomes to be correct in order for it to be consistent. In this simulation, to reflect reality, neither of these models was correct. Similarly, the DDP-GP model was not correct, either. The simulation results are quite striking. Figure 12.7 indicates that, at least in this particular simulated trial of eight DTRs with a moderate sample size of 200, the DDP-GP based posterior estimates of mean survival are both much more accurate and much more reliable than either the IPTW or AIPTW estimates. Given that IPTW and AIPTW are used widely, these simulation results seem quite controversial. Accordingly, the editor of the journal where Xu et al. (2016) was published made it a discussion paper. In their comments on this paper, Qian et al. (2016) repeated the simulations of Xu et al. (2016) described above, but also examined four different accelerated failure time (AFT) models for the log transition times: a normal distribution, an extreme value distribution, a mixture of G-splines, and a semiparametric AFT model with a smooth error distribution. These models gave fits similar to those provided by the DDP-GP, although their simulations reiterated the result that IPTW performed very poorly.

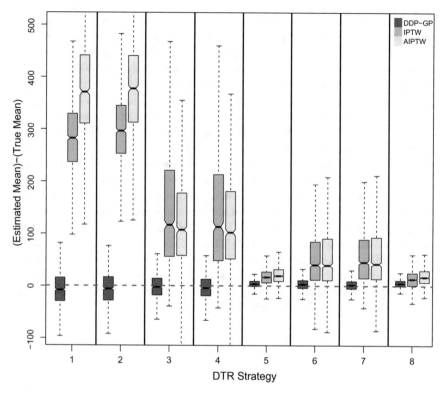

Fig. 12.7 Simulation study results for [Estimated—True] mean survival time, comparing DDP-GP, IPTW, and AIPTW estimation, for each of eight DTRs. In each box-and-whisker plot, the box covers the interquartile range, IQR $= Q_3 - Q_1$, from first quantile Q_1 to third quantile Q_3, the midline is the median, the top whisker goes up to $Q_3 + 1.5$ IQR, and the bottom whisker goes down to $Q_1 - 1.5$ IQR

Given these new simulation results, it is not entirely clear what the future of estimation and bias correction for longitudinal data with embedded adaptive multistage treatment decisions will look like. There are numerous models and methods, cited above, that deal with this broad class of problems to evaluate DTRs, and research in this area is advancing so rapidly that even the excellent books of Chakraborty and Moodie (2013) and Kosorok and Moodie (2015) are only starting points. The inability of IPTW, AIPTW, or similar methods to correct for bias in complex time-to-event data reliably seems clear. The "doubly robust" requirement for AIPTW, that at least one of the models for propensity score or outcome must be correct, actually may be asking too much, since if both models are incorrect then AIPTW seems to perform poorly for time-to-event outcomes. To put this another way, there ain't no Santa Claus and there ain't no Easter Bunny, so don't pretend that at least one of them actually exists.

The simulations described above were for a very specific case with a moderate sample size of 200 patients. In general, due to their complexity, it is widely believed that Bayesian nonparametric models require reasonably large sample sizes to provide reliable fits and inferences. In that regard, these simulation results were surprising. One could elaborate the simulations in many different ways, including examining the effects of larger sample sizes. Another point, for both the AML/MDS trial data analysis and the simulations summarized in Fig. 12.7, is that both used the earlier version of the DDP-GP, without the refinement of the variance–covariance matrix described in formula (11.4), which was used in the AUC-CR-survival data analysis presented in Chap. 11. One important question is whether the DDP-GP model with the refined variance–covariance matrix may do a better job of bias correction.

These data analyses and computer simulations lead to many questions about how one should do robust regression, estimation, and bias correction when analyzing complex longitudinal data involving DTRs. In practice, there are numerous different longitudinal data structures that include DTRs, which may be either prespecified or only identifiable by analyzing the data. To be useful, statistical models and methods must be tailored to accommodate the data at hand and address one's inferential goals. As the bartender Moustache, played by Lou Jacobi in the classic 1963 film "Irma La Douce" would have put it, "But that's another story."

References

Alatrash, G., Kidwell, K. M., Thall, P. F., Di Stasi, A., Chen, J., Zope, M., et al. (2019). Reduced intensity vs. myeloablative conditioning with udarabine in combination with PK-guided busulfan in patients with AML/MDS. *Bone Marrow Transplantation, 54*, 1245–1253.

Albert, J. (2009). *Bayesian computation with R* (2nd ed.). New York: Springer.

Almirall, D., Compton, S., Gunlicks-Stoessel, M., Duan, N., & Murphy, S. A. (2012). Designing a pilot sequential multiple assignment randomized trial for developing an adaptive treatment strategy. *Statistics in Medicine, 31*, 1878–1902.

Almirall, D., Nahum-Shani, I., Sherwood, N., & Murphy, S. A. (2014). Introduction to smart designs for the development of adaptive interventions: With application to weight loss research. *Translational Behavioral Medicine, 66*, 260–274.

Almirall, D., Ten Have, T., & Murphy, S. A. (2010). Structural nested mean models for assessing time-varying effect moderation. *Biometrics, 66*, 131–139.

Altman, D., & Bland, J. (1995). Absence of evidence is not evidence of absence. *British Medical Journal, 311*, 485.

Andersson, B., Thall, P., Valdez, B., Milton, D., Al-Atrash, G., Chen, J., et al. (2017). Fludarabine with pharmacokinetically guided IV busulfan is superior to fixed-dose delivery in pretransplant conditioning of AML/MDS patients. *Bone Marrow Transplantation, 52*, 580–587.

Andersson, B. S., Madden, T., Tran, H. T., Hu, W. W., Blume, K. G., Chow, D. S.-L., et al. (2000). Acute safety and pharmacokinetics of intravenous busulfan when used with oral busulfan and cyclophosphamide as pretransplantation conditioning therapy: A phase I study. *Biology of Blood and Marrow Transplantation, 6*, 548–554.

Andersson, B. S., Thall, P. F., Madden, T., Couriel, D., Wang, X., Tran, H. T., et al. (2002). Busulfan systemic exposure relative to regimen-related toxicity and acute graft-versus-host disease: Defining a therapeutic window for IV BuCy2 in chronic myelogenous leukemia. *Biology of Blood and Marrow Transplantation, 8*, 477–485.

Armitage, P. (2003). Fisher, bradford hill, and randomization. *International Journal of Epidemiology, 32*, 925–928.

Armitage, P., McPherson, C., & Rowe, B. (1969). Repeated significance tests on accumulating data. *Journal of the Royal Statistical Society, Series A, 132*, 235–234.

Armstrong, A., Garrett-Mayer, E., & Eisenberger, M. (2008). Re: Adaptive therapy for androgen-independent prostate cancer: A randomized selection trial of four regimens. *Journal of the National Cancer Institute, 100*, 681–682.

© Springer Nature Switzerland AG 2020

P. F. Thall, *Statistical Remedies for Medical Researchers*, Springer Series in Pharmaceutical Statistics, https://doi.org/10.1007/978-3-030-43714-5

Atkinson, A., & Riani, M. (2000). *Robust diagnostic regression analysis*. New York: Springer.

Austin, P. (2016). Bayesian analysis of binary and polytomous response data. *Statistics in Medicine*, *35*, 5642–5655.

Baggerly, K., & Coombes, K. (2009). Deriving chemosensitivity from cell lines: Forensic bioinformatics and reproducible research in high-throughput biology. *Annals of Applied Statistics*, *3*, 1309–1334.

Barrientos, A. F., Jara, A., & Quintana, F. A. (2012). On the support of MacEachern's dependent Dirichlet processes and extensions. *Bayesian Analysis*, *7*, 277–310.

Bartelink, I. H., Lalmohamed, A., van Reij, E. M., Dvorak, C. C., Savic, R. M., Zwaveling, J., et al. (2016). Association of busulfan exposure with survival and toxicity after haemopoietic cell transplantation in children and young adults: A multicentre, retrospective cohort analysis. *The Lancet Haematology*, *3*(11), e526–e536.

Bauer, P., & Kieser, M. (1999). Combining different phases in the development of medical treatments within a single trial. *Statistics in Medicine*, *18*, 1833–1848.

Bechhofer, R., Santner, T., & Goldsman, D. (1995). *Design and analysis of experiments for statistical selection, screening and multiple comparisons*. Wiley.

Begley, C., & Ellis, L. (2012). Drug development: Raise standards for preclinical cancer research. *Nature*, *483*, 531–533.

Bembom, O., & van der Laan, M. (2007). Statistical methods for analyzing sequentially randomized trials. *Journal of the National Cancer Institute*, *99*, 1577–21582.

Benjamini, Y., & Hochberg, Y. (1995). Controlling the false discovery rate. *Journal of the Royal Statistical Society, Series B*, *57*, 289–300.

Berger, J. O., & Wolpert, R. (1988). *The likelihood principle* (2nd ed.). IMS lecture notes, monograph series. Hayward, California.

Berry, D. (2011). Adaptive clinical trials: The promise and the caution. *Journal of Clinical Oncology*, *29*, 606–609.

Blichert-Toft, L., Rose, C., Andersen, J., Overgaard, M., Axelsson, C. K., Andersen, K. W., et al. (1992). Danish randomized trial comparing breast conservation therapy with mastectomy: Six years of life-table analysis. *Journal of the National Cancer Institute. Monographs*, *11*(11), 19–25.

Bolstad, W., & Curran, J. (2016). *Introduction to Bayesian statistics* (3rd ed.). New York: Wiley.

Boonstra, P., Shen, J., Taylor, J., Braun, T., Griffith, K. A., Daignault, S., et al. (2015). A statistical evaluation of dose expansion cohorts in phase I clinical trials. *Journal of the National Cancer Institute*, *107*, 357–363.

Braun, T. M. (2002). The bivariate continual reassessment method: Extending the CRM to phase I trials of two competing outcomes. *Controlled Clinical Trials*, *23*, 240–256.

Bredeson, C., LeRademacher, J., Kato, K., DiPersio, J. F., Agura, E., Devine, S. M., et al. (2013). Prospective cohort study comparing intravenous busulfan to total body irradiation in hematopoietic cell transplantation. *Blood*, *122*, 3871–3878.

Breslin, F., Sobell, M., Sobell, L., Cunningham, J., Sdao-Jarvie, K., & Borsoi, D. (1999). Problem drinkers: Evaluation of a stepped-care approach. *Journal of Substance Abuse*, *10*, 217–232.

Breslow, N., & Clayton, D. (1993). Approximate inference in generalized linear mixed models. *Journal of the American Statistical Association*, *88*, 9–25.

Bretz, F., Hothorn, T., & Westfall, P. (2010). *Multiple comparisons using R*. New York: Chapman and Hall/CRC.

Bretz, F., Schmidli, H., König, F., Racine, A., & Maurer, W. (2006). Confirmatory seamless phase II/III clinical trials with hypotheses selection at interim: General concepts. *Biometrical Journal*, *48*(4), 623–634.

Brooks, S., Gelman, A., Jones, G., & Meng, X.-L. (2011). *Handbook of markov chain Monte Carlo*. Boca Raton, FL: Chapman & Hall/CRC.

Brown, S., Gregory, W., Twelves, C., & Brown, J. (2014). *A practical guide to designing phase II trials in oncology*. Boca Raton, FL: Wiley.

Carroll, L. (1898). *Alice's adventures in Wonderland*. New York: The Macmillan Company.

Chakraborty, B., & Moodie, E. E. (2013). *Statistical methods for dynamic treatment regimes: Reinforcement learning, causal inference, and personalized medicine.* New York: Springer.

Chappell, R., & Karrison, T. (2007). Letter to the editor. *Statistics in Medicine, 26,* 3046–3056.

Cheung, Y. K., & Chappell, R. (2000). Sequential designs for phase I clinical trials with late-onset toxicities. *Biometrics, 56,* 1177–1182.

Chevret, S. (2006). *Statistical methods for dose-finding experiments.* England: Wiley.

Chipman, H., George, E., & McCulloch, R. (1998). Bayesian cart model search. *Journal of the American Statistical Association, 93,* 935–948.

Christen, J. A., Muller, P., Wathen, K., & Wolf, J. (2004). Bayesian randomized clinical trials: A decision-theoretical sequential design. *Canadian Journal of Statistics, 32,* 387–402.

Claeskens, G., & Hjort, N. L. (2008). *Model selection and model averaging.* Cambridge: Cambridge University Press.

Clavien, P. A., Sanabria, J. R., & Strasberg, S. M. (1992). Proposed classification of complications of surgery with examples of utility in cholecystectomy. *Annals of Surgery, 111,* 518–526.

Cleveland, W. (1979). Robust locally weighted regression and smoothing scatterplots. *Journal of the American Statistical Association, 74,* 829–836.

Collins, L. M., Murphy, S. A., Nair, V. N., & Strecher, V. J. (2005). A strategy for optimizing and evaluating behavioral interventions. *Annals of Behavioral Medicine, 30*(1), 65–73.

Combes, A., Hajage, D., Capellier, G., Demoule, A., Lavoué, S., Guervilly, C., et al. (2018). Extracorporeal membrane oxygenation for severe acute respiratory distress syndrome. *The New England Journal of Medicine, 378,* 1965–1975.

Cook, R. (1998). *Regression graphics: Ideas for studying regressions through graphics.* New York: Wiley.

Cornfield, J. (1969). The Bayesian outlook and its application (with discussion). *Biometrics, 25,* 18–23.

Cox, D. R. (1972). Regression models and life-tables (with discussion). *Journal of the Royal Statistical Society, 34,* 187–220.

Dawson, R., & Lavori, P. (2010). Sample size calculations for evaluating treatment policies in multi-stage design. *Clinical Trials, 7,* 643–6562.

De Iorio, M., Johnson, W. O., Müller, P., & Rosner, G. L. (2009). Bayesian nonparametric nonproportional hazards survival modeling. *Biometrics, 65*(3), 762–771.

de Lima, M., Champlin, R. E., Thall, P. F., Wang, X., Martin, T. G., I. I. I., Cook, J. D., et al. (2008). Phase I/II study of gemtuzumab ozogamicin added to fludarabine, melphalan and allogeneic hematopoietic stem cell transplantation for high-risk CD33 positive myeloid leukemias and myelodysplastic syndrome. *Leukemia, 22,* 258–264.

de Moivre, A. (1718). *The doctrine of chances.* Pearson.

de Moor, C. A., Hilsenbeck, S. G., & Clark, G. M. (1995). Incorporating toxicity grade information in the continual reassessment method for phase I cancer clinical trials. *Controlled Clinical Trials, 16,* 54S–55S.

Dey, D., Ghosh, S., & Mallick, B. (2000). *Generalized linear models: A Bayesian perspective.* New York: Marcel Dekker.

Dindo, D., Demartines, N., & Clavien, P. (2004). Classification of surgical complications: A new proposal with evaluation in a cohort of 6336 patients and results of a survey. *Annals of Surgery, 240,* 205–213.

Dobson, A., & Barnett, A. (2008). *An introduction to generalized linear models* (3rd ed.). CRC Press.

Dragalin, V., & Fedorov, V. (2006). Adaptive designs for dose-finding based on efficacy-toxicity response. *Journal of Statistical Planning and Inference, 136,* 1800–1823.

Dupuy, A., & Simon, R. (2007). Critical review of published microarray studies for cancer outcome and guidelines on statistical analysis and reporting. *Journal of the National Cancer Institute, 99*(2), 147–157.

Efron, B. (1979). Bootstrap methods: Another look at the jackknife. *Anals of Statistics, 7,* 1–26.

Ellenberg, S. S., & Eisenberger, M. A. (1985). An efficient design for phase III studies of combination chemotherapies. *Cancer Treatment Reports, 69*(10), 1147–1154.

Estey, E. H., Thall, P. F., Pierce, S., Cortes, J., Beran, M., Kantarjian, H., et al. (1999). Randomized phase II study of fludarabine + cytosine arabinoside + idarubicin ± all-trans retinoic acid ± granulocyte colony-stimulating factor in poor prognosis newly diagnosed acute myeloid leukemia and myelodysplastic syndrome. *Blood, 93,* 2478–2484.

Faries, D. (1972). Practical modifications of the continual reassessment method for phase I cancer clinical trials. *Journal of Biopharmaceutical Statistics, 4,* 147–164.

Felder, S., & Mayrhofer, T. (2017). *Medical decision making: A health economic primer*. Berlin: Springer.

Feng, W., & Wahed, A. (2009). Sample size for two-stage studies with maintenance therapy. *Statistics in Medicine, 28,* 2028–2041.

Ferguson, T. S. (1973). A Bayesian analysis of some nonparametric problems. *The Annals of Statistics, 1*(2), 209–230.

Fisher, R. A. (1925). *Statistical methods for research workers*. Edinburgh: Oliver and Boyd.

Fisher, R. A. (1926). The arrangement of field experiments. *The Journal of the Ministry of Agriculture, 33,* 503–513.

Fisher, R. A. (1935). *The design of experiments*. Edinburgh: Oliver and Boyd.

Fisher, R. A. (1958). *The nature of probability. Centennial Review, 2,* 261–274.

Fleming, T., Harrington, D., & O'Brien, P. (1984). Designs for group sequential tests. *Contemporary Clinical Trials, 5,* 348–361.

Fowlkes, E. (1987). Some diagnostics for binary logistic regression via smoothing. *Biometrika, 74,* 503–515.

Gamerman, D., & Lopes, H. (2006). *Markov chain Monte Carlo* (2nd ed.). Boca Raton, FL: Chapman & Hall/CRC.

Gelman, A., Carlin, J. B., Stern, H. S., Dunson, D., Vehtari, A., & Rubin, D. (2013). *Bayesian data analysis* (3rd ed.). New York: Taylor & Francis, CRC Press.

Gelman, A., & Stern, H. (2006). The difference between 'significant' and 'not significant' is not itself statistically significant. *The American Statistician, 60,* 328–331.

Gelman, A., & Tuerlinckx, F. (2000). Type S error rates for classical and Bayesian single and multiple comparison procedures. *Computational Statistics, 15,* 373–390.

Giles, F., Kantarjian, H., Cortes, J. E., Garcia-Manero, G., Verstovsek, S., Faderl, S., et al. (2003). Adaptive randomized study of idarubicin and cytarabine versus troxacitabine and cytarabine versus troxacitabine and idarubicin in untreated patients 50 years or older with adverse karyotype acute myeloid leukemia. *Journal of Clinical Oncology, 21,* 1722–1727.

Goodman, S. (2008). A dirty dozen: Twelve p-value misconceptions. *Seminars in Hematology, 45,* 135–140.

Goodman, S., Zahurak, M., & Piantadosi, S. (1995). Some practical improvements in the continual reassessment method for phase I studies. *Statistics in Medicine, 14,* 1149–1161.

Gooley, T. A., Martin, P. J., Fisher, L. D., & Pettinger, M. (1994). Simulation as a design tool for phase I/II clinical trials: An example from bone marrow transplantation. *Controlled Clinical Trials, 15,* 450–462.

Grambsch, P., & Therneau, T. (1994). Proportional hazards tests and diagnostics based on weighted residuals. *Biometrika, 81,* 515–526.

Greenland, A., Senn, S., Rothman, K., Carlin, J., Poole, C., Goodman, S., et al. (2016). Statistical tests, p values, confidence intervals, and power: A guide to misinterpretations. *European Journal of Epidemiology, 31,* 337–350.

Hampson, L., & Jennison, C. (2015). Optimizing the data combination rule for seamless phase II/III clinical trials. *Statistics in Medicine, 34,* 39–58.

Harrell, F. (2001). *Regression modeling strategies*. New York: Springer.

Hernan, M. A., Brumback, B., & Robins, J. M. (2000). Marginal structural models to estimate the causal effect of zidovudine on the survival of HIV-positive men. *Epidemiology, 11*(5), 561–570.

Hey, S., & Kimmellman, J. (2015). Are outcome-adaptive allocation trials ethical? *Clinical Trials*, *12*, 102–106.

Ho, D., Imai, K., King, G., & Stuart, E. (2007). Matching as nonparametric preprocessing for reducing model dependence in parametric causal inference. *Political Analysis*, *15*, 199–236.

Ho, D., Imai, K., King, G., & Stuart, E. (2011). Matchit: Nonparametric preprocessing for parametric causal inference. *Journal of Statistical Software*, *42*, 1–28.

Holland, P. (1986). Statistics and causal inference. *Journal of the American Statistical Association*, *81*, 945–960.

Holm, S. (1979). A simple sequentially rejective multiple test procedure. *Scandinavian Journal of Statistics*, *6*, 65–70.

Hsu, J. (1996). *Multiple comparisons: Theory and methods*. London: Chapman and Hall, CRC Press.

Ibrahim, J. G., Chen, M.-H., & Sinha, D. (2001). *Bayesian survival analysis*. New York: Springer.

Imbens, G., & Rubin, D. (2015). *Causal inference for statistics, social, and biomedical sciences: An introduction*. Cambridge, UK: Cambridge University Press.

Inoue, L., Thall, P., & Berry, D. (2002). Seamlessly expanding a randomized phase II trial to phase III. *Biometrics*, *58*, 823–831.

Ioannidis, J. (2005). Contradicted and initially stronger effects in highly cited clinical research. *Journal of the American Medical Association*, *294*, 218–228.

Ioannidis, J., Allison, D. B., Ball, C. A., Coulibaly, I., Cui, X., Culhane, A. C., et al. (2009). Repeatability of published microarray gene expression analyses. *Nature Genetics*, *41*, 149–155.

Jaki, T., Clive, S., & Weir, C. (2013). Principles of dose finding studies in cancer: A comparison of trial designs. *Cancer Chemotherapy and Pharmacology*, *58*, 1107–1114.

James, G., Witten, D., Hastie, G., & Tibshirani, R. (2013). *An introduction to statistical learning with applications*. New York: Springer.

Jatoi, I., & Proschan, M. (2005). Randomized trials of breast-conserving therapy versus mastectomy for primary breast cancer: A pooled analysis of updated results. *American Journal of Clinical Oncology*, *26*, 289–294.

Jeffreys, H. (1961). *Theory of probability* (3rd ed.). Oxford: Clarendon Press.

Jennison, C., & Turnbull, B. (2007). Adaptive seamless designs: Selection and prospective testing of hypotheses. *Journal of Biopharmaceutical Statistics*, *17*, 1135–1161.

Jin, I. H., Liu, S., Thall, P. F., & Yuan, Y. (2014). Using data augmentation to facilitate conduct of phase I-II clinical trials with delayed outcomes. *Journal of the American Statistical Association*, *109*(506), 525–536.

Jung, S.-H. (2013). *Randomized phase II cancer clinical trials*. Boca Raton, FL: CRC Press, Taylor and Francis Group.

Kaplan, E. L., & Meier, P. (1958). Nonparametric estimation from incomplete observations. *Journal of the American Statistical Association*, *53*, 457–481.

Karrison, T., Huo, D., & Chappell, R. (2003). A group sequential, response-adaptive design for randomized clinical trials. *Controlled Clinical Trials*, *24*, 506–522.

Kass, R., & Raftery, A. (1995). Bayes factors. *Journal of the American Statistical Association*, *90*, 773–795.

Kelly, P. J., Stallard, N., & Todd, S. (2005). An adaptive group sequential design for phase II/III clinical trials that select a single treatment from several. *Journal of Biopharmaceutical Statistics*, *15*, 641–658.

Kim, E., Herbst, R., Wistuba, I., Lee, J., Blumenschein, G., Tsao, A., et al. (2011). The battle trial: Personalizing therapy for lung cancer. *Cancer Discovery*, *1*, 44–53.

Kleinberg, S. (2013). *Causality, probability, and time*. Cambridge, UK: Cambridge University Press.

Konopleva, M., Thall, P. F., Yi, C. A., Borthakur, G., Coveler, A., Bueso-Ramos, C., et al. (2015). Phase I/II study of the hypoxia-activated prodrug PR104 in refractory/relapsed acute myeloid leukemia and acute lymphoblastic leukemia. *Haematologica*, *100*(7), 927–934.

Korn, E. L., & Freidlin, B. (2011). Outcome-adaptive randomization: Is it useful? *Journal of Clinical Oncology*, *29*, 771–776.

Korn, E. L., Freidlin, B., Abrams, J. S., & Halabi, S. (2012). Design issues in randomized phase II/III trials. *Journal of Clinical Oncology, 30*, 667–671.

Korn, E. L., Midthune, D., Chen, T. T., Rubinstein, L., Christian, M. C., & Simon, R. M. (1994). A comparison of two phase I trial designs. *Statistics in Medicine, 13*, 1799–1806.

Kosorok, M., & Moodie, E. (Eds.). (2015). Dynamic treatment regimes in practice: Planning trials and analyzing data for personalized medicine. SIAM.

Kruschke, J. (2015). *Doing Bayesian data analysis, second edition: A tutorial with R, JAGS, and Stan* (2nd ed.). New York: Elsevier.

Lan, K. K. G., & DeMets, D. L. (1983). Discrete sequential boundaries for clinical trials. *Biometrika, 70*(3), 659–663.

Laplace, P.-S. (1902). *A philosophical essay on probabilities*. London: Wiley/Chapman and Hall. (translated from the 6th French edition).

Lavori, P. W., & Dawson, R. (2004). Dynamic treatment regimes: Practical design considerations. *Clinical Trials, 1*(1), 9–20.

Lee, J., Thall, P. F., Ji, Y., & Müller, P. (2015). Bayesian dose-finding in two treatment cycles based on the joint utility of efficacy and toxicity. *Journal of the American Statistical Association, 110*(510), 711–722.

Leek, J., & Peng, R. D. (2015). Reproducible research still can be wrong: Adopting a prevention approach. *Proceedings of the National Academy of Sciences, 112*, 1645–1646.

Lei, H., Nahum-Shani, I., Lynch, K., Oslin, D., & Murphy, S. (2012). A "SMART" design for building individualized treatment sequences. *Annual Review of Clinical Psychology, 8*, 1–14.

Li, Z., & Murphy, S. (2011). Sample size formulae for two-stage randomized trials survival outcomes. *Biometrika, 98*, 503–518.

Liang, F., Liu, C., & Carroll, R. (2010). *Advanced markov chain Monte Carlo methods: Learning from past samples*. New York: Wiley.

Lin, S., Wang, L., Myles, B., Thall, P., Hoffstetter, W., Swisher, S., et al. (2012). Propensity score-based comparison of long-term outcomes with 3-dimensional conformal radiotherapy vs intensity-modulated radiotherapy for esophageal cancer. *Journal of Radiation Oncology, Biology, Physics, 84*, 1078–1085.

Lin, Y., & Shih, W. (2001). Statistical properties of the traditional algorithm-based designs for phase I cancer clinical trials. *Biostatistics, 2*, 203–215.

Little, R. J. A., & Rubin, D. B. (2002). *Statistical analysis with missing data* (2nd ed.). New York: Wiley.

Localio, R., Berlin, J., Ten Have, T., & Kimmel, S. (2001). Adjustments for center in multicenter studies: An overview. *Annals of Internal Medicine, 135*, 112–123.

Love, S., Brown, S., Weir, C., Harbron, C., Yap, C., Gaschler-Markefski, B., et al. (2017). Embracing model-based designs for dose-finding trials. *British Journal of Cancer, 117*, 332–339.

Lunceford, J. K., Davidian, M., & Tsiatis, A. (2002). Estimation of survival distributions of treatment policies in two-stage randomization designs in clinical trials. *Biometrics, 58*, 48–57.

MacEachern, S. N. (1999). Dependent nonparametric processes. In *ASA Proceedings of the Section on Bayesian Statistical Science* (pp. 50–55). Alexandria, VA: American Statistical Association.

MacEachern, S. N., & Müller, P. (1998). Estimating mixture of Dirichlet process models. *Journal of Computational and Graphical Statistics, 7*(2), 223–238.

Maki, R., Wathen, J., Hensley, M., Patel, S., Priebat, D., Okuno, S., et al. (2007). An adaptively randomized phase III study of gemcitabine and docetaxel versus gemcitabine alone in patients with metastatic soft tissue sarcomas. *Journal of Clinical Oncology, 25*, 2755–2763.

Mandrekar, S., Qin, R., & Sargent, D. (2010). Model-based phase I designs incorporating toxicity and efficacy for single and dual agent drug combinations: Methods and challenges. *Statistics in Medicine, 29*, 1077–1083.

May, G. S., DeMets, D. L., Friedman, L. M., Furberg, C., & Passamani, E. (1981). The randomized clinical trial: Bias in analysis. *Circulation, 64*(4), 669–673.

Mc Grayne, S. B. (2011). *The theory that would not die*. New Haven & London: Yale University Press.

McCullagh, P. (1980). Regression models for ordinal data (with discussion). *Journal of the Royal Statistical Society*, 42, 109–142.

McCullagh, P., & Nelder, J. A. (1989). *Generalized linear models* (2nd ed.). New York: Chapman and Hall.

McShane, B., & Gal, D. (2017). Statistical significance and the dichotomization of evidence. *Journal of the American Statistical Association*, 12, 885–895.

Miller, R. (1981). *Simultaneous statistical inference* (2nd ed.). New York: Springer.

Millikan, R., Logothetis, C., Thall, P.Response to comments on "Adaptive therapy for androgen independent prostate cancer: A randomized selection trial including four regimens" by P.F. Thall, et al. (2008). *Journal of the National Cancer Institute*, 100, 682–683.

Mitra, R., & Mueller, P. (2015). *Nonparametric Bayesian inference in biostatistics*. New York: Springer.

Moodie, E. E. M., Richardson, T. S., & Stephens, D. A. (2007). Demystifying optimal dynamic treatment regimes. *Biometrics*, 63(2), 447–455.

Morgan, S., & Winthrop, C. (2015). *Counterfactuals and causal inference: Methods and principles for social research* (2nd ed.). Cambridge, UK: Cambridge University Press.

Müller, P., & Mitra, R. (2013). Bayesian nonparametric inference-Why and how. *Bayesian Analysis*, 8(2), 269–302.

Müller, P., & Mitra, R. (2013). Bayesian nonparametric inference-Why and how. *Bayesian Analysis*, 8, 269–302.

Müller, P., Parmigiani, G., & Rice, K. (2006). FDR and Bayesian multiple comparisons rules. *Johns Hopkins University, Department of Biostatistics Working Papers*.

Müller, P., & Rodriguez, A. (2013). Nonparametric Bayesian inference. *IMS-CBMS lecture notes* (p. 270). IMS.

Murphy, S. (2005a). An experimental design for the development of adaptive treatment strategies. *Statistics in Medicine*, 24, 1455–1481.

Murphy, S. (2005b). Generalization error for q-learning. *Journal of Machine Learning*, 6, 1073–1097.

Murphy, S., Lynch, K., McKay, J., & TenHave, T. (2005). Developing adaptive treatment strategies in substance abuse research. *Drug and Alcohol Dependence*, 88, 524–530.

Murphy, S., van der Laan, M., & Robins, J. (2001). Marginal mean models for dynamic regimes. *Journal of the American Statistical Association*, 96, 1410–1423.

Murphy, S. A. (2003). Optimal dynamic treatment regimes. *Journal of the Royal Statistical Society: Series B*, 65(2), 331–355.

Murray, T., Yuan, Y., & Thall, P. (2018a). A Bayesian machine learning approach for optimizing dynamic treatment regimes. *Journal of the American Statistical Association*, 113, 1255–1267.

Murray, T., Yuan, Y., Thall, P., Elizondo, J., & Hofstetter, W. (2018b). A utility-based design for randomized comparative trials with ordinal outcomes and prognostic subgroups. *Biometrics*, 74, 1095–1103.

Nelder, J. A., & Wedderburn, R. (1972). Generalized linear models. *Journal of the Royal Statistical Society, Series A*, 135, 370–384.

Neyman, J., & Pearson, E. (1933). On the problem of the most efficient tests of statistical hypotheses. *Philosophical Transactions of the Royal Society, A*, 231, 289–337.

O'Brien, P., & Fleming, T. (1979). A multiple testing procedure for clinical trials. *Biometrics*, 35, 549–556.

O'Quigley, J., & Chevret, S. (1992). Methods for dose-finding in cancer clinical trials: A review and results of a Monte-Carlo study. *Statistics in Medicine*, 34, 1647–1664.

O'Quigley, J., Pepe, M., & Fisher, L. (1990). Continual reassessment method: A practical design for phase I clinical trials in cancer. *Biometrics*, 46, 33–48.

O'Quigley, J., & Shen, L. (1996). Continual reassessment method: A likelihood approach. *Biometrics*, 52, 673–684.

Parmigiani, G. (2002). *Modeling in medical decision making: A Bayesian approach*. New York: Wiley.

Pearl, J., Glymour, M., & Jewell, N. (2016). *Causal inference in statistics: A primer*. New York, NY, United States: Wiley.

Piret, A., & Amado, C. (2008). Interval estimators for a binomial proportion: Comparison of twenty methods. *Statistical Journal, 6*, 165–197.

Plotkin, S., Gerber, J., & Offit, P. (2009). Vaccines and autism: A tale of shifting hypotheses. *Clinical Infectious Diseases, 48*, 456–461.

Pocock, S. (1977). Group sequential methods in the design and analysis of clinical trials. *Biometrika, 64*, 191–199.

Pregibon, D. (1980). Goodness of link tests for generalized linear models. *Applied Statistics, 29*, 15–24.

Pregibon, D. (1981). Logistic regression diagnostics. *Annals of Statistics, 9*, 705–724.

Qian, G., Laber, E., & Reich, B. (2016). Comment on 'Bayesian nonparametric estimation for dynamic treatment regimes with sequential transition times'. *Journal of the American Statistical Association, 111*, 936–942.

Ratain, M. J., & Karrison, T. G. (2007). Testing the wrong hypothesis in phase II oncology trials: there is a better alternative. *Clinical Cancer Research, 13*(3), 781–782.

Robert, C. P., & Cassella, G. (1999). *Monte Carlo statistical methods*. New York: Springer.

Robins, J. M. (1986). A new approach to causal inference in mortality studies with a sustained exposure period application to control of the healthy worker survivor effect. *Mathematical Modeling, 7*(9), 1393–1512.

Robins, J. M. (1997). *Causal inference from complex longitudinal data*. Springer.

Robins, J. M. (1998). Marginal structural models. In *Proceedings of the American Statistical Association Section on Bayesian Statistics* (pp. 1–10).

Robins, J. M., & Rotnitzky, A. (1992). Recovery of information and adjustment for dependent censoring using surrogate markers. In N. Jewell, K. Dietz, & V. Farewell (Eds.), *AIDS epidemiology, methodological issues* (pp. 24–33).

Rosenbaum, P. (2001). *Observational studies*. Springer.

Rosenbaum, P. (2010). *Design of observational studies. Springer series in statistics*.

Rosenbaum, P., & Rubin, D. B. (1983). The central role of the propensity score in observational studies for causal effects. *Biometrika, 70*, 41–55.

Rosenbaum, P., & Rubin, D. B. (1985). Constructing a control group using multivariate matched sampling methods that incorporate the propensity score. *The American Statistician, 39*, 33–38.

Rosenberger, W., & Lachin, J. (2004). *Randomization in clinical trials: Theory and practice*. Wiley.

Royson, P., Parmar, M., & Qian, W. (2003). Novel designs for multi-arm clinical trials with survival outcomes with an application in ovarian cancer. *Statistics in Medicine, 22*, 2239–2256.

Royston, P., & Sauerbrei, W. (2008). *Multivariable model building*. Chichester, West Sussex, England: Wiley.

Rubin, D. (1973). Matching to remove bias in observational studies. *Biometrics, 29*, 159–183.

Rubin, D. (1978). Bayesian inference for causal effects: The role of randomization. *Annals of Statistics, 6*, 34–58.

Rubinstein, L. V., Korn, E. L., Freidlin, B., Hunsberger, S., Ivy, S. P., & Smith, M. A. (2005). Design issues of randomized phase II trials and a proposal for phase II screening trials. *Journal of Clinical Oncology, 23*, 7199–7206.

Rush, A. J., Trivedi, M., & Fava, M. (2003). Depression, IV: STAR*D treatment trial for depression. *American Journal of Psychiatry, 160*, 237.

Schaid, D. J., Wieand, S., & Therneau, T. M. (1990). Optimal two stage screening designs for survival comparisons. *Biometrika, 77*, 507–513.

Schmidli, H., Bretz, F., Racine, A., & Maurer, W. (2006). Confirmatory seamless phase II/III clinical trials with hypothesis selection at interim: Applications and practical considerations. *Biometrical Journal, 48*, 635–643.

Senn, S. (2013). Seven myths of randomisation in clinical trials. *Statistics in Medicine, 32*, 1439–1450.

Sethuraman, J. (1994). A constructive definition of Dirichlet priors. *Statistica Sinica, 4*, 639–650.

Shafer, D., & Grant, S. (2016). Update on rational targeted therapy in AML. *Blood Reviews, 30*, 275–283.

Sharma, M. R., Stadler, W. M., & Ratain, M. J. (2011). Randomized phase II trials: A long-term investment with promising returns. *Journal of the National Cancer Institute, 103*(14), 1093–1100.

Simes, R. (1986). An improved Bonferroni procedure for multiple tests of significance. *Biometrika, 73*, 751–754.

Simmons, J., Nelson, J., & Simonsohn, U. (2011). False-positive psychology: Undisclosed flexibility in data collection and analysis allows presenting anything as significant. *Psychological Science, 22*, 1359–1366.

Simon, R., Wittes, R. E., & Ellenberg, S. S. (1985). Randomized phase II clinical trials. *Cancer Treatment Reports, 69*, 1375–1381.

Simon, R. M. (1989). Optimal two-stage designs for phase II clinical trials. *Controlled Clinical Trials, 10*, 1–10.

Sox, H. C., Higgins, M. C., & Owens, D. K. (2013). *Medical decision making* (2nd ed.). Wiley.

Stallard, N., & Todd, S. (2003). Sequential designs for phase III clinical trials incorporating treatment selection. *Statistics in Medicine, 22*, 689–703.

Stigler, S. (1986). Laplace's 1774 memoir on inverse probability. *Statistical Science, 1*, 359–378.

Sugihara, M. (2009). Survival analysis using inverse probability of treatment weighted methods based on the generalized propensity score. *Pharmaceutical Statistics, 9*, 21–34.

Sutton, R. S., & Barto, A. G. (1998). *Reinforcement learning: An introduction.* Cambridge, MA: MIT Press.

Sverdlov, O. (2015). *Modern adaptive randomized clinical trials: Statistical and practical aspects.* Boca Raton: CRC Press/Taylor and Francis.

Tannock, I., de Wit, R., Berry, W., Horti, J., Pluzanska, A., Chi, K. N., et al. (2004). Docetaxel plus prednisone or mitoxantrone plus prednisone for advanced prostate cancer. *New England Journal of Medicine, 351*, 1502–1512.

Thall, P., Fox, P., & Wathen, J. (2015). Statistical controversies in clinical research: Scientific and ethical problems with adaptive randomization in comparative clinical trials. *Annals of Oncology, 26*(8), 1621–1628.

Thall, P., Logothetis, C., Pagliaro, L., Wen, S., Brown, M., Williams, D., et al. (2007). Adaptive therapy for androgen independent prostate cancer: A randomized selection trial including four regimens. *Journal of the National Cancer Institute, 99*, 1613–1622.

Thall, P. F., & Cook, J. D. (2004). Dose-finding based on efficacy-toxicity trade-offs. *Biometrics, 60*, 684–693.

Thall, P. F., Herrick, R., Nguyen, H., Venier, J., & Norris, J. (2014a). Using effective sample size for prior calibration in Bayesian phase I-II dose-finding. *Clinical Trials, 11*, 657–666.

Thall, P. F., Millikan, R. E., & Sung, H.-G. (2000). Evaluating multiple treatment courses in clinical trials. *Statistics in Medicine, 19*(8), 1011–1028.

Thall, P. F., Mueller, P., Xu, Y., & Guindani, M. (2017). Bayesian nonparametric statistics: A new toolkit for discovery in cancer research. *Pharmaceutical Statistics, 16*, 414–423.

Thall, P. F., & Nguyen, H. Q. (2012). Adaptive randomization to improve utility-based dose-finding with bivariate ordinal outcomes. *Journal of Biopharmaceutical Statistics, 22*(4), 785–801.

Thall, P. F., Nguyen, H. Q., Braun, T. M., & Qazilbash, M. H. (2013). Using joint utilities of the times to response and toxicity to adaptively optimize schedule-dose regimes. *Biometrics, 69*, 673–682.

Thall, P. F., Nguyen, H. Q., & Estey, E. H. (2008). Patient-specific dose finding based on bivariate outcomes and covariates. *Biometrics, 64*, 1126–1136.

Thall, P. F., Nguyen, H. Q., Zohar, S., & Maton, P. (2014b). Optimizing sedative dose in preterm infants undergoing treatment for respiratory distress syndrome. *Journal of the American Statistical Association, 109*(507), 931–943.

Thall, P. F., & Russell, K. E. (1998). A strategy for dose-finding and safety monitoring based on efficacy and adverse outcomes in phase I/II clinical trials. *Biometrics, 54*, 251–264.

Thall, P. F., Simon, R., & Ellenberg, S. S. (1988). Two-stage selection and testing designs for comparative clinical trials. *Biometrika, 75*(2), 303–310.

Thall, P. F., Simon, R. M., & Estey, E. H. (1995). Bayesian sequential monitoring designs for single-arm clinical trials with multiple outcomes. *Statistics in Medicine, 14*, 357–379.

Thall, P. F., & Sung, H.-G. (1998). Some extensions and applications of a Bayesian strategy for monitoring multiple outcomes in clinical trials. *Statistics in Medicine, 17*, 1563–1580.

Thall, P. F., Sung, H.-G., & Estey, E. H. (2002). Selecting therapeutic strategies based on efficacy and death in multicourse clinical trials. *Journal of the American Statistical Association, 97*(457), 29–39.

Thall, P. F., & Wathen, J. K. (2005). Covariate-adjusted adaptive randomization in a sarcoma trial with multi-stage treatments. *Statistics in Medicine, 24*(13), 1947–1964.

Thall, P. F., & Wathen, J. K. (2007). Practical Bayesian adaptive randomisation in clinical trials. *European Journal of Cancer, 43*(5), 859–866.

Thall, P. F., Wooten, L., & Tannir, N. (2005). Monitoring event times in early phase clinical trials: Some practical issues. *Clinical Trials, 2*, 467–478.

Therneau, T., & Grambsch, P. (2000). *Modeling survival data: Extending the Cox model.* New York: Springer.

Thompson, W. (1933). On the likelihood that one unknown probability exceeds another in view of the evidence of the two samples. *Biometrika, 25*, 285–294.

Ting, N., Chen, D.-G., Ho, S., & Cappelleri, J. (2017). *Phase II clinical development of new drugs.* Singapore: Springer.

Tsiatis, A. (2006). *Semiparametric theory and missing data.* New York: Springer.

Vittinghoff, E., Glidden, D., Shiboski, S., & McCulloch, C. (2005). *Regression methods in biostatistics.* New York: Springer.

Wahed, A. S., & Thall, P. F. (2013). Evaluating joint effects of induction-salvage treatment regimes on overall survival in acute leukaemia. *Journal of the Royal Statistical Society: Series C (Applied Statistics), 62*(1), 67–83.

Wahed, A. S., & Tsiatis, A. A. (2004). Optimal estimator for the survival distribution and related quantities for treatment policies in two-stage randomization designs in clinical trials. *Biometrics, 60*, 124–133.

Wang, L., Rotnitzky, A., Lin, X., Millikan, R. E., & Thall, P. F. (2012). Evaluation of viable dynamic treatment regimes in a sequentially randomized trial of advanced prostate cancer. *Journal of the American Statistical Association, 107*, 493–508.

Wang, S., & Tsiatis, A. (1987). Approximately optimal one-parameter boundaries for trials. *Biometrics, 43*, 193–200.

Wasserstein, R., & Lazar, N. (2016). The ASA's statement on p-values: Context, process, and purpose. *The American Statistician, 70*, 129–133.

Wathen, J. K., & Thall, P. F. (2017). A simulation study of outcome adaptive randomization in multi-arm clinical trials. *Clinical Trials, 14*, 432–440.

Watkins, C. J. C. H. (1989). *Learning from delayed rewards.* Ph.D. thesis, Cambridge University.

Whelan, H. T., Cook, J. D., Amlie-Lefond, C. M., Hovinga, C. A., Chan, A. K., Ichord, R. N., et al. (2008). Practical model-based dose finding in early-phase clinical trials-Optimizing tissue plasminogen activator dose for treatment of ischemic stroke in children. *Stroke, 39*, 2627–2636.

Wittes, J. (2012). Jerome Cornfield's contributions to early large randomized clinical trials and some reminiscences from the years of the slippery doorknobs. *Statistics in Medicine, 31*, 2791–2797.

Xie, J., & Liu, C. (2005). Adjusted Kaplan-Meier estimator and log-rank test with inverse probability of treatment weighting for survival data. *Statistics in Medicine, 24*, 3089–3110.

Xu, Y., Mueller, P., Wahed, A., & Thall, P. (2016). Bayesian nonparametric estimation for dynamic treatment regimes with sequential transition times. *Journal of the American Statistical Association (with discussion), 111*, 921–950.

Xu, Y., Thall, P., Hua, W., & Andersson, B. (2019). Bayesian nonparametric survival regression for optimizing precision dosing of intravenous busulfan in allogeneic stem cell transplantation. *Journal of the Royal Statistical Society, Series C, 68*, 809–828.

Yan, F., Thall, P. F., Lu, K., Gilbert, M., & Yuan, Y. (2018). Phase I-II clinical trial design: A state-of-the-art paradigm for dose finding with novel agents. *Annals of Oncology, 29*, 694–699.

Yates, F. (1964). Sir Ronald Fisher and the design of experiments. *Biometrics, 20*, 307–321.

Yin, G., Li, Y., & Ji, Y. (2006). Bayesian dose-finding in phase I/II clinical trials using toxicity and efficacy odds ratios. *Biometrics, 62*, 777–787.

Yin, G., & Yuan, Y. (2009). Bayesian model averaging continual reassessment method in phase I clinical trials. *Journal of the American Statistical Association, 104*(487), 954–968.

Yuan, Y., Nguyen, H., & Thall, P. (2016). *Bayesian designs for phase I-II clinical trials*. Boca Raton, FL: CRC Press.

Yuan, Y., & Yin, G. (2009). Bayesian dose-finding by jointly modeling toxicity and efficacy as time-to-event outcomes. *Journal of the Royal Statistical Society: Series C (Applied Statistics), 58*, 719–736.

Yuan, Y., & Yin, G. (2011). On the usefulness of outcome adaptive randomization. *Journal of Clinical Oncology, 29*, 771–776.

Zheng, L., & Zelen, M. (2008). Multi-center clinical trials: Randomization and ancillary statistics. *Annals of Applied Statistics, 2*, 582–600.

Zhou, Y.-Q., Zeng, C., Laber, E., & Kosorok, M. (2015). New statistical learning methods for estimating optimal dynamic treatment regimes. *Journal of the Statistical Association, 110*, 583–598.

Index

© Springer Nature Switzerland AG 2020
P. F. Thall, *Statistical Remedies for Medical Researchers*, Springer Series
in Pharmaceutical Statistics, https://doi.org/10.1007/978-3-030-43714-5